RIGHT-ORDERED GROUPS

SIBERIAN SCHOOL OF ALGEBRA AND LOGIC

Definability and Computability • *Yuri L. Ershov*

Right-Ordered Groups • *Valeriĭ M. Kopytov and Nikolaĭ Ya. Medvedev*

RIGHT-ORDERED GROUPS

Valeriĭ M. Kopytov
Institute of Mathematics
Novosibirsk, Russia

and

Nikolaĭ Ya. Medvedev
Altaĭ University
Barnaul, Russia

Consultants Bureau • New York, London, and Moscow

Library of Congress Cataloging-in-Publication Data

On file

Siberian School of Algebra and Logic is a simultaneous translation
of the book series Sibirskaya Shkola Algebry i Logiki,
which is published in and translated by
Scientific Books (RIMIBE NGU)
2, Pirogova Street, Novosibirsk 630090, Russia

ISBN 0-306-11060-1

A Division of Plenum Publishing Corporation
233 Spring Street, New York, N.Y. 10013

10 9 8 7 6 5 4 3 2 1

Printed in the United States of America

Preface

The theory of right-ordered groups is a branch of contemporary algebra that is naturally connected with the abstract group theory and with the theory of ordered algebraic systems. Originating in the early century, this theory concerns the works of Dedekind, Hölder, and Hilbert. By now, the theory is well developed; it is useful for both the group theory and other fields of mathematics. The main purpose of this book is to give a complete review of methods and results of this theory.

In the first three chapters of the book, we recall classical theorems and methods of the theory of partially ordered groups as a whole, with special emphasis on right partial orders. In Chapter 4, we discuss the theory of groups of automorphisms of totally ordered sets, which is a powerful, although rather complicated, instrument for studying right-ordered groups.

In Chapters 5–8, we describe properties and structures of right-ordered groups. Much attention is focused on extensions of orders, conditions under which right orders can be introduced on groups, and the definition of operations on the class of all right-ordered groups and on its most important subclasses. The most significant results obtained recently are given. Perhaps some of them will serve as a basis for new research and pave the way for further development of the theory of right-ordered groups. We hope that presentation of the current status of the right-ordered theory will be a starting point.

The reader does not need any preliminary knowledge other than that of elements of the group theory, the set theory, and topology. All notions and terms specific for the right-ordered groups are defined here.

We are grateful to the representatives of the School of algebraists and logicians in Novosibirsk. Our interest in right-ordered groups was aroused by the works of the founder of this school, A. I. Mal'cev, and the well-known Siberian researchers M. I. Kargapolov, A. I. Kokorin, and D. M. Smirnov. The works of the American mathematicians P. F. Conrad, C. W. Holland, and S. H. McCleary on the groups of automorphisms of totally ordered sets and on lattice ordered groups also influenced us greatly.

Valeriĭ M. Kopytov
Nikolaĭ Ya. Medvedev

Contents

Chapter 1. Introduction 1
1.1. Partially Ordered Sets 1
1.2. Lattices 4
1.3. Properties of Lattices 7
1.4. Orders on Groups 12
1.5. Positive Cones 15
1.6. Basic Notions 19

Chapter 2. Systems of Convex Subgroups 29
2.1. General Properties 29
2.2. Archimedean Groups 33
2.3. Totally Ordered Groups 35
2.4. Conradian Groups 37

Chapter 3. Orderability Conditions 45
3.1. Semigroup Conditions 45
3.2. Sufficient Conditions 49
3.3. Group Conditions for Total Orderability 52
3.4. Groups of Automorphisms 57
3.5. Connection with Totally Ordered Groups 59

3.6. Fully Orderable Groups 62

Chapter 4. Groups of Order Automorphisms 73

4.1. Preliminaries 73
4.2. Wreath Products 85
4.3. Order Types of Right-Ordered Groups 97
4.4. The Chehata Groups and the Dlab Groups 100
4.5. Embeddings 113

Chapter 5. Relatively Convex Subgroups 121

5.1. Orderable Representations 121
5.2. A Finite Number of Right Orders 132
5.3. A Finite Number of Total Orders 142
5.4. Center of Right-Ordered Groups 148
5.5. Criteria for Orderability 153
5.6. Centers of "Small" Subgroups 157

Chapter 6. Orders on Free Products 165

6.1. The Vinogradov Theorem 165
6.2. Free Products with Amalgamation 169
6.3. Right-Orderable Groups with Amalgamation 178

Chapter 7. Quasivarieties 181

7.1. Properties of Quasivarieties 181
7.2. Model Theory 190
7.3. Axiomatic Rank 194
7.4. Locally Indicable Groups 200
7.5. The Local Indicability of Extensions 205
7.6. Lattice of Quasivarieties 207

Chapter 8. Semilinearly Ordered Groups 217

8.1. Definitions 218
8.2. Basic Properties 223
8.3. Convex Subgroups 226
8.4. Constructions 232

References 239

Subject Index 245

Author Index 249

Notation

$(G, \Omega)\mathrm{Wr}\,(H, T)$	the Cartesian wreath product
$(G, \Omega)\mathrm{wr}\,(H, T)$	the direct wreath product
$[A, B]$	the commutator subgroup of A and B
$[a, b]$	the commutator of a and b
$\overline{\Omega}$	the Dedekind completion of Ω
$\boldsymbol{c}(H)$	the centralizer of H
$\boldsymbol{n}(H)$	the normalizer of H
$\mathcal{A}(G)$	the group of all automorphisms of G
$\mathfrak{C}(G)$	a system of subgroups of G
$A \prec B$	a jump of groups A and B
$\mathrm{gp}\,(M)$	the subgroup generated by M
$\mathrm{sgr}\,(M)$	the subsemigroup generated by M
$\mathrm{supp}\,(\alpha)$	the support of α

Chapter 1

Introduction

1.1. Partially Ordered Sets

1.1.1. In this section, we recall some definitions.

- A nonempty set M is called *partially ordered* (*po-set*) if a binary relation $\leqslant$ (called a *partial order relation* or simply *partial order*) is defined on M such that the following axioms hold:

$$x \leqslant x \qquad \text{(reflexivity)}$$

$$x \leqslant x,\ y \leqslant x \ \Rightarrow\ x = y \qquad \text{(antisymmetricity)}$$

$$x \leqslant y,\ y \leqslant z \ \Rightarrow\ x \leqslant z \qquad \text{(transitivity)}$$

A partially ordered set M with a partial order $\leqslant$ is denoted by $\{M; \leqslant\}$.

If $x \leqslant y$, then we say "x is less than or equal to y" or "y is greater than or equal to x" and write $y \geqslant x$. If $x \leqslant y$ and $x \neq y$, then we say that "x is less (strictly less) than y" or "y is greater (strictly greater) than x" and write $x < y$ or $y > x$.

- Elements x and y are called *comparable* if $x \leqslant y$ or $y \leqslant x$ and *incomparable* otherwise.

- A partially ordered set M is called *totally ordered* if any two elements of M are comparable.

- A totally ordered set J is called *well-ordered* if any nonempty subset of M has the least element.

In the sequel, we suppose that the Zermelo theorem is true. As is known, this theorem is equivalent to the axiom of choice. In other words, we assume that any set can be made well-ordered.

A subset A of a partially ordered set $\{M;\ \leqslant\}$ is also a partially ordered set with respect to the induced order $\leqslant_{|A}$, where $x \leqslant_{|A} y$ for $x, y \in A$ if $x \leqslant y$ in $\{M;\ \leqslant\}$. The restriction on $A \subset M$ of an order $\leqslant$ defined on M will be denoted by the same symbol $\leqslant$.

As the simplest examples of partially ordered sets, we indicate the set $\mathbb{R}$ of real numbers and its subsets with the natural order.

EXAMPLE 1.1.1. Let X be arbitrary and Y be totally ordered sets. We denote by $\mathcal{F}(X, Y)$ the set of all functions defined on X with values in Y. On $\mathcal{F}(X, Y)$ we introduce the partial order $\leqslant$ by setting $f \leqslant g$ for $f, g \in \mathcal{F}(X, Y)$ if $f(x) \leqslant g(x)$ in Y for each $x \in X$. It is easy to check immediately that the relation just defined is a partial order. The latter is called the *coordinate-wise order*.

EXAMPLE 1.1.2. On the set $\mathcal{F}(X, Y)$ we introduce another order. First, we well order somehow the set X by a relation $\prec$:

$$x_1 \prec x_2 \prec \ldots \prec x_\alpha \prec x_{\alpha+1} \prec \cdots$$

Then for $f, g \in \mathcal{F}(X, Y)$ we set $f \leqslant g$ if $f(x_\alpha) \leqslant g(x_\alpha)$, where the element $x_\alpha \in X$ is the least element with respect to $\prec$ among x_ξ such that $f(x_\xi) \neq g(x_\xi)$. This order is called *lexicographic with respect to the order* $\prec$ (or simply *lexicographic* if it is clear from the context what order we mean).

Let two partial orders $\leqslant$ and $\leqslant'$ be defined on a set M. If $x \leqslant' y$ for any $x, y \in M$ such that $x \leqslant y$, then the order $\leqslant'$ is called an *extension of the order* $\leqslant$. It is easy to see that the order $\leqslant'$ on $\mathcal{F}(X, Y)$ from Example 1.1.2 is an extension of the order $\leqslant$ on $\mathcal{F}(X, Y)$ from Example 1.1.1.

1.1.2. Example 1.1.1 is a special case of a general algebraic construction. Let $\{M_\alpha \mid \alpha \in I\}$ be a collection of partially ordered sets and $\bar{\prod\limits_{\alpha\in I}} M_\alpha$ be the set of all functions f defined on I with values in $\bigcup\limits_{\alpha\in I} M_\alpha$: $f(\alpha) \in M_\alpha$ for all $\alpha \in I$. On the set $\bar{\prod\limits_{\alpha\in I}} M_\alpha$ we introduce the partial order $\leqslant$ by setting $f \leqslant g$ for $f, g \in \bar{\prod\limits_{\alpha\in I}} M_\alpha$ if $f(\alpha) \leqslant g(\alpha)$ for all $\alpha \in I$.

The partially ordered set $\bar{\prod\limits_{\alpha\in I}} M_\alpha$ is called the *Cartesian product* (*cardinal product*) of partially ordered sets M_α, $\alpha \in I$.

On the set $\bar{\prod\limits_{\alpha\in I}} M_\alpha$ it is possible to introduce another order generalizing the order from Example 1.1.2. Let I be a well-ordered set with an order $\prec$:

$$\alpha_1 \prec \alpha_2 \prec \ldots \prec \alpha_\xi \prec \alpha_{\xi+1} \prec \ldots$$

For $f, g \in \bar{\prod\limits_{\alpha\in I}} M_\alpha$ we set $f < g$ if $f(\beta) < g(\beta)$, where $\beta \in I$ is the least element of the set $\{\alpha_\xi \in I \mid f(\alpha_\xi) \neq g(\alpha_\xi)\}$ with respect to the order $\prec$. The set $\bar{\prod\limits_{\alpha\in I}} M_\alpha$ with the order just introduced is called the *lexicographic product* of M_α, $\alpha \in I$, and is denoted by $\overrightarrow{\prod\limits_{\alpha\in I}} M_\alpha$.

Let $\prec$ be a dual order on a well-ordered set I (i.e., with respect to the order $\prec$ the set I is totally ordered and any nonempty subset of I has the maximal element):

$$\ldots \prec \alpha_{\xi-1} \prec \alpha_\xi \prec \ldots \prec \alpha_{-1} \prec \alpha_0$$

For $f, g \in \bar{\prod\limits_{\alpha\in I}} M_\alpha$ we set $f < g$ if $f(\beta) < g(\beta)$, where β is the least element of the set $\{\alpha_\xi \in I \mid f(\alpha_\xi) \neq g(\alpha_\xi)\}$ with respect to the order $\prec$. The set $\bar{\prod\limits_{\alpha\in I}} M_\alpha$ with this order is called the *antilexicographic product* of M_α, $\alpha \in I$, and is denoted by $\overleftarrow{\prod\limits_{\alpha\in I}} M_\alpha$.

In the case of the finite set $I = \{1, 2, \ldots, n\}$ for partially ordered sets M_i, $i \in I$, we use the following notation:

$M_1 \times M_2 \times \ldots \times M_n$ is the Cartesian product,

$$M_1 \overrightarrow{\times} M_2 \overrightarrow{\times} \dots \overrightarrow{\times} M_n \text{ or } \overrightarrow{\prod_{1 \leqslant i \leqslant n}} M_\alpha \text{ is the lexicographic product,}$$

$$M_1 \overleftarrow{\times} M_2 \overleftarrow{\times} \dots \overleftarrow{\times} M_n \text{ or } \overleftarrow{\prod_{1 \leqslant i \leqslant n}} M_\alpha \text{ is the antilexicographic product.}$$

1.2. Lattices

1.2.1. Elements u and v of a partially ordered set M are called an *upper bound* and a *lower bound* of elements $x, y \in M$ if $x \leqslant u$, $y \leqslant u$ and $v \leqslant x$, $v \leqslant y$.

- A partially ordered set M is called
  - — *connected* if for any $a, b \in M$ there is a finite sequence $x_1 = a, x_2, \dots, x_n = b$ such that for any i, $1 \leqslant i \leqslant n-1$, the elements x_i and x_{i+1} are comparable,
  - — *directed up* or simply *directed* if each pair of elements of the set M has an upper bound in M,
  - — *directed down* if each pair of elements of the set M has a lower bound in M.

The Cartesian and lexicographic products of directed sets are directed. Therefore, $\mathcal{F}(X,Y)$ from Examples 1.1.1 and 1.1.2 are directed sets.

If, for elements x and y of a partially ordered set M, there is an upper bound a such that $a \leqslant z$ for any upper bound z of x, y, then a is called the *least upper bound* or the *join* of x, y and is denoted by $x \vee y$.

If for elements x, y there is a lower bound b such that $t \leqslant b$ for any lower bound t of x, y, then b is called the *greatest lower bound* or the *meet* of x, y and is denoted by $x \wedge y$.

It is clear that the join $x \vee y$ is unique provided that it exists. The same is true for the meet.

- A partially ordered set L is called a *lattice* if for any elements $x, y \in L$ there exist the join $x \vee y$ and the meet $x \wedge y$.

Any totally ordered set is a lattice.

EXAMPLE 1.2.1. The partially ordered set $\mathcal{F}(X,Y)$ of functions $f\colon X \to Y$ transforming a set X into a totally ordered set Y is a lattice with the coordinate-wise order on $\mathcal{F}(X, Y)$ and

$$(f \vee g)(x) = \begin{cases} g(x) & \text{if} \quad f(x) \leqslant g(x) \\ f(x) & \text{if} \quad g(x) \leqslant f(x) \end{cases}$$

$$(f \wedge g)(x) = \begin{cases} f(x) & \text{if} \quad f(x) \leqslant g(x) \\ g(x) & \text{if} \quad g(x) \leqslant f(x) \end{cases}$$

EXAMPLE 1.2.2. We denote by $\mathcal{C}(0,1)$ the set of all real-valued functions continuous on the interval $[\![0,1]\!]$. We note that $\mathcal{C}(0,1) \subseteq \mathcal{F}([\![0,1]\!], \mathbb{R})$. It is evident that $\mathcal{C}(0,1)$ is a lattice with order induced by the coordinate-wise order on the set $\mathcal{F}([\![0,1]\!], \mathbb{R})$ and $f \vee g$ and $f \wedge g$ (with respect to the order on $\mathcal{F}([\![0,1]\!], \mathbb{R})$) are continuous functions, i.e., $f, g \in \mathcal{C}(0,1)$ for any $x, y \in \mathcal{C}(0,1)$.

EXAMPLE 1.2.3. Let Ω be a totally ordered set and $\mathcal{A}(\Omega)$ be a set of all *automorphisms* (or *order permutations*) of Ω, i.e., the set of one-to-one mappings $f \in \mathcal{A}(\Omega)$ of Ω onto itself such that $\alpha f < \beta f$ for $\alpha, \beta \in \Omega$, $\alpha < \beta$. On $\mathcal{A}(\Omega)$ we introduce the relation $\leqslant$ by setting $f \leqslant g$ for $f, g \in \mathcal{A}(\Omega)$ if $\alpha f \leqslant \alpha g$ for any $\alpha \in \Omega$. We show that $\leqslant$ is a partial order on $\mathcal{A}(\Omega)$. Moreover, $\mathcal{A}(\Omega)$ is a lattice with the order $\leqslant$. Indeed, for arbitrary elements $f, g \in \mathcal{A}(\Omega)$ we construct a mapping a of Ω into Ω such that αa is the greatest element among αf and αg in Ω. This means that a is a mapping of Ω in Ω. If $\alpha < \beta$ in Ω, then $\alpha f < \beta f$, $\alpha g < \beta g$, and $\max\{\alpha f, \alpha g\} < \max\{\beta f, \beta g\}$, i.e., the mapping a is one-to-one and preserves the order on Ω. Further, if $\gamma \in \Omega$, then there are elements $\alpha, \beta \in \Omega$ such that $\alpha f = \gamma$ and $\beta g = \gamma$. Without loss of generality, we can assume that $\alpha \leqslant \beta$. Then $\alpha g \leqslant \beta g = \gamma = \alpha f$ and $\alpha a = \gamma$. This means that a is a mapping of Ω onto Ω, i.e., a is an automorphism of Ω. It is easy to verify that $a = f \vee g$ in $\mathcal{A}(\Omega)$. Arguing in a similar way, we can prove that the $b \colon \Omega \to \Omega$ defined by the formula $\alpha b = \min\{\alpha f, \alpha g\}$ is an element of $\mathcal{A}(\Omega)$ and $b = f \wedge g$.

Let A be an arbitrary subset of a partially ordered set M. By the *least upper bound* of A we mean an element $a \in M$ such that $x \leqslant a$ for any $x \in A$ and $a \leqslant u$ for any upper bound u of A. By the *greatest lower bound* of A we mean an element $b \in M$ such that $b \leqslant x$ for any $x \in A$ and $v \leqslant b$ for any lower bound v of A. The greatest lower bound and the least upper bound will be denoted by $a = \sup A = \vee A$ and $b = \inf A = \wedge A$, respectively.

- A lattice L is called
  - *complete* if there exist $\vee A$ and $\wedge A$ in L for any nonempty subset A of L,
  - *conditionally complete* if there exist $\vee A$ and $\wedge A$ for any bounded subset A of L.

1.2.2. We can assume that on a lattice L two binary operations are defined: the join $\vee$ and the meet $\wedge$. Therefore, the lattice L can be regarded as an algebraic system of signature $\{\vee, \wedge\}$. The class of lattices can be characterized by the following system of axioms, each of which is an identity of signature $\{\vee, \wedge\}$:

L.1 $$x \vee x = x, \quad x \wedge x = x$$

L.2 $$x \vee y = y \vee x, \quad x \wedge y = y \wedge x$$

L.3 $$(x \vee y) \vee z = x \vee (y \vee z), \quad (x \wedge y) \wedge z = x \wedge (y \wedge z)$$

L.4 $$(x \vee y) \wedge x = x, \quad (x \wedge y) \vee x = x$$

The relations L.1, L.2, L.3, and L.4 are called the laws of idempotency, commutativity, associativity, and absorption respectively.

Theorem 1.2.1. *The relations* L.1–L.4 *hold for any lattice L. Conversely, if L is an algebraic system of signature $\{\vee, \wedge\}$ such that the relations* L.1–L.4 *hold, then L is a lattice with partial order defined by the following rule:*

$$x \leqslant y \Leftrightarrow x \vee y = y$$

Under such an order $x \vee y$ is the least upper bound and $x \wedge y$ is the greatest lower bound of x and y.

The proof is standard and is left to the reader.

The greatest element of a lattice (if it does exist) is called the *unit* of the lattice and is denoted by $\mathbf{1}$. The least element of a lattice (if it does exist) is called the *null element* of the lattice and is denoted by $\mathbf{0}$.

1.3. Properties of Lattices

1.3.1. A subset A of a lattice L is called a *sublattice* of the lattice L if $x \vee y \in A$ and $x \wedge y \in A$ for any $x, y \in L$. It should be borne in mind that a subset of a lattice L can be a lattice itself with respect to the induced order, but at the same time cannot be a sublattice of the lattice L. It is easy to construct suitable examples.

A subset A of a partially ordered set M is called *convex* if for any elements $x \in M$, $a, b \in A$, the inequality $a \leqslant x \leqslant b$ implies the inclusion $x \in A$. If a convex subset A of a lattice L is directed up and down, then A is a sublattice of the lattice L.

A mapping φ: $M \to N$ of a partially ordered set M into a partially ordered set N is called an *order mapping* if $x\varphi \leqslant y\varphi$ for any elements $x, y \in M$, $x \leqslant y$. A one-to-one order mapping φ of a partially ordered set M onto a partially ordered set N is called an *isomorphism of partially ordered sets* if the inverse mapping φ^{-1} is also an order mapping. In this case partially ordered sets M and N are called *order-isomorphic*. Generally speaking, an order isomorphism φ of a lattice L into a lattice S does not necessarily preserve joins and meets of elements. We can only assert that $x\varphi \vee y\varphi \leqslant (x \vee y)\varphi$ and $x\varphi \wedge y\varphi \geqslant (x \wedge y)\varphi$.

In view of the aforesaid, it is expedient to introduce one more type of mappings of lattices.

- A mapping φ of a lattice L into a lattice S is called a *homomorphism* of L if for any $x, y \in L$

$$(x \vee y)\varphi = (x)\varphi \vee (y)\varphi, \quad (x \wedge y)\varphi = x\varphi \wedge y\varphi$$

Any homomorphism of a lattice is an order mapping. It is easy to see that a homomorphism of a lattice L into a lattice S maps L onto a sublattice of the lattice S.

- A one-to-one homomorphism of a lattice L onto a lattice S is called an *isomorphism* of L onto S. In this case the lattices L and S are called *isomorphic*.

1.3.2. With every index $\alpha \in I$ we associate a lattice L_α. We consider the set $\bar{\prod\limits_{\alpha \in I}} L_\alpha = \bar{L}$ of functions f mapping I into $\bigcup\limits_{\alpha \in I} L_\alpha$ such that

$\alpha f \in L_\alpha$. On $\overline{L}$ we define the operations $\vee$ and $\wedge$ by setting

$$\alpha(f \vee g) = \alpha f \vee \alpha g, \quad \alpha(f \wedge g) = \alpha f \wedge \alpha g$$

for $f, g \in \overline{L}$, $\alpha \in I$. It is easy to show that these operations on $\overline{L}$ satisfy L.1–L.4. Hence $\overline{L}$ is a lattice. We call this lattice the *Cartesian product* of L_α, $\alpha \in I$. We note that the Cartesian product of lattices L_α, $\alpha \in I$, regarded as a partially ordered set is the Cartesian product of the partially ordered sets L_α, $\alpha \in I$, in view of the definition from Section 1.1.2. We single out this assertion because of its importance.

Proposition 1.3.1. *The Cartesian product of lattices is a lattice.*

1.3.3. For lexicographic products (cf. Section 1.1.1) the situation is more complicated. It is not difficult to construct an example demonstrating that the class of lattices is not closed under lexicographic products.

Proposition 1.3.2. *The lexicographic product of lattices L_α, $\alpha \in I$, over a well-ordered set I is a lattice if and only if, for any lattice L_α containing incomparable elements, each lattice L_β, $\alpha < \beta$, contains the null element* **0** *and the unit* **1**.

The proof is left to the reader.

Proposition 1.3.3. *The lexicographic product of totally ordered sets over a well-ordered set is totally ordered.*

1.3.4. A lattice L is called *distributive* if the following relations (distributivity axioms) hold:

L.5 $\quad x \vee (y \wedge z) = (x \vee y) \wedge (x \vee z), \quad x \wedge (y \vee z) = (x \wedge y) \vee (x \wedge z)$

Simple computation shows that each of these identities implies another one. The lattices from Examples 1.2.1–1.2.3 are distributive. It is clear that any totally ordered set is a distributive lattice. The class of distributive lattices is closed under the Cartesian products, homomorphisms, and sublattices.

A lattice L is called *modular* if the following relation (modularity axiom) holds:

L.6 $\quad x \vee (y \wedge z) = (x \vee y) \wedge (x \vee z)$ for $x, y, z \in L$, $x \leqslant y$

It is easy to verify that a lattice is modular if and only if the following relation holds:

L.7 $$x \vee ((x \vee y) \wedge z) = (x \vee y) \wedge (x \vee z)$$

It is evident that any distributive lattice is modular.

The following criterion for distributivity of a lattice is often used.

Theorem 1.3.1. *A lattice L is distributive if and only if for any of its elements x, y, z the equalities $x \vee y = x \vee z$ and $x \wedge y = x \wedge z$ imply the equality $y = z$.*

Lattices, especially finite ones, are prescribed sometimes by means of graphs: elements of a lattice are represented as points and $x < y$ is marked as follows: the point x is placed below the point y and the points x and y are joined by a segment.

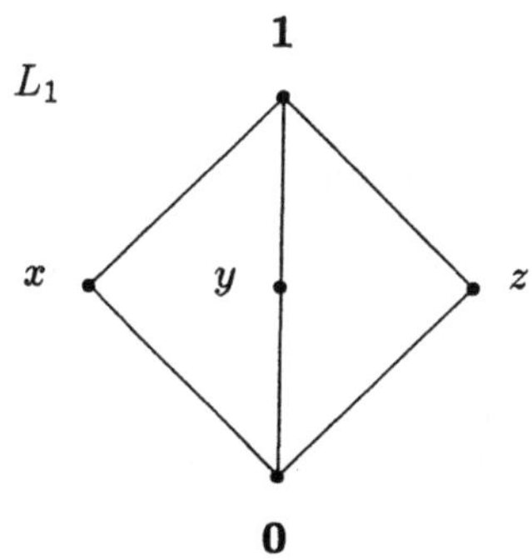

Fig. 1.3.1.

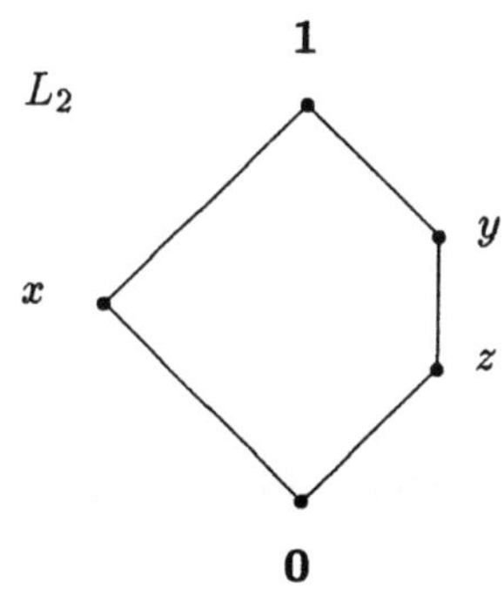

Fig. 1.3.2.

EXAMPLE 1.3.1. The lattice L_1 represented in Fig. 1.3.1 is not distributive.

EXAMPLE 1.3.2. The lattice L_2 represented in Fig. 1.3.2 is not distributive and not modular.

The following statement often turns out to be useful for verifying the distributivity and modularity of a lattice.

Theorem 1.3.2. *A lattice L is not distributive if and only if it contains a sublattice isomorphic to the lattice L_1 or L_2* (see Figs. 1.3.1 and 1.3.2).

A lattice L is not modular if and only if it contains a sublattice isomorphic to the lattice L_2 (see Fig. 1.3.2).

The significance of modular lattices is demonstrated by the following Dedekind theorem, which is important for the group theory.

Theorem 1.3.3. *The lattice of normal subgroups of a group G is modular.*

PROOF. By L.7, it suffices to establish the equality $XY \cap XZ = (XY \cap Z)X$ for arbitrary normal subgroups X, Y, and Z of a group G. If $a \in XY \cap XZ$ and $a = uy = vz$ for $u, v \in X$, $y \in Y$, $z \in Z$, then $z = v^{-1}uy$, $z \in XY \cap Z$, and $a = vv^{-1}y \in X(XY \cap Z)$, i.e., $XY \cap XZ \subseteq (XY \cap Z)X$. Conversely, let $a \in X(XY \cap Z)$, $a = xh$, where $x \in X$ and $h \in XY \cap Z$. Then $h \in Z$ and $a \in XY$. Since $h \in XY$, we have $h = uv$, where $u \in X$, and $v \in Y$, i.e., $a = xuv \in XY$. This means that the following relations hold: $a \in XY \cap XZ$, $X(XY \cap Z) \subseteq XY \cap XZ$, and $X(XY \cap Z) = XY \cap XZ$. □

1.3.5. In this section, we give some definitions.

- A lattice L is called
  - *Brouwerian* if it is complete and for any $x, y_\alpha \in L$, $\alpha \in I$, the infinite distributivity identity is fulfilled:

$$x \wedge \left(\bigvee_{\alpha \in I} y_\alpha \right) = \bigvee_{\alpha \in I} (x \wedge y_\alpha)$$

  - *dually Brouwerian* if it is complete and the duality identity is fulfilled:

$$x \vee \left(\bigwedge_{\alpha \in I} y_\alpha \right) = \bigwedge_{\alpha \in I} (x \vee y_\alpha)$$

  - *completely distributive* if

$$\bigvee_{\alpha \in A} \left(\bigwedge_{\beta \in B} x_{\alpha,\beta} \right) = \bigwedge_{\rho \in A^B} (\bigvee_{\alpha \in A} x_{\alpha,\rho})$$

    where A and B stand for sets of indices and A^B denotes the set of all functions from A into B.

1.3.6. Let L be a lattice with null element $\mathbf{0}$ and unit $\mathbf{1}$. An element x' of the lattice L is called the *complement* to an element $x \in L$ if $x \vee x' = \mathbf{1}$ and $x \wedge x' = \mathbf{0}$. A distributive lattice B is called a *Boolean lattice* if B contains $\mathbf{0}$, $\mathbf{1}$, and the complement x' for any element $x \in B$.

By Theorem 1.3.1, the complement x' to the element x is unique in each Boolean lattice. Therefore, on B we can introduce the unary operation $'$ satisfying the following relations:

B.1
$$(x')' = x$$

B.2
$$(x \vee y)' = x' \wedge y', \quad (x \wedge y) = x' \vee y'$$

B.3
$$(x \vee x') \wedge y = y, \quad (x \wedge x') \vee y = y$$

A Boolean lattice B can be regarded as an algebraic system of signature $\{\vee, \wedge, '\}$ such that L.1–L.5 and B.1–B.3 hold. The algebraic system is called a *Boolean algebra*. A subset S of a Boolean algebra B is called a *subalgebra* of the algebra B if it is closed under the operations $\vee$, $\wedge$, and $'$. A mapping φ: $B \to C$ is called a *homomorphism* of the Boolean algebra B into the Boolean algebra C if for any $x, y \in B$

$$(x \vee y)\varphi = x\varphi \vee y\varphi, \quad (x \wedge y)\varphi = x\varphi \wedge y\varphi, \quad (x')\varphi = (x\varphi)'$$

If φ is a one-to-one homomorphism of a Boolean algebra B onto a Boolean algebra C, then φ is called a *Boolean isomorphism.*

EXAMPLE 1.3.3. The set $\mathcal{S}(M)$ of all subsets of a set M is a Boolean algebra with the set-theoretical operations of union, intersection, and complement. As the null element in $\mathcal{S}(M)$ we can take the empty set $\varnothing$ and as the unit we can take the set M.

The following theorem shows the importance of the notion of the Boolean algebra $\mathcal{S}(M)$ for the lattice theory.

Theorem 1.3.4 (Birkhoff [1] and Stone [2]). *Any distributive lattice (Boolean algebra) is isomorphically embedded into the Boolean lattice (Boolean algebra) $\mathcal{S}(M)$ of a set M. If the Boolean algebra is finite, then it is isomorphic to the Boolean algebra of subsets of a finite set.*

The proof can be found, for example, in [1, 2].

1.4. Orders on Groups

1.4.1. By *a partially right-ordered group* (*pro-group*) we mean a nonempty set G with two binary operations $\cdot$ and $\leqslant$ such that $\{G; \cdot\}$ is a group, $\{G; \leqslant\}$ is a partially ordered set, and the following relation holds:

M.1 $$x \leqslant y \Rightarrow xz \leqslant yz \text{ for any } x, y, z \in G$$

A group G is called *partially left-ordered* if a partial order $\leqslant$ is given on G and the following relation holds:

M.2 $$x \leqslant y \Rightarrow zx \leqslant zy \text{ for any } x, y, z \in G$$

If a group G is partially right-ordered and partially left-ordered with respect to some partial order simultaneously, then it is simply called *partially ordered* (*po-group*).

- A partially ordered (partially right-ordered) group G is called
  - — *connected* if the order is a connected partially ordered set,
  - — *directed* if the order is a directed set,
  - — *lattice-ordered* or an *l-group* if the order is a lattice,
  - — *totally ordered* or *o-group* (*right-ordered* or *ro-group*) if the order is a totally ordered set.

Any l-group G is an algebraic system of signature $l = \{\cdot, ^{-1}, e, \vee, \wedge\}$, where $x \vee y$ is the least upper bound and $x \wedge y$ is the greatest lower bound of the elements $x, y \in G$. Henceforth, we denote by e the identity element of a group.

Proposition 1.4.1. *Let G be a partially ordered group. Then for each of its elements x, y*

$$x \leqslant e \Rightarrow e \leqslant x^{-1} \tag{1.4.1}$$

$$x \leqslant y \Rightarrow xy^{-1}x^{-1} \leqslant x^{-1} \tag{1.4.2}$$

If G is partially ordered, then for all $x, y, z, t \in G$ the following relations also hold:

$$x \leqslant y \Rightarrow y^{-1} \leqslant x^{-1} \tag{1.4.3}$$

$$x \leqslant y, \quad z \leqslant t \Rightarrow xz \leqslant yt \tag{1.4.4}$$

Moreover, if the group G is lattice-ordered, then the following identities hold:

$$x(y \vee z)t = xyt \vee xzt \tag{1.4.5}$$

$$x(y \wedge z)t = xyt \wedge xzt \tag{1.4.6}$$

$$(x \vee y)^{-1} = x^{-1} \wedge y^{-1} \tag{1.4.7}$$

$$(x \wedge y)^{-1} = x^{-1} \vee y^{-1} \tag{1.4.8}$$

$$x \vee (y \wedge z) = (x \vee y) \wedge (x \vee z) \tag{1.4.9}$$

$$x \wedge (y \vee z) = (x \wedge y) \vee (x \wedge z) \tag{1.4.10}$$

PROOF. If $x \leqslant e$ in a partially right-ordered group G, then we multiply the last inequality from the right by x^{-1} and obtain $e \leqslant x^{-1}$. The relation (1.4.1) is proved. If $x \leqslant y$ in G, then $xy^{-1} = xy^{-1}x^{-1}x \leqslant e$. Hence $xy^{-1}x^{-1} \leqslant x^{-1}$ and the relation (1.4.2) is proved. The relation (1.4.3) is obtained by multiplication of the inequality $x \leqslant y$ from the left by x^{-1} and from the right by y^{-1}. The relation (1.4.4) is obtained from the following obvious property: if $x \leqslant y$ and $z \leqslant t$, then $xz \leqslant yz$ and $yz \leqslant yt$. Consequently, $xz \leqslant yt$. The proof of (1.4.5)–(1.4.10) is left to the reader as an exercise. □

Proposition 1.4.2. *Each positive element of a partially right-ordered group generates an infinite cyclic group. Each right-ordered group is a torsion-free group.*

PROOF. Let x be a positive element of a partially right-ordered group G and $x > e$. We assume that for some natural number n the inequality $x^n > e$ holds. Multiplying it from the right by x, we obtain $x^{n+1} \geqslant x > e$, which proves the proposition. □

1.4.2. The additive groups $\mathbb{Z}$ (integers), $\mathbb{Q}$ (rational numbers), and $\mathbb{R}$ (real numbers) are the simplest examples of totally ordered groups. As usual, we denote by $\mathbb{N}$ the set of natural numbers.

EXAMPLE 1.4.1. Let $\{\mathbb{C}; +\}$ be the additive group of complex numbers $x + iy$, where $x, y \in \mathbb{R}$. On $\{\mathbb{C}; +\}$ we introduce the following partial order: $x + iy \leqslant u + iv$ if $x \leqslant u$ and $y \leqslant v$ with respect to the standard order on $\mathbb{R}$. Then $\{\mathbb{C}; +, \leqslant\}$ is an l-group in which

$(x + iy) \vee (u + iv) = a + ib$ and $(x + iy) \wedge (u + iv) = c + id$, where $a = \max(x, u)$, $b = \max(y, v)$, $c = \min(x, u)$, and $d = \min(y, v)$ in $\mathbb{R}$.

EXAMPLE 1.4.2. Let $\{\mathbb{C}; +\}$ be the additive group of complex numbers. On $\{\mathbb{C}; +\}$ we introduce the following total order: $x + iy \leqslant u + iv\}$ if $x < u$ or $x = u$, $y \leqslant v$. Then $\{\mathbb{C}; +, \leqslant\}$ is a totally ordered group.

EXAMPLE 1.4.3. Let $\{\mathbb{Q}^+; \cdot\}$ be the multiplicative group of positive rational numbers. On $\{\mathbb{Q}^+; \cdot\}$ we define the following order: $r \leqslant s$ if $r^{-1}s \in \mathbb{Z}$. Then $\{\mathbb{Q}^+; \cdot, \leqslant\}$ is an l-group.

EXAMPLE 1.4.4. For a totally ordered set Γ we denote by $\mathcal{A}(\Gamma)$ the set of all one-to-one mappings $f\colon \Gamma \to \Gamma$ such that $\alpha f \leqslant \beta f$ for any $\alpha, \beta \in \Gamma$, $\alpha \leqslant \beta$. It is easy to see that $\mathcal{A}(\Gamma)$ is a group with the composition $\alpha fg = (\alpha f)g$ for $\alpha \in \Gamma$ and $f, g \in \mathcal{A}(\Gamma)$. In Example 1.2.3, it is shown that $\mathcal{A}(\Gamma)$ is a lattice with the following partial order: $f \leqslant g$ if $\alpha f \leqslant \alpha g$ for any $\alpha \in \Gamma$. We now show that $\mathcal{A}(\Gamma)$ is an l-group with respect to the order just introduced.

Let $f, g, h \in \mathcal{A}(\Gamma)$, $g \leqslant f$. Then $\alpha g \leqslant \alpha f$ and $\alpha gh \leqslant \alpha fh$ for any $\alpha \in \Gamma$, i.e., $gh \leqslant fh$. On the other hand, $\alpha h = \alpha h$, $\alpha hg \leqslant \alpha hf$, and $hg \leqslant hf$. Therefore, $\mathcal{A}(\Gamma)$ is an l-group.

EXAMPLE 1.4.5. Let $\mathcal{A}(\Gamma)$ be the group of all automorphisms of a totally ordered set Γ with the order $\leqslant$. We well order somehow the set Γ by the relation $\prec$. On $\mathcal{A}(\Gamma)$ we introduce the following order: $f < g$ if $\alpha f < \alpha g$, where $\alpha \in \Gamma$ is such that $\xi f = \xi g$ for any $\xi \in \Gamma$, $\xi \prec \alpha$. Then $\mathcal{A}(\Gamma)$ is a right-ordered group. Indeed, as was shown in Example 1.1.2, the lexicographic order defined on $\mathcal{A}(\Gamma)$ is total. Let $g, h \in \mathcal{A}(\Gamma)$, $g < h$, and let the least element α of the set Γ with respect to $\prec$ be such that $\alpha g \neq \alpha h$. Then $\alpha g < \alpha h$ and $\xi g = \xi h$ for any $\xi \in \Gamma$, $\xi \prec \alpha$. Therefore, for $f \in \mathcal{A}(\Gamma)$ we obtain $\alpha gf < \alpha hf$ and $\xi gf = \xi hf$ for any $\xi \in \Gamma$, $\xi \prec \alpha$, i.e., the relation M.1 holds on the group $\mathcal{A}(\Gamma)$. On $\mathcal{A}(\Gamma)$ the relation M.2 holds not for every set Γ. For example, as Γ we take the set $\mathbb{Q}$ and well order it so that the relation $\alpha_0 = 0 \prec \alpha_1 = 1$ is valid in $\mathbb{Q}$. Then we order the remainder of Γ in an arbitrary way. We consider the total right order on the group $\mathcal{A}(\mathbb{Q})$ corresponding to the introduced relation $\prec$. For elements f and g such that $f(x) = x/3 + 1/3$ and $g(x) = x + 1$ we have $f > e$ and $g \cdot f < g$, i.e., the constructed right order on $\mathcal{A}(\mathbb{Q})$ is not two-sided.

EXAMPLE 1.4.6. Let a group G be defined by generators and relations:

$$G = \text{gp}\,\{a, b, c \mid ab = ba,\ c^{-1}ac = b,\ c^{-1}bc = a\}$$

Then any element $x \in G$ is uniquely represented as $x = c^m a^n b^k$, where $m, n, k \in \mathbb{Z}$. On G we introduce a partial order as follows: $c^m a^n b^k \leqslant c^p a^q b^r$ if $m < p$ or $m = p$, $n \leqslant q$, $k \leqslant r$. The group $\{G; \cdot, \leqslant\}$ is an l-group.

EXAMPLE 1.4.7. Let X be a set and $G = \mathcal{F}(X, \mathbb{R})$ be the set of all functions $f\colon X \to \mathbb{R}$ with the coordinate-wise order (cf. Example 1.1.1). Then G is an l-group with the coordinate-wise order and the addition operation $(f + g)(x) = f(x) + g(x)$.

EXAMPLE 1.4.8. The additive group of all differentiable real-valued functions on the interval $[\![0, 1]\!]$ is a directed partially ordered group but not an l-group with respect to the coordinate-wise order.

1.5. Positive Cones

1.5.1. Let G be a partially right-ordered group. Recall that the letter e stands for the identity element of G.

- An element $x \in G$ is called
  - *positive* (*strictly positive*) if $x \geqslant e$ $(x > e)$,
  - *negative* (*strictly negative*) if $x \leqslant e$ $(x < e)$.
- The set of positive elements of a partially right-ordered group G is called the *positive cone*.

Parallel with the positive cone of a partially right-ordered group it is sometimes useful to utilize the notion of negative cone, i.e., the set of all negative elements of the group.

Theorem 1.5.1. *For a partially right-ordered group G with a positive cone P, the following relations hold:*

O.1 $\qquad P{\cdot}P \subseteq P \quad$ (*P is a semigroup*)

O.2 $\qquad P \cap P^{-1} = \{e\} \quad$ (*P is a pure set*)

If G is a connected partially right-ordered group, then

O.3 $$G \text{ is generated by its positive cone } P$$

If G is a directed partially right-ordered group, then

O.4 $$G = P^{-1} \cdot P$$

If G is a right-ordered group, then

O.5 $$G = P \cup P^{-1}$$

Conversely, if in a group G there is a subset P satisfying O.1, O.2, *then it is possible to introduce a partial order $\leqslant$ on G such that $\{G; \cdot, \leqslant\}$ is a partially right-ordered group with the positive cone P. If P also satisfies* O.3 (O.4 *or* O.5), *then G is a connected partially right-ordered group (directed partially right-ordered group or right-ordered group).*

PROOF. Let G be a partially right-ordered group and P be its positive cone. If $x, y \in P$, then $x \geqslant e$, $y \geqslant e$, and $xy \geqslant e$, i.e., $xy \in P$, $P \cdot P \subseteq P$. If $x \in P$, $x^{-1} \in P$, then $x \geqslant e$, $x^{-1} \geqslant e$. Hence $x = e$ and $P \cap P^{-1} = \{e\}$.

If G is a connected right-ordered group and $a \in G$, then there are elements $x_1 = a, x_2, \ldots, x_n = e$ such that $x_i \leqslant x_{i+1}$ or $x_i \geqslant x_{i+1}$ for any i, $1 \leqslant i \leqslant n-1$. Therefore,

$$a = (x_1 x_2^{-1}) \cdot (x_2 x_3^{-1}) \ldots (x_{n-1} x_n^{-1}) \cdot x_n$$

where all elements $x_i x_{i+1}$, $i = 1, 2, \ldots, n-1$, are comparable with e and $x_n = e$. Consequently, the element a lies in the subgroup generated by the cone P and the relation O.3 holds.

If G is a directed right-ordered group, then for any element $x \in G$ there is an element $y \in G$ such that $x \leqslant y$ and $e \leqslant y$. Therefore, $xy^{-1} \leqslant e$ and $x = xy^{-1}y$, where $xy^{-1} \in P^{-1}$, $y \in P$, i.e., the relation O.4 holds.

It is clear that the relation O.5 holds for the positive cone P of the right-ordered group G.

Conversely, let G be a group and P be its subset satisfying O.1 and O.2. We define the relation $\leqslant$ on G by setting $x \leqslant y$ for $x, y \in G$ if $yx^{-1} \in P$. Since $e \in P$, we have $e = x^{-1}x$ and $x \leqslant x$. If $x \leqslant y$ and $y \leqslant x$, then $yx^{-1} \in P$ and $xy^{-1} = (yx^{-1})^{-1} \in P$, i.e., $yx^{-1} \in P \cap P^{-1}$, $yx^{-1} = e$, $x = y$. Finally, if $x \leqslant y$ and $y \leqslant z$, then $yx^{-1} \in P$ and $zy^{-1} \in P$. By O.1, we obtain $zy^{-1}yx^{-1} \in P$, i.e., $zx^{-1} \in P$ and $x \leqslant z$. Hence $\leqslant$ is a partial order.

Let us show that the relation $\leqslant$ satisfies M.1. If $x, y, z \in G$, $x \leqslant y$, then $yx^{-1} \in P$ and $yz \cdot z^{-1}x^{-1} \in P$, $yz \cdot (xz)^{-1} \in P$, $xz \leqslant yz$. This means that G is a partially right-ordered group with the positive cone P.

Let the group G be generated by its positive cone P, i.e., the relation O.3 holds. We show that the partial order on G defined by the positive cone P is connected. Let $a, b \in G$. Then the element ab^{-1} is the product of elements comparable with the identity element e: $ab^{-1} = y_1 y_2 \dots y_{n-1}$. Thus, $a = y_1 y_2 \dots y_n b$. Let $x_n = b, x_{n-1} = y_{n-1}x_n, \dots, x_i = y_i x_{i+1}, \dots, x_1 = y_1 y_2$. It is evident that $x_1 = y_1 y_2 \dots y_{n-1} b = a$ and $x_i x_{i+1}^{-1} = y_i \in P \cup P^{-1}$, $i = 1, 2, \dots n$, which indicates the connectedness of the partially ordered set G.

Let the positive cone P of the partially right-ordered group G satisfy O.4. First, we establish that for any element $x \in G$ there is an element y such that $x \leqslant y$ and $e \leqslant y$. In fact, since $G = P^{-1} \cdot P$, we have $x = t^{-1}z$ for some elements $z, t \in P$. Let $y = tx = z$. Then $y \geqslant e$ and $yx^{-1} = t \in P$, i.e., $yx^{-1} \geqslant e$ and $y \geqslant x$. We now show that for any elements $x, y \in G$ there is an element z such that $x \leqslant z$ and $y \leqslant z$. By the property of the element xy^{-1} we find an element $z_1 \in P$ such that $xy^{-1} \leqslant z_1$ and $e \leqslant z_1$. Multiplying the inequalities $e \leqslant z_1$ and $xy^{-1} \leqslant z_1$ from the right by y, we obtain $y \leqslant z_1 \cdot y = z$ and $x \leqslant z_1 \cdot y = z$. Thus, we have proved that the partial right order on the group G with the positive cone is directed.

The verification of the fact that the relation O.5 is equivalent to the totality property of the partial right order on a group (this order must not necessarily be a two-sided one) is left to the reader. □

1.5.2. In view of the particular importance of the notion of a partially ordered group, we single out the properties of positive cones of partially ordered groups into a special statement.

Theorem 1.5.2. *A partially right-ordered group G with positive cone P is a partially ordered group if and only if P satisfies* O.1, O.2, *and the following relation:*

O.6 $$x^{-1} \cdot P \cdot x \subseteq P \quad \text{for any } x \in G$$

(A subset of a group satisfying O.6 is called *normal.*)

PROOF. We show that the positive cone P of a partially ordered group G is a normal set. Indeed, if $a \in P$ and $x \in G$, then, by M.1

and M.2, the following inequalities hold:

$$a \geqslant e, \quad ax \geqslant x, \quad x^{-1}ax \geqslant e$$

which proves the validity of O.6. Conversely, if in the group G there is a subset P satisfying O.1, O.2, and O.6, then, by Theorem 1.5.1, the set P defines a partial right order $\leqslant$ on G. We show that this partially right-ordered group satisfies the relation M.2. Indeed, if $x \leqslant y$ and $x, y, z \in G$, then $yx^{-1} \in P$. By O.6, we have $z \cdot yx^{-1} \cdot z^{-1} \in P$, i.e., $zy \cdot x^{-1}z^{-1} = zy \cdot (zx)^{-1} \in P$, $zx \leqslant zy$. □

It is clear that the connection between the properties of a partial order and the relations O.3–O.6 for the positive cone, which was established in Theorem 1.5.1, takes place for a partially ordered group. It is also obvious that any connected partially ordered group is directed up and down. The assertion is false for partially right-ordered groups: there exist partially right-ordered groups directed up but not directed down. The corresponding examples are given in Chapter 8.

The following assertion is useful in proving the fact that a partial order on a group is a lattice order.

Proposition 1.5.1. *A partially ordered group G with positive cone P is a lattice-ordered group if and only if it is directed and P is a lattice with respect to the induced order.*

PROOF. Let G be an l-group, P be its positive cone, and $x, y \in P$. Then $x \geqslant e$, $y \geqslant e$ and $x \vee y \geqslant e$, $x \wedge y \geqslant e$, i.e., $x \vee y$, $x \wedge y \in P$. Hence P is a sublattice of the lattice G. Conversely, let G be a partially ordered group and its positive cone P be a lattice with respect to the induced order. Then the lattice operations $\wedge$ and $\vee$ can be introduced in P. We now show that in this case for any elements $x, y \in G$ their least upper and greatest lower bounds exist. Since G is a directed partially ordered group, it is directed down in view of (1.4.3). Therefore, for any elements $x, y \in G$ there is an element $z \in G$ such that $x^{-1} \leqslant z^{-1}$, $y^{-1} \leqslant z^{-1}$, $e \leqslant z^{-1}x$, and $e \leqslant z^{-1}y$. Since P is a lattice, the elements $u = z^{-1}x \vee z^{-1}y$ and $v = z^{-1}x \wedge z^{-1}y$ belong to P. Let us prove that zu is the least upper bound and zv is the greatest lower bound of the elements $x, y \in G$. Indeed, since $z^{-1}x \leqslant u$ and $z^{-1}y \leqslant u$, we have $x \leqslant zu$ and $y \leqslant zu$, i.e., zu is the least upper bound of x and y in G. Furthermore, if $x \leqslant t$ and $y \leqslant t$ in G, then $e \leqslant z^{-1}x \leqslant z^{-1}t$ and $e \leqslant z^{-1}y \leqslant z^{-1}t$, i.e., $z^{-1}t$ is an upper bound of $z^{-1}x$ and $z^{-1}y$ in P. Since u is the least upper bound of $z^{-1}x$ and $z^{-1}y$ in P, we have

$u \leqslant z^{-1}t$ and $zu \leqslant t$. Consequently, $zu = x \vee y$ in G. The proof that $zv = x \wedge y$ in G is similar. □

According to Theorem 1.5.1, we identify a partial right order $\leqslant$ on a group G with its positive cone P and denote by (G, P) the partially right-ordered group G with positive cone P and by $P(G)$ the positive cone.

Among all partial right orders on a group we distinguish the *trivial* partial order, i.e., the partial order such that $x \leqslant y$ if and only if $x = y$. It is clear that the positive cone $P(G)$ of a trivially ordered group G consists of only the identity element of the group: $P(G) = \{e\}$.

1.6. Basic Notions

1.6.1. It is clear that a subgroup of a partially right-ordered group is itself a partially right-ordered group with respect to the induced order. If G is a partially ordered group, then each of its subgroups is also a partially ordered group. If G is a right-ordered or totally ordered group, then any of its subgroups is right-ordered or totally ordered respectively. Other properties of the order are not always transferable to a subgroup. For example, a subgroup of a directed partially right-ordered group can be not directed and a subgroup of an l-group can be not an l-group.

EXAMPLE 1.6.1. Let $G = A \times B$ be the direct product of nontrivial totally ordered groups A and B. If on G we introduce the partial order

$$P(G) = \{x = ab \mid a \in A, b \in B, a \geqslant e \text{ in } A, \ \ b \geqslant e \text{ in } B\}$$

then G becomes an l-group, which can be easily verified. It is not hard to see that the subgroup $H = \operatorname{gp}(ab^{-1})$ of G generated by the element ab^{-1}, where $a > e$ in A and $b > e$ in B, is a trivially ordered group and, consequently, is neither a directed group nor an l-group.

In an l-group it is natural to single out subgroups which are sublattices of the l-group. Such subgroups are called l-subgroups. In other words, a subgroup H of an l-group G is called an *l-subgroup* of the group G if for any elements $x, y \in H$ their join $x \vee y$ and meet $x \wedge y$ belong to H. Each l-subgroup of an l-group is a subalgebra of the l-group regarded as an algebraic system of signature $l = \{\cdot, {}^{-1}, e, \vee, \wedge\}$.

Proposition 1.6.1. *A subgroup H of an l-group G is an l-subgroup if and only if $x \vee e \in H$ for any $x \in H$.*

PROOF. If H is an l-subgroup of an l-group G, then $x \vee e \in H$ for any $x \in H$ by definition. Conversely, let H be a subgroup of an l-group G such that $x \vee e \in H$ for any $x \in H$. We choose elements $a, b \in H$ and note that $c = ab^{-1} \vee e \in H$. Therefore, $a \vee b = cb \in H$. The aforesaid and (1.4.7), (1.4.8) mean that the subgroup H is closed under the meet. □

1.6.2. In the theory of partially right-ordered groups, the notion of a convex subgroup is crucial.

- A subgroup H of a partially right-ordered group G is called *convex* if it is a convex subset of the partially ordered set $\{G; \leqslant\}$, i.e., for any $x, y, z \in G$ such that $x, y \in H$ and $x \leqslant z \leqslant y$, we have $z \in H$.

It is easy to see that the intersection of any set of convex subgroups of a partially right-ordered group is a convex subgroup.

- A convex, normal, directed up, and directed down subgroup of a directed partially right-ordered group is called an *ideal.*

From this definition it follows immediately that a subgroup H of an l-group G is an ideal if and only if H is a convex normal subgroup of the group G and a sublattice simultaneously.

1.6.3. If H is a convex normal subgroup of a partially right-ordered group G, then on the quotient group G/H we can introduce the partial order $\leqslant$ by setting $Hx \leqslant Hy$ for cosets Hx and Hy if there exists an element $h \in H$ such that $hx \leqslant y$. We now show that this relation is a partial order on G/H.

It is clear that $Hx \leqslant Hx$ for any $x \in H$. If $Hx \leqslant Hy$ and $Hy \leqslant Hx$, then there are elements $u, v \in H$ such that $ux \leqslant y$ and $vy \leqslant x$. Therefore, the inequalities $u \leqslant yx^{-1}$ and $vyx^{-1} \leqslant e$ are valid. From the second inequality we obtain

$$vyx^{-1} \leqslant e$$

$$vux^{-1} = yx^{-1}(yx^{-1})^{-1}v(yx^{-1}) = yx^{-1} \cdot w \leqslant e$$

where $w \in H$, because H is a normal subgroup. However, this means that $yx^{-1} \leqslant w^{-1}$. Taking into account the above inequality, we get $u \leqslant yx^{-1} \leqslant w^{-1}$. The convexity of H guarantees the validity of the inclusion $xy^{-1} \in H$ and equality $Hx = Hy$.

Further, if $Hx \leqslant Hy$ and $Hy \leqslant Hz$, then in H there are elements u and v such that $ux \leqslant y$ and $vy \leqslant z$. Taking into account the equalities $vy = y \cdot y^{-1}vy = yw$, where $w = y^{-1}vy \in H$, and multiplying the inequality $ux \leqslant y$ from the right by w, we obtain the following relation:

$$uxw = uxy^{-1}vy = u \cdot xy^{-1}vyx^{-1} \cdot x \leqslant yw = y \cdot y^{-1}vy = vy \leqslant z$$

In view of the inclusion $u \cdot xy^{-1}vyx^{-1} \in H$ we have $Hx \leqslant Hz$. Hence $\leqslant$ is a partial order on the quotient group G/H.

Proposition 1.6.2. *If H is a convex normal subgroup of a partially right-ordered group G, then on the quotient group G/H we can introduce a partial order induced by the partial order on G and such that G/H becomes a partially right-ordered group. Moreover, if the group G is partially ordered, then the group G/H is also partially ordered.*

PROOF. The partial order on the quotient group G/H turns out to be stable under multiplication from the right. Indeed, if $Hx \leqslant Hy$ and $Hx, Hy, Hz \in G/H$, then $hx \leqslant y$ for some element $h \in H$, $hxz \leqslant hyz$. Consequently, $Hx{\cdot}Hz = Hxz \leqslant Hyz = Hy{\cdot}Hz$.

In a similar way, we can show that G/H is a partially ordered group provided that the group G is partially ordered. □

The partially right-ordered group constructed in Proposition 1.6.1 is called the *quotient group of the partially right-ordered group G by the convex subgroup H.*

We emphasize that without additional assumptions on the convex subgroup H some properties of the group G disappear after passage to the quotient group G/H. For example, the quotient group of an l-group by a convex normal subgroup can be not an l-group.

Proposition 1.6.3. *The quotient group of an l-group G by an ideal H is an l-group.*

PROOF. Let $Hx, Hy \in G/H$. We consider the element $x \vee y$ of the group G, where x and y are representatives of the cosets Hx and Hy respectively. It is evident that $Hx \leqslant H(x \vee y)$ and $Hy \leqslant H(x \vee y)$. We assume that for the coset Hz the inequalities $Hx \leqslant Hz$ and $Hy \leqslant Hz$

hold in G/H. Then in H there are elements u and v such that $ux \leqslant z$ and $vy \leqslant z$. Since H is an l-subgroup of the group G, there is an element $h \in H$ such that $h \leqslant u$ and $h \leqslant v$. Since the order on an l-group is two-sided, we deduce that $hx \leqslant z$ and $hy \leqslant z$. Hence $h(x \vee y) = hx \vee hy \leqslant z$, i.e., $H(x \vee y) \leqslant Hz$ in G/H. This means that $H(x \vee y)$ is the least upper bound of Hx and Hy.

Arguing in a similar way, we can prove that $H(x \wedge y)$ is the greatest lower bound of Hx and Hy in G/H. □

1.6.4. A mapping φ of a partially right-ordered group G into a partially right-ordered group H is called an *order homomorphism* (*o-homomorphism*) if φ is a homomorphism of G into H and preserves the partial order. In other words, a mapping φ: $G \to H$ is an o-homomorphism if for any $x, y \in G$

$$(xy)\varphi = x\varphi \cdot y\varphi \tag{1.6.1}$$

$$x \leqslant y \Rightarrow x\varphi \leqslant y\varphi \tag{1.6.2}$$

If G and H are partially right-ordered groups, φ is an isomorphism of the group G into the group H, and φ is an isomorphism of the partially ordered sets G and $G\varphi$, then φ is called an *order isomorphism* (*o-isomorphism*). The fact that G and H are isomorphic partially right-ordered groups will be denoted as follows: $G \cong H$.

Proposition 1.6.4. *A homomorphism φ of a partially right-ordered group (G, P) into a partially right-ordered group (H, Q) is an o-homomorphism if and only if $P\varphi \subseteq Q$.*

PROOF. For an o-homomorphism φ the inclusion $P\varphi \subseteq Q$ is obvious. Conversely, let φ be a homomorphism of a group G into a group H such that $P\varphi \subseteq Q$ and $x \leqslant y$ in G. Then $yx^{-1} \in P$, $(yx^{-1})\varphi \in Q$, and $(yx^{-1})\varphi = (y\varphi)(x\varphi)^{-1} \in Q$, $x\varphi \leqslant y\varphi$ in H. □

EXAMPLE 1.6.2. Let A be the additive group of integers with the trivial order and B be the additive group of integers with the natural total order. If φ: $A \to B$ is the natural homomorphism of A onto B, then φ is an o-homomorphism, but not an o-isomorphism.

An order homomorphism φ: $(G, P) \to (H, Q)$ of partially ordered groups (G, P) and (H, Q) is called *strict* if $P\varphi = Q \cap G\varphi$. It is clear that every o-isomorphism is strict. The converse assertion is also true:

every strict o-homomorphism of a partially right-ordered group is an o-isomorphism, provided that it is an isomorphism of the groups.

To study l-groups we need one more notion.

- A homomorphism $\varphi\colon G \to H$ of an l-group G into an l-group H such that

$$(x \vee y)\varphi = x\varphi \vee y\varphi \qquad (1.6.3)$$

$$(x \wedge y)\varphi = x\varphi \wedge y\varphi \qquad (1.6.4)$$

  for all $x, y \in G$ is called an *l-homomorphism* (*homomorphism of l-groups*).

In a similar way, by an *l-isomorphism* (*isomorphism of l-groups*) we mean an l-homomorphism that is an isomorphism of groups. Any l-homomorphism is strict.

- A partially right-ordered group is called *order-simple* if any of its o-homomorphisms is either an o-isomorphism or an o-homomorphism onto the trivial partially right-ordered group E.

- A lattice-ordered group G is called *l-simple* (a *simple l-group*) if any l-homomorphism of G is either an l-isomorphism or an l-homomorphism of G onto E.

If H is a convex normal subgroup of a partially right-ordered group G, then the natural homomorphism of G onto the introduced partially right-ordered group G/H is a strict order homomorphism. The assertion follows immediately from Proposition 1.6.4. Such an o-homomorphism is called *natural.*

For partially right-ordered groups the following theorem on homomorphisms holds.

Theorem 1.6.1. *Let $\varphi\colon G \to H$ be an order homomorphism of a partially right-ordered group G into a partially right-ordered group H. Then the kernel N of the homomorphism φ is a normal convex subgroup of the group G and there exists an order isomorphism ψ of the quotient group G/N with the natural order onto H such that $x\varphi = xN\psi$ for all $x \in G$.*

PROOF. The fact that N is a normal subgroup of G and the existence of a unique group homomorphism ψ of G/N into H such that $x\varphi = xN\psi$ for any $x \in G$ is exactly the theorem on isomorphisms for groups. It remains to prove that N is a convex subgroup of G and ψ is an order homomorphism. If $x \leqslant y \leqslant z$ in G for some elements $x, z \in N$ and $y \in G$, then $e = x\varphi \leqslant y\varphi \leqslant z\varphi = e$, $y\varphi = e$, $y \in N$, i.e., the subgroup N is convex. To complete the proof it suffices to show that, for any coset Nx such that $Nx \geqslant N$ in the naturally ordered quotient group G/N, the inequality $Nx\psi \geqslant e$ holds in H. However, if $Nx \geqslant N$ then, in accordance with the definition of the order on the quotient group G/N, in the group G the inequality $ax \geqslant e$ holds for some $a \in N$. Therefore, $Nx\psi = x\varphi = ax\varphi \geqslant e$. □

Note that the theorem on homomorphisms also remains true for l-groups. Moreover, in this case the theorem has two aspects. If we deal with the order homomorphism φ, then we cannot guarantee additional properties of N and ψ. However, under the assumption that φ is an l-homomorphism, it turns out that N is an ideal and the mapping ψ is an l-isomorphism. The proof of this assertion is not complicated.

1.6.5. We now describe the simplest methods of constructing partially right-ordered groups. Let $\{A_\alpha \mid \alpha \in I\}$ be a family of partially right-ordered groups.

By the *Cartesian product* $\overline{A} = \overline{\prod_{\alpha \in I}} A_\alpha$ of partially right-ordered groups A_α, $\alpha \in I$, we mean a partially right-ordered group which is the Cartesian product of the groups A_α, $\alpha \in I$, and the Cartesian product of the partially ordered sets A_α, $\alpha \in I$. This means that $\overline{A}$ is the set of all functions $f\colon I \to \bigcup_{\alpha \in I} A_\alpha$ such that $f(\alpha) \in A_\alpha$, $fg(\alpha) = f(\alpha)g(\alpha)$, and $f \leqslant g$ if $f(\alpha) \leqslant g(\alpha)$ for all $\alpha \in I$. It is evident that $\overline{A}$ is a partially ordered group or an l-group if and only if all groups A_α, $\alpha \in I$, are partially ordered or l-groups respectively.

By the *direct product* $A = \prod_{\alpha \in I} A_\alpha$ of partially right-ordered groups A_α, $\alpha \in I$, we mean a subgroup of the group $\overline{A}$ such that $f \in A$ if and only if $f(\alpha) \neq e$ for some finite set of indices. We will write $A = A_1 \times A_2 \times \ldots \times A_n$, in the case $I = \{1, 2, \ldots, n\}$.

If each group A_α is an l-group, then A is called the *direct product* of the l-groups A_α, $\alpha \in I$. Note that the direct product of l-groups is an l-subgroup of the group $\overline{A}$.

Proposition 1.6.5. *Let $\{N_\alpha \mid \alpha \in I\}$ be the set of convex normal subgroups of a partially right-ordered group G and $\bigcap_{\alpha\in I} N_\alpha = E$. Then there is an isomorphism of G into the Cartesian product $\overline{G} = \overline{\prod_{\alpha\in I}} G_\alpha$ of the partially right-ordered groups $G_\alpha = G/N_\alpha$ which is an order mapping.*

If G is an l-group and all subgroups N_α, $\alpha \in I$, are ideals in G, then G is l-isomorphic to an l-subgroup of the l-group $\overline{G}$.

PROOF. We introduce the mapping $f\colon G \to \overline{G}$ such that $f(g)(\alpha) = N_\alpha g$, $g \in G$, $\alpha \in I$. It is clear that f is an isomorphism of G into $\overline{G}$. If $x \in G$, $x \geqslant e$, then $N_\alpha x \geqslant N_\alpha$ in G/N_α for any $\alpha \in I$. Therefore, $f(x) \geqslant e$ in $\overline{G}$. The first assertion of Proposition 1.6.5 is proved.

If G is an l-group and N_α, $\alpha \in I$, are ideals in G, then

$$f(x \vee y)(\alpha) = N_\alpha(x \vee y) = N_\alpha x \vee N_\alpha y = (f(x)(\alpha) \vee f(y)(\alpha)$$

for $x, y \in G$, i.e., $f(x \vee y) = f(x) \vee f(y)$. In a similar way, we can prove that $f(x \wedge y) = f(x) \wedge f(y)$, i.e., f is an l-isomorphism. □

The following example shows that under the hypotheses of Proposition 1.6.5 it is impossible to assert, in general, that the constructed isomorphism of the group G is an o-isomorphism.

EXAMPLE 1.6.3. Let $G = (a) \times (b)$ be the direct product of infinite cyclic groups (a) and (b) with partial order $P = \{a^k b^k \mid k \geqslant 0\}$. Then the subgroups $A = (a)$ and $B = (b)$ are convex normal subgroups of the group G and $A \cap B = E$. Furthermore, the naturally ordered quotient groups G/A and G/B are order-isomorphic to the totally ordered group of integers and the Cartesian product $\overline{G}$ is order-isomorphic to the l-group $(a) \times (b)$ with order $Q = \{a^m b^n \mid m \geqslant 0,\ n \geqslant 0\}$. It is evident that there exists no isomorphism between the partially ordered group G and the l-group $\overline{G}$.

On the Cartesian product of partially right-ordered groups one can introduce another order. Let $\{A_\alpha \mid \alpha \in I\}$ be a family of partially right-ordered groups, and let the set of indices I be well-ordered with the relation $\prec$. We denote by $\overline{A} = \overline{\prod_{\alpha\in I}} A_\alpha$ the Cartesian product of the groups A_α, $\alpha \in I$, and set $f > e$ in $\overline{A}$ if $f \neq e$ and $f(\alpha) > e$ in A_α, where α is the least element with respect to $\prec$ such that $f(\alpha) \neq e$. The group just constructed is called the *lexicographic product* of the partially right-ordered groups A_α, $\alpha \in I$, by the well-ordered set I. This partially right-ordered group is denoted by $\vec{A} = \vec{\prod\limits_{\alpha \in I}} A_\alpha$. It is clear that $\vec{A}$ is the lexicographic product of the partially ordered sets A_α, $\alpha \in I$ (cf. 1.1.2).

The lexicographic order on the direct product of partially right-ordered groups over a well-ordered set of indices and the antilexicographic product of partially right-ordered groups A_α, $\alpha \in I$, over a well-ordered set of indices can be introduced in a similar way.

The following proposition is central in practically all branches of the ordered group theory. It will be proved with the help of the notion just introduced.

Proposition 1.6.6. *Any torsion-free Abelian group admits a total order.*

PROOF. Let A be a torsion-free Abelian group and V be its divisible closure, i.e., V is the minimal torsion-free divisible Abelian group containing A. Then V is the direct product $V \cong \prod\limits_{\alpha \in I} Q_\alpha$, where each group Q_α is isomorphic to the additive group $\mathbb{Q}$ of rational numbers. We totally order each group Q_α, well order somehow the set I, and consider the lexicographic product $\vec{A} = \vec{\prod\limits_{\alpha \in I}} Q_\alpha$. It is evident that the group $\vec{A}$ is totally ordered. Therefore, the group A is also totally ordered. □

1.6.6. If N is a convex normal subgroup of a partially right-ordered group G, then the quotient group G/N is also partially right-ordered. In this case G is said to be an *extension of the partially right-ordered group N by the partially right-ordered group G/N*. Moreover, if the inequality $gx > e$ holds for any $g \in G$ and $x \in N$ such that $gN > N$ in G/N, then G is called the *lexicographic extension*. In the sequel, we use the standard notation of the Schreier extension theory (cf., for example, [3]).

An element g of the extension G of a group N by $\overline{G}$ will be written as the pair $(\overline{g}, x)$, where $\overline{g} \in \overline{G}$ and $x \in N$. A partially right-ordered group G is the lexicographic extension of the partially right-ordered group (N, P_N) by the partially right-ordered group $(\overline{G}, \overline{P})$ provided that G is the extension of N by $\overline{G}$ and $(\overline{g}, x) > (\overline{h}, y)$ if and only if $\overline{g} > \overline{h}$ in $\overline{G}$ or $\overline{g} = \overline{h}$, and $x > y$ in N.

Theorem 1.6.2 (Levi [4]). *Let N be a normal subgroup of a group G, P_N be a partial right order on the group N, and $\overline{P}$ be a partial right order on the quotient group $\overline{G} = G/N$. Then there is a partial right order P on the group G such that (G, P) is the lexicographic extension of (N, P_N) by $(\overline{G}, \overline{P})$. If the groups (N, P_N) and $(\overline{G}, \overline{P})$ are partially ordered and $g^{-1}P_N g = P_N$ for any $g \in G$, then the group (G, P) is also partially ordered.*

The theorem is proved by direct verification.

We finish the chapter with a theorem that turns out to be useful when we need to construct a new right order starting from some prescribed right order. We first give the following definition. A group G is called the *semidirect product* of its normal subgroup H and the complement subgroup N if G is their product and $N \cap H = E$.

Theorem 1.6.3 (Kopytov). *Let a group G be the semidirect product of its normal subgroup H and the complement subgroup N. If right orders P_N and P_H are given on N and H such that $x^{-1}P_H x \subseteq P_H$ for any $x \in P_N$, then it is possible to introduce a right order Q on G such that the subgroup N is convex with respect to Q and $Q \cap N = P_N$.*

PROOF. Since G is the semidirect product of H and N, each element $g \in G$ is uniquely represented as $g = h \cdot n$, where $h \in H$ and $n \in N$. In G we consider a subset Q of elements g of the form $g = h \cdot n$, $h \in H$, $n \in N$, such that $h \in P_H$, $h \neq e$, or $g = n \in P_N$. We now show that Q defines a right order on G.

First, we verify that Q is a semigroup. Let $g_1 = h_1 n_1 \in Q$ and $g_2 = h_2 n_2 \in Q$. Then $g_1 g_2 = (h_1 \cdot n_1 h_2 n_1^{-1}) \cdot n_1 n_2$. According to the construction of Q, the elements h_1 and h_2 are contained in P_H. Moreover, if at least one of them is different from e, then $g_1 g_2 \in Q$. In the case $h_1 = h_2 = e$ we have $g_1 = n_1 \in P_N$ and $g_2 = n_2 \in P_N$. Hence $g_1 \cdot g_2 = n_1 n_2 \in P_N$, i.e., Q is a semigroup.

Moreover, if $g \in G$, $g \neq e$, $g = hn$, where $h \in H$, $n \in N$, and $h \in P_H \setminus \{e\}$, then $g \in Q$. If $h = e$, then either the equality $g = n$ holds or the element $g^{-1} = n^{-1}$ is contained in P_N, i.e., in this case we also have $g \in Q$ or $g^{-1} \in Q$.

Finally, if $h \in P_H^{-1} \setminus \{e\}$, then $g^{-1} = (hn)^{-1} = n^{-1}h^{-1} = n^{-1}h^{-1}n \cdot n^{-1}$ and $n^{-1}h^{-1}n \in P_H$ by the above assumption. Therefore, $g^{-1} \in Q$. Thus, Q is a linear semigroup.

We can verify that Q is a pure set, i.e., $Q \cap Q^{-1} = E$ in a similar way. Consequently, the set Q satisfies O.1, O.2, and O.5 (cf. Theorem 1.5.1). Hence Q is a positive cone corresponding to a right order on the group G.

We now prove that N is convex with respect to this order. Let $e \leqslant g \leqslant m$ for $g \in G$ and $m \in N$ with respect to the order Q. Let $g = hn$, where $h \in H$ and $n \in N$. If $h \neq e$, then $h \in P_H$ because of the inclusion $hn \in Q$. On the other hand, $hn \leqslant m$ with respect to the order Q, i.e., $hn \cdot m^{-1} \in Q^{-1}$. Hence $h \in P_H^{-1}$, which contradicts the inclusion $hn \in Q$. Thus, $h = e$, i.e., $g \in N$. Consequently, the subgroup N is convex with respect to the order Q. □

Chapter 2

Systems of Convex Subgroups

2.1. General Properties

2.1.1. We consider right-ordered groups, i.e., groups on which there is a total order relation $\leqslant$ stable under multiplication from the right by elements of these groups. Such groups are sometimes called *ro*-groups. In this chapter, the properties of the system of convex subgroups of a right-ordered group are studied.

- A subset H of a partially right-ordered group G is called *convex* if for any $x, z \in H$ and $y \in G$ the inequalities $x \leqslant y \leqslant z$ imply $y \in H$.

- A subset H of a group G is called *isolated* if the inclusion $x^n \in H$, where $x \in G$ and $n \in \mathbb{N}$, implies the inclusion $x \in H$.

Proposition 2.1.1. *Every convex subgroup of a right-ordered group is isolated.*

PROOF. Let H be a convex subgroup of a right-ordered group G, $x \in G$, and $x^n \in H$ for some $n \in \mathbb{N}$. If $e \leqslant x$, then $e \leqslant x \leqslant \ldots \leqslant x^n$.

Since $x^n \in H$ and the subgroup H is convex, we obtain $x \in H$. The case $x \leqslant e$ can be studied in a similar way. □

Proposition 2.1.2. *If H is a convex subgroup of a right-ordered group G, then every right coset Hx of G by H is a convex set.*

PROOF. Let $a, b, c \in G$, $a \leqslant b \leqslant c$, and $a, c \in Hx$ for some $x \in G$. Then $e = aa^{-1} \leqslant ba^{-1} \leqslant ca^{-1}$. Since $e \in H$ and H is a convex subgroup of G, we get $ba^{-1} \in H$, $b \in H$. □

As in Section 1.6, we introduce an order relation on cosets of a right-ordered group G by its convex subgroup H as follows. We set $Hx \leqslant Hy$ if there is an element $h \in H$ such that $hx \leqslant y$.

We denote by $R(G : H)$ the collection of right cosets of a right-ordered group G by its convex subgroup H.

Proposition 2.1.3. *The set $R(G : H)$ is totally ordered.*

PROOF. From Proposition 2.1.2 it follows that a totally ordered set $\{G; \leqslant\}$ can be represented as a union of its disjoint subsets Hx which are convex. Therefore, if $Hx \neq Hy$ for $x, y \in G$ and $x < y$, then $ux < vy$ for any $u, v \in H$, which proves the correctness of the definition of $\leqslant$ on $R(G : H)$. It is clear that the relation $\leqslant$ is a total order. □

Proposition 2.1.4. *Let G be a right-ordered group and H be its convex subgroup. Then for any $a \in G$ the mapping $R(a)$: $Hx \to Hxa$ is an order isomorphism of $R(G : H)$ and R: $a \to R(a)$ is a homomorphism of G into the group of order automorphisms of the totally ordered set $R(G : H)$. The kernel of this homomorphism is the greatest normal subgroup of the group G contained in H.*

PROOF. The mapping $R(a)$ is an order automorphism of the set $R(G : H)$ since the inequality $x < y$ in G implies the inequality $xa < ya$. The rest of the assertions are proved directly. □

We note that the mapping R from Proposition 2.1.4 can be not an order homomorphism of the right-ordered group G.

2.1.2. In what follows, we need the following definitions.

- A system $\mathfrak{C}(G)$ is called

  — *normal* if any $H \in \mathfrak{C}(G)$ is a normal subgroup of G,

— *complete* if $E \in \mathfrak{C}(G)$, $G \in \mathfrak{C}(G)$, and the union and intersection of any collection of subgroups belonging to $\mathfrak{C}(G)$ are contained in $\mathfrak{C}(G)$.

- A pair of subgroups $A, B \in \mathfrak{C}(G)$ is called a *jump in* $\mathfrak{C}(G)$ (denoted by $A \prec B$) provided that $A \subseteq B$, $A \neq B$, and if $A \subseteq C \subseteq B$ for some subgroup $C \in \mathfrak{C}(G)$, then either $A = C$ or $C = B$.

In the case of a complete system $\mathfrak{C}(G)$, for any nonidentity element $x \in G$ there is a jump $A \prec B$ in $\mathfrak{C}(G)$ such that $x \in B \setminus A$. As B we can take the intersection of all subgroups of $\mathfrak{C}(G)$ containing x and as A we can take the union of all subgroups of $\mathfrak{C}(G)$ which do not contain x.

- A complete system $\mathfrak{C}(G)$ is called

  — *subnormal* if for any jump $A \prec B$ in $\mathfrak{C}(G)$ the subgroup A is a normal subgroup of B,

  — *infrainvariant* if $x^{-1}Hx \in \mathfrak{C}(G)$ for any subgroup $H \in \mathfrak{C}(G)$ and element $x \in G$,

  — *soluble* if $\mathfrak{C}(G)$ is subnormal and the quotient group B/A is Abelian for any jump $A \prec B$ in $\mathfrak{C}(G)$.

Theorem 2.1.1. *Convex subgroups of a right-ordered group form a complete totally ordered system.*

PROOF. Let A and B be nontrivial convex subgroups of a right-ordered group G. We assume that A and B are incomparable with respect to the set-theoretical inclusion. Then there are elements $a \in A \setminus B$ and $b \in B \setminus A$. We can assume that $a > e$ and $b > e$. Since the order on G is total, we have $e < a < b$ or $e < b < a$. Each of these inequalities contradicts the assumption that the subgroups A and B are convex. □

It should be noted that the system of convex subgroups of a right-ordered group cannot be soluble, nor normal, nor infrainvariant.

EXAMPLE 2.1.1 (Smirnov [5]). Let G be a group of matrices of the form $\left[\begin{smallmatrix} k & a \\ 0 & 1 \end{smallmatrix}\right]$, where k, a are rational numbers and $k > 0$. Such matrices will be denoted by (k, a). We fix a positive irrational number ε and denote by P_ε the set of matrices (k, a) of G such that $k + a\varepsilon \geqslant 1$. If

$(k_1, a_1), (k_2, a_2) \in P_\varepsilon$, then

$$(k_1, a_1)(k_2, a_2) = (k_1k_2, k_1a_2 + a_1)$$
$$k_1k_2 + \varepsilon(k_1a_2 + a_1) = k_1(k_2 + \varepsilon a_2) + \varepsilon a_1 \geqslant 1$$

Consequently, P_ε is a semigroup. It is evident that

$$P_\varepsilon \cap P_\varepsilon^{-1} = \{(1, 0)\}$$
$$P_\varepsilon \cup P_\varepsilon^{-1} = G$$

i.e., P_ε determines a right total order on G; moreover,

$$(k_1, a_1) \leqslant (k_2, a_2) \Leftrightarrow (k_2, a_2)(k_1, a_1)^{-1} \in P_\varepsilon$$

We now prove that the group G does not have nontrivial convex subgroups. Let A be a subgroup of the group G formed by the matrices $(1, a)$, $a \in \mathbb{Q}$. On A the right order P_ε induces the total order $(1, a) > (1, 0) \Leftrightarrow a > 0$ with respect to which A does not have nontrivial convex subgroups. For any matrix $g \in G$, $g = (k, a) > (1, 0)$, there are rational numbers x and y such that

$$x > (1 - (k^{-1} - k^{-1}a\varepsilon))\varepsilon^{-1}$$
$$0 < y < (k + a\varepsilon^{-1})k^{-1}\varepsilon^{-1}$$

Therefore, for the matrices $(1, x)$ and $(1, y)$ the inequalities $(1, x) \geqslant (k, a)$ and $(1, y) > (1, 0)$ hold. If C is a convex subgroup of the group G, then $C \cap A$ is a convex subgroup of the totally ordered group A. Consequently, $C \cap A = A$ or $C \cap A = E$. If $A \subseteq C$, then there is an element $g > (1, 0)$ such that $g \notin A$. In accordance with what has been said above, there exists an element $(1, x)$ of A such that $(1, x) > g > (1, 0)$, which contradicts the above assumption. Hence $C = G$. If $A \cap C = E$, then for any element $c \in C$, $c > (1, 0)$, the matrix $(1, y)$ satisfies the relation $c > (1, y) > (1, 0)$. Since the subgroup C is convex, we obtain $(1, y) \in C \cap A$, which is impossible.

In particular, Example 2.1.1 shows that there exist right-ordered groups such that the system of their convex subgroups is subnormal but not soluble. We note that McCleary [6] and Droste [7] have proved the existence of a free right-ordered group which has no nontrivial convex subgroup. Right-ordered groups with soluble systems of their convex subgroups and with torsion-free factors were studied by Conrad [8], Smirnov [5], and Zaĭtseva [9]. Such groups are called *Conradian.*

2.2. Archimedean Groups

To study the structure of the system of convex subgroups of a right-ordered group we first consider the simplest situation and then we try to extend it to more general cases.

- A right-ordered group G is called *Archimedean* if the following *Archimedean principle* holds:

  (A) *For any strictly positive elements $a, b \in G$ there is a natural number n such that $a^n > b$.*

Archimedean totally ordered groups are described by the following classical Hölder theorem.

Theorem 2.2.1 (Hölder [10]). *Every Archimedean totally ordered group is order-isomorphic to a subgroup of the additive group of real numbers with the natural order.*

PROOF. Let G be an Archimedean totally ordered group. Two cases are possible.

CASE 1: in G there is the smallest positive element a. In this case for any element $b \in G$ there is a real number n such that $a^n \leqslant b < a^{n+1}$ or $e \leqslant ba^{-n} < a$. Since a is the smallest positive element of the group G, we have $ba^{-n} = e$ and $b = a^n$. Hence G is order-isomorphic to the additive group of integers with the natural order.

CASE 2: in G there is not the smallest positive element. Then for any positive element $a > e$ there is an element $c > e$ such that $c^2 \leqslant a$. Indeed, starting from a, we find an element $b \in G$ such that $e < b < a$. If $b^2 \leqslant a$, we set $c = b$. If $b^2 > a > b$, then $e > b^{-1}ab > b^{-1}$ and $a > ab^{-1} \cdot ab^{-1} > ab^{-1} > e$, $c = ab^{-1}$.

We now prove that the group G is commutative. We assume that there are noncommutative positive elements $x, y \in G$. Then $z = [x, y] = x^1y^{-1}xy \neq e$. For the sake of definiteness, let $z > e$. From what has been proved, it follows that we can find an element $t \in G$, $t > e$, such that $t^2 < z$. Since G is an Archimedean totally ordered group, there are integers n and m such that $t^n \leqslant x < t^{n+1}$ and $t^m \leqslant y < t^{m+1}$. Consequently, $xy < t^{n+m+2}$ and $x^{-1}y^{-1} \leqslant t^{-n-m}$. Therefore, $z = x^{-1}y^{-1}xy < t^2$, which contradicts the choice of t. Thus, the group G is Abelian.

To complete the proof of the theorem, we embed the group G into the minimal divisible Abelian group A and consider the following subset of A:

$$P(A) = \{x \in A \mid x^n \in G,\ x^n \geqslant e \text{ for some } n \in \mathbb{N}\}$$

It is clear that $P(A)$ is a pure linear subsemigroup of the group A. Therefore, on A the set $P(A)$ determines a total order that is an extension of the original total order on the group G. It is evident that $(A, P(A))$ is also an Archimedean totally ordered group. Since the group A is torsion-free divisible Abelian, there exists its subgroup Q which is order-isomorphic to the totally ordered additive group $\mathbb{Q}$. Consequently, in the totally ordered set Q each element $g \in A$ determines a Dedekind cut with which, in turn, a real number α_g is associated. It is easy to show that the constructed mapping α: $g \to \alpha_g$ is an order isomorphism of the totally ordered group G into the naturally ordered additive group $\mathbb{R}$. □

Theorem 2.2.2 (Conrad [8]). *An Archimedean right-ordered group is commutative and, consequently, order-isomorphic to a subgroup of the additive group of real numbers with the natural order.*

PROOF. By Theorems 2.2.1 and 1.5.2, it suffices to prove that a right order P introduced on an Archimedean right-ordered group G is stable under its inner automorphisms. Let $x, y \in G$, $x > e$, $y > e$. By the above assumption, there is the smallest natural number n such that $x < y^n$. Therefore, $xyx^{-1} > e$ because $xyx^{-1} \leqslant e$ implies $xy < x < y^n$. But in this case $x < y^{n-1}$, which contradicts the fact that n is minimal. This means that $xyx^{-1} > e$ for any elements x and y such that $x > e$ and $y > e$. We assume that $x^{-1}yx \leqslant e$ for some elements x and y such that $x > e$ and $y > e$. Then $x^{-1}y^{-1}x > e$. Consequently, $x(x^{-1}y^{-1}x)x^{-1} = y^{-1} > e$, which contradicts the choice of y. Thus, $x^{-1}Px \subseteq P$ for any $x \in G$, i.e., an Archimedean right-ordered group is totally ordered, which proves the theorem. □

In the sequel, we often use the following assertion concerning Archimedean totally ordered groups.

Proposition 2.2.1. *If A and B are subgroups of the additive group of real numbers with the natural order and φ: $A \to B$ is an order isomorphism of A onto B, then there is a positive real number r such that $x\varphi = rx$ for each $x \in G$.*

PROOF. For the group $\mathbb{R}$ we use here the additive denotation of the group operation and the natural multiplication operation. We prove the assertion by contradiction. Let $(a\varphi)\cdot(b\varphi)^{-1} \neq ab^{-1}$ for some elements $a, b \in A$ such that $a, b > 0$ in $\mathbb{R}$. For the sake of definiteness, we set $(a\varphi)(b\varphi)^{-1} < ab^{-1}$. Then there exists a rational number mn^{-1} $(m, n \in \mathbb{N})$ such that the relation $(a\varphi)(a\varphi)^{-1} < mn^{-1} < ab^{-1}$ holds, which leads to the contradictory inequalities $mb < na$ and $m(b\varphi) > n(a\varphi)$. □

Corollary 2.2.1. *The group of order automorphisms of an Archimedean right-ordered group is isomorphic to a subgroup of the multiplicative group of positive real numbers.*

2.3. Totally Ordered Groups

Among classes of right-ordered groups, the class of totally ordered groups is the most important. For this class the structure of the system of convex subgroups has been studied fairly well.

Proposition 2.3.1. *For the smallest convex subgroup A of a totally ordered group G containing the nonidentity element a the following equality holds*:

$$A = \{x \in G \mid |x| \leqslant |a|^n \quad \textit{for some } n \in \mathbb{N}\}$$

where, as usual, $|x| = x$ for $x \geqslant e$ and $|x| = x^{-1}$ for $x \leqslant e$.

PROOF. In view of (1.4.3) and (1.4.4), elements $x \in G$ such that $|x|^n \leqslant |a|$ for some $n \in \mathbb{N}$ form a subgroup of the group G. By definition, this subgroup is convex. It is clear that it is the smallest subgroup among the convex subgroups of the group G containing the element a. □

The simplest situation where a totally ordered group has no convex subgroup is completely described in the following assertion.

Proposition 2.3.2. *A totally ordered group has no proper convex subgroup if and only if it is Archimedean.*

PROOF. The sufficiency is obvious. To prove the necessity we consider a totally ordered group G which has no proper convex subgroup. Let $a, b \in G$, $a, b > e$. The smallest convex subgroup A of the group G containing the element a coincides with G. Hence $b \in A$. By

Proposition 2.3.1, we have $b \leqslant a^n$ for some $n \in \mathbb{N}$, which proves the proposition. □

By Proposition 2.3.2, a totally ordered group is Archimedean if and only if it has no proper convex subgroups. The Smirnov example (cf. Example 2.1.1) shows that for right-ordered groups this assertion fails: there are no Archimedean right-ordered groups without proper convex subgroups.

To describe the structure of the system of convex subgroups we use the standard set-theoretical notions. Let G be a group and H be its subgroup. We introduce the following notation:

$$\boldsymbol{n}(H) = N_G(H) = \{x \in G \mid x^{-1}Hx = H\}$$

$$\boldsymbol{c}(H) = C_G(H) = \{x \in G \mid x^{-1}hx = h \quad \text{for all } h \in H\}$$

It is clear that $\boldsymbol{n}(H)$ and $\boldsymbol{c}(H)$ are subgroups of the group G and $\boldsymbol{c}(H)$ is a normal subgroup of $\boldsymbol{n}(H)$. The subgroup $\boldsymbol{n}(H)$ is called the *normalizer* and $\boldsymbol{c}(H)$ is called the *centralizer* of the subgroup H.

If $A \prec B$ is a jump in the infrainvariant system U of subgroups of G, then $\boldsymbol{n}(A) = \boldsymbol{n}(B)$. It is obvious that each element $x \in \boldsymbol{n}(B)$ induces an automorphism of the quotient group B/A: $bA \to x^{-1}bxA$. The group of all such automorphisms of the group B/A is isomorphic to the quotient group $\boldsymbol{n}(A)/\boldsymbol{c}(B/A)$; here $\boldsymbol{c}(B/A) = \{x \in \boldsymbol{n}(B) \mid x^{-1}bxA = bA\}$ is a subgroup of G and is called the *centralizer of the jump* $A \prec B$.

Theorem 2.3.1 (Iwasawa [11] and Mal'cev [12]). *The system $\mathfrak{C}(G)$ of convex subgroups of a totally ordered group G is infrainvariant. If $A \prec B$ is a jump in $\mathfrak{C}(G)$, then the quotient group B/A is order-isomorphic to a subgroup of the totally ordered additive group of real numbers. The group of automorphisms of the quotient group B/A induced by the inner automorphisms of the normalizer $\boldsymbol{n}(A)$ of the subgroup A of G is isomorphic to a subgroup of the multiplicative group of positive real numbers.*

Proof. Let G be a totally ordered group. By Theorem 2.1.1, the system $\mathfrak{C}(G)$ is complete and totally ordered by inclusion. If $H \in \mathfrak{C}(G)$, then $g^{-1}Hg$ is a convex subgroup of the group G for any $g \in G$ because an inner automorphism of a totally ordered group preserves the order. Consequently, the system $\mathfrak{C}(G)$ is infrainvariant. If $A \prec B$ is a jump in $\mathfrak{C}(G)$, then A is a normal subgroup of B. By Proposition 2.3.2, the quotient group B/A is Archimedean. By the Hölder theorem (cf. Theorem 2.2.1), it is order-isomorphic to a subgroup of the naturally ordered group $\mathbb{R}$, and, by Proposition 2.2.1, the action of the inner automorphisms prescribed by the elements of $\boldsymbol{n}(A)$ on B/A is the multiplication by positive real numbers. □

In Chapter 3 (cf. Theorem 3.3.1) we give a description of the system of convex subgroups of a totally ordered group which is close to the above description but will be expressed in terms of the abstract group theory.

2.4. Conradian Groups

2.4.1. The class of right-ordered groups such that the system of its convex subgroups is subnormal and the quotient groups of jumps are Archimedean is rather close to the class of totally ordered groups. It is natural that a right-ordered group has the subnormal system of convex subgroups with the Archimedean quotient groups of jumps if and only if it is Conradian. The last assertion follows immediately from Theorem 2.2.2.

We recall that, in Section 2.1, by a Conradian group we mean a group such that the system of its convex subgroups is subnormal and the quotient groups of jumps are torsion-free Abelian groups. Such right-ordered groups are also called *right-ordered groups with Archimedean jumps.* The description of such groups was given by Conrad [8].

Lemma 2.4.1 (Conrad [8]). *The following properties of a right-ordered group G are equivalent:*

(a) *for any $a, b \in G$ such that $a > e$ and $b > e$ there is a number $n \in \mathbb{N}$ such that $(ab)^n > ba$,*

(b) *for any $a, b \in G$ such that $e < a < b$ there is a number $n \in \mathbb{N}$ such that $ab^n a^{-1} > b$,*

(c) *for any $a, b \in G$ such that $a > e$ and $b > e$ there is a number $n \in \mathbb{N}$ such that $a^n b > a$.*

Proof. Let a right-ordered group G have the property (a) and $e < a < b$. Then $ba^{-1} > e$. Starting from the elements a and ba^{-1}, we find a number n such that $ab^n a^{-1} = (a \cdot ba^{-1})^n > ba^{-1} \cdot a = b$, i.e., G has the property (b). Conversely, if G has the property (b) and $e < a, b$, then $b < ab$. By the above assumption, we obtain $(ba)^n = b(ab)^n b^{-1} > ab$ for some positive number n. Hence the properties (a) and (b) are equivalent.

We now show the equivalence of (a) and (c). We suppose that G has the property (a) and $e < a \leqslant b$. If $a \leqslant b$, then $e < a \leqslant b \leqslant ab$ and we can assume that $n = 1$. If $a > b$, then $a = cb$, where $c = ab^{-1} > e$, and there is a number $n \in \mathbb{N}$ such that $(cb)^n > bc$. Therefore, $a^n b = (cb)^n b > bcb = ba > a$. We assume that G has the property (c) but the assertion (a) is false for some positive elements $a, b \in G$. Then $(ab)^n < ba$ for all $n \in \mathbb{N}$. Since $a > e$, we have $(ba)^n b < a(ba)^n b = (ab)^{n+1} < ba$, which contradicts (c). Thus, the equivalence of the properties (a) and (c) is proved. □

Theorem 2.4.1 (Conrad [8]). *A right-ordered group G is Conradian if and only if G possesses one of the properties* (a)–(c) *indicated in Lemma* 2.4.1.

PROOF. We assume that a right-ordered group G has one of the properties (a)–(c). We consider a jump $A \prec B$ in the system of convex subgroups of the group G. We first show that for any $x, y \in B \setminus A$ such that $x, y > e$ there is a positive integer n such that $x^n > y$. We assume by way of contradiction that there are elements $a, b \in B \setminus A$ such that $a, b > e$ and $e < a^n < b$ for any $n \in \mathbb{N}$. Let S denote the set of elements $x \in G$ such that $x > e$ and $x < a^n$ for some $n \in \mathbb{N}$. Let $N = S \cup S^{-1} \cup \{e\}$. We now show that N is a subgroup of the group G. It is clear that $N = N^{-1}$. Therefore, it suffices to note that the subset N is closed under multiplication. If $x, y \in N$ and $x = e$ or $y = e$, then $xy \in N$. Let $x, y \in N$ and $x \neq e$, $y \neq e$, $xy \neq e$. If $x, y \in S$, then $e < x < a^n$ and $e < y < a^n$ for a suitable $n \in \mathbb{N}$. We now show that there is a number r such that $a^n \leqslant a^r y^{-1}$. Indeed, in the opposite case we have $a^r y^{-1} < a^n$ for each integer r. Consequently, $a^{n+q} y^{-1} < a^n$ for each integer q. But $y < a^n$ and $ya^q < a^{n+q}$. Hence $ya^q y^{-1} < a^{n+q} < a^n$ for each $q \in \mathbb{N}$, which contradicts the assertion (b). Thus, there is a number r such that $a^n \leqslant a^r y^{-1}$ and $a^n y \leqslant a^r$, i.e., the inequalities $e < x < a^n$ imply the inequalities $e < xy < a^n y \leqslant a^r$, from which we have the following inclusions: $xy \in S$ and $xy \in N$.

In the case $x, y \in S^{-1}$ the inclusions $xy \in S^{-1}$ and $xy \in N$ can be proved in a similar way.

If $x \in S$ and $y \in S^{-1}$, then $e < x < a^n$ and $e < y < a^n$ for some $n \in \mathbb{N}$. We first consider the case $e < xy$ and $xy \notin S$. Then $xy > a^r$

and $x > a^r y^{-1}$ for any integer r. In view of (a), starting from the number n, we can find an integer q such that $a^{nq} y^{-1} > a^n$. Therefore, for $r = nq$ we obtain $x > a^{nq} y^{-1} > a^n$, which contradicts the choice of n. In the case $xy < e$ we have to prove that $y^{-1}x^{-1} = (xy)^{-1} \in S$. But $y^{-1} \in S$, $x^{-1} \in S^{-1}$, $e < y^{-1}x^{-1}$. From the above argument it follows that $y^{-1}x^{-1} \in S$ and $xy \in N$.

Let $x \in S^{-1}$, $y \in S$, $e < x^{-1} < a^n$, and $e < y < a^n$. If $xy > e$ and $xy \notin S$, then $xy > a^r$, $x > a^r y^{-1}$ for any integer r. Since $y < a^n$, we obtain $x > a^n y^{-1} > e$, which contradicts the assumption $x < e$. If $xy < e$, then $y^{-1} \in S^{-1}$, $x \in S$, $y^{-1}x^{-1} > e$. By the above argument, $y^{-1}x^{-1} \in S$, $xy \in S$, and $xy \in N$. Thus, we have proved that N is a subgroup of the group G.

We now show that the subgroup N is convex. It is evident that the semigroup S is convex. If $x < y < t$, where $x \in S^{-1}$, then $e < yx^{-1} < x^{-1}$. Since $x^{-1} \in S$ and the semigroup S is convex, we can deduce the inclusions $yx^{-1} \in S$ and $yx^{-1} \in N$. Then $y \in N$, which completes the proof of the fact that the subgroup N is convex.

Owing to the choice of the elements a and b, we have $a \in N$ and $b \notin N$, i.e., $A \subset N$, $N \subset B$, $A \neq N$, $B \neq N$. Thus, we arrive at a contradiction with the fact that $A \prec B$ is a jump in the system of convex subgroups of the group G. Consequently, for any positive elements $x, y \in B \setminus A$ there is a number $n \in \mathbb{N}$ such that $x^n > y > e$.

We now prove that A is a normal subgroup of B. Let $b > e$ and $b \in B \setminus A$. We first show that $bab^{-1} < b$ for any $a \in A$. Indeed, since $a < b$ for each $a \in A$, we have $ba^{-1} > e$ and $ba^{-1} \in B \setminus A$. In accordance with what has been proved above, $(ba^{-1})^n > b$ for some $n \in \mathbb{N}$. Let n be the smallest number with such a property. We note that the following inequality holds: $b \cdot ba^{-1}b^{-1} > e$; otherwise the inequality $b \cdot ba^{-1}b^{-1} \leqslant e$ implies the relation $b \cdot ba^{-1} \leqslant b < (ba^{-1})^n$ from which we obtain the inequality $b < (ba^{-1})^{n-1}$, which contradicts the choice of n. Thus, if $a \in A$, then $b \cdot ba^{-1}b^{-1} > e$, i.e., $bab^{-1} \cdot b^{-1} < e$. Therefore, $bab^{-1} < b$. We note that the simultaneous validity of the inequality $bab^{-1} > e$ and of the inclusion $bab^{-1} \in B \setminus A$ is impossible for any $a \geqslant A$. Indeed, otherwise $(bab^{-1})^n > b$ for some $n \in \mathbb{N}$. However, from what has been proved above, we obtain $(bab^{-1})^n = ba^n b^{-1} < b$ for any n. Hence we have $bab^{-1} \in A$ or $bab^{-1} < e$ for any $a \in A$. In the last case $ba^{-1}b^{-1} > e$ and $a^{-1} \in A$. Therefore, $bAb^{-1} \subseteq A$ for any positive element $b \in B$.

To complete the proof of the normality of A in B it suffices to prove the inclusion $b^{-1}ab \in A$ for any $b \in B \setminus A$, $b > e$, and $a \in A$. We assume that the last assertion is not true, i.e., $b^{-1}ab \in B \setminus A$ for some

$a \in A$ and $b \in B \setminus A$, $b > e$. Without loss of generality, we can assume that $b^{-1}ab > e$. Then for a suitable $n \in \mathbb{N}$ we have $b^{-1}a^n b > b$ and $b^{-1}a^n > e$. Using the property (b), we can find a number $m \in \mathbb{N}$ such that $b(b^{-1}a^n)^m b^{-1} > b^{-1}a^n > e$. Since $b(b^{-1}a^n)^m b^{-1} = (a^n b^{-1})^m > e$, the following inequalities hold: $(a^n b^{-1})^m > e$, $a^n b^{-1} > e$, and $a^n > b$, which contradicts the choice of b. Hence $b^{-1}ab \in A$ for any $a \in A$ and $b \in B$, which implies the normality of A in B.

The quotient group B/A is a right-ordered group such that for any $x, y \in B \setminus A$ with $x > e$ and $y > e$ there is a number $n \in \mathbb{N}$ such that $x^n A > yA$, i.e., the group B/A is Archimedean.

Conversely, let G be a Conradian right-ordered group and $a, b \in G$, $e < a < b$. We consider a jump $A \prec B$ in the system of convex subgroups of the group G which is defined by the element b, i.e., $b \in B \setminus A$. Since the quotient group B/A is Archimedean, we obtain $ab = cba$ and $ab^2a^{-1} = c'b^2$ for some elements $c, c' \in A$. If $c'b^2 < b$, then $c' < b^{-1} < e$, $b \in A$, which contradicts the convexity of the subgroup A and the relation $b \notin A$. This means that the inequality $ab^2a^{-1} > b$ holds and the group G has the property (b). □

2.4.2. It is obvious that an Abelian right-ordered group is Conradian. In this section, we present another important and wide class of Conradian right-ordered groups.

- A group G is called
  - — *nilpotent* if it has a finite *central series* of subgroups, i.e., a system of subgroups
  $$E = H_0 \subseteq H_1 \subseteq \ldots \subseteq H_n = G$$
  such that all H_i, $0 \leqslant i \leqslant n$, are normal subgroups of G and all quotient groups H_i/H_{i-1}, $1 \leqslant i \leqslant n$, are contained in the center of the quotient group G/H_{i-1} (the smallest integer n with this property is called the *nilpotency class* of the group G),
  - — *locally nilpotent* if each of its finitely generated subgroups is nilpotent.

As usual, we denote by $\zeta_k(G)$ the kth term of the upper central series of a group G. More exactly, if $\zeta_1(G)$ is the center of a group G and $\zeta_{k-1}(G)$ is defined, then $\zeta_k(G)$ is the full preimage in G of the center of the quotient group $G/\zeta_{k-1}(G)$. As is known, a group is nilpotent with nilpotency class n if $\zeta_n(G) = G$ and $\zeta_{n-1}(G) \neq G$.

For an arbitrary group we denote by $\mathrm{gp}(M)$ and $\mathrm{sgr}(M)$ its subgroup and subsemigroup which are generated by the set M.

We now give the statement and proof of B. Neumann's result, following Rhemtulla [13].

Lemma 2.4.2 (B. Neumann). *Let G be a locally nilpotent group and S be its subsemigroup. Then for any $u, v \in S$ there are $z, t \in S$ such that $zu = tv$.*

PROOF. We prove the assertion by induction on the nilpotency class of the subgroup $L = \mathrm{gp}(u, v)$ of G. For the Abelian group L we set $z = v$ and $t = u$. We assume that for the groups of nilpotency class $k \geqslant 1$ the assertion of the lemma is true and the group L is of nilpotency class $k+1$. We note that the nilpotency class of the group $M = \mathrm{gp}(uv, vu)$ is at most k because of the equality $vu = uv[v, u]$, where, as usual, $[v, u] = v^{-1}u^{-1}vu$ and $[v, u] \in \zeta_k(G)$. But the quotient group $M/\zeta_k(L) \cap M$ is cyclic and $\zeta_k(M) \supseteq \zeta_k(M) \cap M$. Hence M is a nilpotent group of nilpotency class k. By assumption, there are elements $x, y \in M$, $x, y \in \mathrm{sgr}(u, v)$, such that $xvu = yuv$. Let $z = xv$ and $t = yu$. Since $u, v \in S$, we have $x, y \in S$, which completes the proof. □

Theorem 2.4.2 (Ault [14] and Rhemtulla [13]). *A right-ordered locally nilpotent group is Conradian.*

PROOF. Let G be a right-ordered locally nilpotent group. We now show that G possesses the property (c) (cf. Lemma 2.4.1), which together with Theorem 2.4.1 will prove the required assertion. Let $a, b \in G$, $a, b > e$. We assume that $a^n b \leqslant a$ for each integer n. We denote by S a subsemigroup of the group G generated by the elements ab and a. By Lemma 2.4.2, in S there are elements z and t such that $zab = ta$. It is clear that $z > e$, $t > e$, and $ta > a$. Furthermore, $zab = a^{k_1}ba^{k_2}b \ldots a^{k_r}b$, where $k_1, k_2, \ldots, k_r \geqslant 0$. Since $a^n b < a$ for each $n \in \mathbb{N}$, we have $zab \leqslant a^{k_2+1}b \ldots a^{k_r}b \leqslant a^{k_r+1}b \leqslant a$. However, $ta > a$. The contradiction obtained completes the proof. □

Corollary 2.4.1. *A right-ordered locally nilpotent group without proper convex subgroups is Archimedean and, consequently, isomorphic to a subgroup of the additive group of real numbers with the natural order.*

2.4.3. Following Zaĭtseva [9], Teh [15], Trevisan [16], and Hion [17], we describe all total (obviously, all right) orders on an Abelian group of finite rank.

Let G be a written additively torsion-free Abelian group of finite rank r, and let $d_1, \ldots, d_r$ be a maximal linearly-independent-over-$\mathbb{Q}$ system of elements of the group G. Then each element $g \in G$ can be uniquely represented in the form

$$g = k_1 d_1 + \ldots + k_r d_r$$

where $k_i \in \mathbb{Q}$. We denote by G^* the minimal completion of the group G with respect to solving equations $nx = g$, where $n \in \mathbb{N}$. Then the group G can be regarded as a vector space over the field $\mathbb{Q}$ with dimension r and basis $d_1, \ldots, d_r$. For each $x \in G$ there is a number $n(x) \in \mathbb{N}$ such that $n(x) \cdot x \in G$. It is easy to verify that every total order on the group G can be uniquely extended to a total order on the group G^*. Therefore, the description of all total orders on G is equivalent to that of all total orders on G^*.

We now fix a total order on the group G^*. The finiteness of the rank of the group G^* implies the finiteness of the number of its convex subgroups, and the divisibility of the group G^* implies that the group G^* is the lexicographic product of the quotient groups of its convex subgroups:

$$G^* = A_1 \vec{\oplus} A_2 \ldots \vec{\oplus} A_n$$

moreover, on every group A_i, $1 \leqslant i \leqslant n$, an Archimedean order is induced and the equality $x = a_1 + a_2 + \ldots + a_n > 0$ holds if and only if $a_k > 0$ in the totally ordered group A_k, where k is the smallest index such that $a_i \neq 0$. By the Hölder theorem (Theorem 2.2.1), every totally ordered group A_i is order-isomorphic to a subgroup of the naturally ordered additive group $\mathbb{R}$.

Let φ_i be a fixed order isomorphism of the group A_i into $\mathbb{R}$ and $a_{i1}, \ldots, a_{ik(i)}$ be a basis of the vector space A_i. Every total order on the group G^* determines a decomposition of G^* into the direct sum of subgroups $A_1, \ldots, A_n$ and gives a collection of real numbers $\{c_{ij} = a_{ij}\varphi_i,\ 1 \leqslant i \leqslant n,\ 1 \leqslant j \leqslant k(i)\}$ such that for any i, $1 \leqslant i \leqslant n$, the numbers $c_{i1}, \ldots, c_{ik(i)}$ are linearly-independent-over-$\mathbb{Q}$. It is evident that such a decomposition of G^* is unique up to an isomorphism and the subgroups $B_2 = A_2 \oplus \ldots \oplus A_n, \ldots,\ B_s = A_s \oplus \ldots \oplus A_n,\ B_n = A_n$ are uniquely determined. The mappings $\varphi_i\colon A_i \to \mathbb{R}$ are uniquely determined (for a fixed basis $a_{i1}, \ldots, a_{ik(i)}$ of the vector space A_i) up to a positive real factor. Namely, if $c_{i1}, \ldots, c_{ik(i)}$ and $c'_{i1}, \ldots, c'_{ik(i)}$

correspond to the same order on the group A_i with the same basis of the vector space A_i, then, by Proposition 2.2.1, there is a positive real number c such that $c'(ij) = c \cdot c_{ij}$. Therefore, under the assumption that $a_{i1}\varphi_i = 1$ we conclude that the sequence

$$B_1 = G \supset B_2 \supset B_3 \ldots \supset B_n = E$$

and φ_i uniquely (up to the choice of the basis $a_{i1}, \ldots, a_{ik(i)}$ of the group B_i over B_{i+1}) determine an order on the group G^*.

We now fix an arbitrary basis $a_1, \ldots, a_r$ of the group G^* and represent the number r as a sum of natural numbers in all possible ways. With every decomposition of the number $r = k(1)+\ldots+k(s)$ we associate the decomposition of the group G^* into the direct sum of its subgroups $C_1, \ldots C_s$, where C_i is a subspace of the space G^* generated by the elements

$$u_{i1} = a_{k(1)+\ldots k(i-1)+1}, \ldots, u_{ik(i)} = a_{k(1)+\ldots+k(i)}$$

After that we choose embeddings $\varphi_i\colon C_i \to \mathbb{R}$ such that $u_{i1}\varphi_i = 1, u_{ij}\varphi_i = c_{ij}$ and linearly-independent-over-$\mathbb{Q}$ numbers $c_{i1}, \ldots, c_{ik(i)}$. We consider the set

$$P = \{x \mid x \in G^*, \ \ x = \sum_{i,j} t_{ij}u_{ij}\}$$

where t_{ij} are integers, $t_{i1}c_{i1} + \ldots t_{ik(i)}c_{ik(i)} > 0$ in $\mathbb{R}$, and i is the smallest index such that $\sum_{j=1}^{k(i)} t_{ij}c_{ij} \neq 0$. It is clear that P is a pure linear subsemigroup of G^* and, consequently, determines a total order on G^* with respect to which all subgroups $C_k \oplus C_{k+1} \oplus \ldots C_s$, $k = 1, 2, \ldots, s$, are convex and the mappings φ_i determine Archimedean orders on the factors of this system. It is evident that the constructed system of orders is a complete (up to the automorphisms of the group G^*) system of representatives of the set of all orders on the group G^*.

Corollary 2.4.2. *A finitely generated totally ordered Abelian group is the lexicographic product of Archimedean totally ordered groups.*

Corollary 2.4.3 (Matsusita [18]). *A free group with a finite number $n \geqslant 2$ of free generators admits an uncountable set of distinct total orders.*

Corollary 2.4.4. *A free Abelian group with a finite number $n \geqslant 2$ of free generators admits an uncountable set of distinct total orders.*

Chapter 3

Orderability Conditions

3.1. Semigroup Conditions

One of the principal purposes of the theory of right-orderable groups is to characterize a class of groups admitting right orders in terms of the abstract group theory. The language of the semigroup theory is frequently and successfully employed for such a description. In the sequel, sgr (M) denotes the subsemigroup of a group G generated by a given subset M of G. We recall that a partial right order $\leqslant$ on a group G is identified with the semigroup $P = \{x \in G \mid x \leqslant e\}$ of the group G (cf. Chapter 1).

3.1.1. We begin with definitions.

- A group G is called

  — *orderable* if a total two-sided order can be introduced on G,

  — *right-orderable* if a total right order can be introduced on G.

Lemma 3.1.1. *A partial right order P on a group G can be extended to a total right order on G if and only if it satisfies the following property:*

EO.1 *for every finite set of nonidentity elements of the group G, there is a set of integers $\varepsilon_1, \dots, \varepsilon_n$ ($\varepsilon_i = \pm 1$) such that the semigroup $\operatorname{sgr}(P \setminus \{e\}, x_1^{\varepsilon_1}, \dots, x_n^{\varepsilon_n})$ does not contain e.*

PROOF. If an order P is extended to a total right order Q, then, starting from elements $x_1, \dots, x_n$, $x_i \neq e$, we choose integers $\varepsilon_1, \dots, \varepsilon_n$ ($\varepsilon_i = \pm 1$) such that $x^{\varepsilon_i} \in Q$. Then

$$\operatorname{sgr}(P \setminus \{e\}, x_1^{\varepsilon_1}, \dots, x_n^{\varepsilon_n}) \subseteq Q \setminus \{e\}$$

To prove the converse assertion we first show that if the semigroup P satisfies the condition EO.1, then for every element $x \in G$, $x \neq e$, there is an integer $\varepsilon = \pm 1$ such that the subsemigroup $\operatorname{sgr}(P \setminus \{e\}, x^{\varepsilon})$ satisfies the condition EO.1. Let us assume the contrary. Then semigroups $\operatorname{sgr}(P \setminus \{e\}, x)$ and $\operatorname{sgr}(P \setminus \{e\}, x^{-1})$ do not satisfy the condition EO.1, i.e., there exist elements $y_1, \dots, y_n$ and $z_1, \dots, z_m$ of the set $G \setminus \{e\}$ such that $e \in \operatorname{sgr}(P \setminus \{e\}, x, y_1^{\varepsilon_1}, \dots, y_n^{\varepsilon_n})$ and $e \in \operatorname{sgr}(P \setminus \{e\}, x^{-1}, z_1^{\delta_1}, \dots, z_m^{\delta_m})$ for any integers ε_i, δ_j, $1 \leqslant i \leqslant n$, $1 \leqslant j \leqslant m$, equal to ± 1. Consequently,

$$e \in \operatorname{sgr}(P \setminus \{e\}, x^{\varepsilon}, y_1^{\varepsilon_1}, \dots, y_n^{\varepsilon_n}, z_1^{\delta_1}, \dots, z_n^{\delta_m})$$

for integers $\varepsilon, \varepsilon_1, \dots, \varepsilon_n, \delta_1, \dots, \delta_m$, which contradicts the condition EO.1 for P. Thus, if P satisfies the condition EO.1, then $\operatorname{sgr}(P \setminus \{e\}, x)$ or $\operatorname{sgr}(P \setminus \{e\}, x^{-1})$ satisfies this condition too.

Let a maximal right order Q be an extension of the order P and satisfy the condition EO.1. If the order Q is not total, there is an element x such that $x \notin Q \cup Q^{-1}$. Hence there is an integer $\varepsilon = \pm 1$ such that the semigroup $\operatorname{sgr}(Q \setminus \{e\}, x^{\varepsilon})$ satisfies the condition EO.1. At the same time, $Q \subset \operatorname{sgr}(Q \setminus \{e\}, x^{\varepsilon}) \cup \{e\}$, which contradicts the choice of Q. □

Theorem 3.1.1 (Conrad [8]). *A group G admits a total right order if and only if the following condition is fulfilled:*

EO.2 *for any finite set $x_1, \dots, x_n$ of nonidentity elements of a group G there are $\varepsilon_1, \dots, \varepsilon_n$ ($\varepsilon_i = \pm 1$) such that $e \notin \operatorname{sgr}(x_1^{\varepsilon_1}, \dots, x_n^{\varepsilon_n})$.*

Proof. The assertion is established by the direct application of Lemma 3.1.1 to the partial right order defined by the identity semigroup of G. □

Corollary 3.1.1. *If every finitely generated subgroup of G admits a total right order, then the group G admits a total right order too.*

3.1.2. We now show necessary and sufficient conditions for the orderability of groups in terms of semigroups. The following lemma is obvious.

Lemma 3.1.2. *If P and Q are pure normal subsemigroups of a group G and $P \cap Q = \{e\}$, then the subset PQ^{-1} is also a pure normal subsemigroup of the group G.*

Henceforth, $S(a_1, \dots, a_n)$ stands for the normal subsemigroup of G generated by $a_1, \dots, a_n$.

Theorem 3.1.2 (Fuchs [19]). *A partial order P on a group G can be extended to a total order if and only if the following conditions are fulfilled:*

EO.3 *for any finite set $a_1, \dots, a_n$ of nonidentity elements of the group G there are integers $\varepsilon_1, \dots, \varepsilon_n$ ($\varepsilon_i = \pm 1$) such that*

$$P \cap S(a_1^{\varepsilon_1}, \dots, a_n^{\varepsilon_n}) = \varnothing$$

If P satisfies the condition EO.3 *and $a \in G$, $a \neq e$, then either the semigroup $P \cdot S(a, e)$ or semigroup $P \cdot S(a^{-1}, e)$ determines a partial order P' satisfying the condition* EO.3.

Proof. Let a total order Q be the extension of a partial order P on a group G and $a_1, \dots, a_n$ be nonidentity elements of the group G. We choose integers $\varepsilon_1, \dots, \varepsilon_n$ ($\varepsilon_i = \pm 1$) such that $a_i^{\varepsilon_i} \notin Q$ for $1 \leqslant i \leqslant n$. Then $S(a_1^{\varepsilon_1}, \dots, a_n^{\varepsilon_n}) \subseteq Q^{-1}$. Consequently, $P \cap S(a_1^{\varepsilon_1}, \dots, a_n^{\varepsilon_n}) = \varnothing$.

To prove the converse assertion, we assume that a partial order P satisfies the condition EO.3 and $a \in G$, $a \neq e$. We first show that the semigroup $P \cdot S(a, e)$ or the semigroup $P \cdot S(a^{-1}, e)$ is a partial order on the group G and satisfies the condition EO.3. Indeed, for the element a there is an integer $\varepsilon = \pm 1$ (without loss of generality, we can assume that $\varepsilon = 1$) such that $P \cap S(a^{\varepsilon}, e) = \varnothing$. By Lemma 3.1.2, the semigroup $P \cdot S^{-1}(a, e)$ equal to $P \cdot S(a^{-1}, e)$ is a partial order on the group G.

We consider two cases.

CASE 1: $P \cap S(a^{-1}, e) \neq \varnothing$. Then the semigroup $P \cdot S(a^{-1}, e)$ satisfies the condition EO.3. Indeed, otherwise there are elements $a_1, \ldots, a_n$, $a_i \neq e$, $1 \leqslant i \leqslant n$, such that for any integers $\varepsilon_i = \pm 1$ the following relation holds: $P \cdot S(a^{-1}, e) \cap S(a_1^{\varepsilon_1}, \ldots, a_n^{\varepsilon_n}) \neq \varnothing$. Hence

$$P \cap S(a, a_1^{\varepsilon_1}, \ldots, a_n^{\varepsilon_n}) \neq \varnothing, \quad P \cap S(a, a_1^{\varepsilon_1}, \ldots, a_n^{\varepsilon_n}) \neq \varnothing$$

which contradicts the condition EO.3 for the partial order P.

CASE 2: $P \cap S(a^{-1}, e) = \varnothing$. Then the semigroups $P \cdot S(a, e)$ and $P \cdot S(a^{-1}, e)$ are partial orders. We show that at least one of them satisfies the condition EO.3. Otherwise in the group G there are elements $a_1, \ldots, a_n, b_1, \ldots, b_m$ different from e and such that, for any integers ε_i, δ_j, $1 \leqslant i \leqslant n$, $1 \leqslant j \leqslant m$, equal to ± 1, the following relations are fulfilled:

$$P \cdot S(a, e) \cap S(a_1^{\varepsilon_1}, \ldots, a_n^{\varepsilon_n}) \neq \varnothing$$

$$P \cdot S(a^{-1}, e) \cap S(b_1^{\delta_1}, \ldots, b_m^{\delta_m}) \neq \varnothing$$

Consequently,

$$P \cap S(a^{\delta}, a_1^{\varepsilon_1}, \ldots, a_n^{\varepsilon_n}, b_1^{\delta_1}, \ldots, b_m^{\delta_m}) \neq \varnothing$$

for any integers $\delta = \pm 1$ $\varepsilon_i = \pm 1$, $\delta_j = \pm 1$, which contradicts the condition EO.3 for the partial order P. The contradiction shows that either $P \cdot S(a, e)$ or $P \cdot S(a^{-1}, e)$ satisfies the condition EO.3.

To complete the proof we assume that Q is a maximal partial order on the group G satisfying the condition EO.3 and $a \in G$, $a \neq e$. In accordance with what has been proved above, the semigroup $Q \cdot S(a, e)$ or the semigroup $Q \cdot S(a^{-1}, e)$ determines a partial order on the group G which satisfies the condition EO.3. Since the order Q is maximal, we conclude that either $Q = Q \cdot S(a, e)$ and $a \in Q$ or $Q = Q \cdot S(a^{-1}, e)$ and $a^{-1} \in Q$. Thus, Q is a total order on the group G. □

By setting $P = \{e\}$ in Theorem 3.1.2, we obtain the following theorem.

Theorem 3.1.3 (Ohnishi [20] and Loś [21]). *A group G admits a total order if and only if it satisfies the following property:*

EO.4 *for any finite set $a_1, \ldots, a_n$ of nonidentity elements of the group G there are integers $\varepsilon_1, \ldots, \varepsilon_n$ ($\varepsilon_i = \pm 1$) such that $e \notin S(a_1^{\varepsilon_1}, \ldots, a_n^{\varepsilon_n})$.*

Theorem 3.1.4 (Lorenzen [22]). *A group G is orderable if and only if for any finite set $a_1 \dots, a_n$ of its nonidentity elements the following equality holds:*

$$\cap S(a_1^{\varepsilon_1}, \dots, a_n^{\varepsilon_n}) = \varnothing$$

where the intersection is taken over all sets of integers $\varepsilon_i = \pm 1$, $1 \leqslant i \leqslant n$.

Corollary 3.1.2 (Mal'cev [23] and B. Neumann [24]). *A group G is orderable if and only if every finitely generated subgroup is orderable.*

3.2. Sufficient Conditions

3.2.1. In a sense, it is simpler to utilize right-ordered groups than totally ordered ones. In particular, this observation is confirmed by the Levi theorem (cf. Theorem 1.6.2). In this section, we list the assertions that follow immediately from the Levi theorem.

Proposition 3.2.1. *Let $\mathfrak{C}(G)$ be a subnormal linearly-ordered-by-inclusion system of subgroups of a group G which is complete under taking unions and intersections. We suppose that all factors of the system $\mathfrak{C}(G)$ admit right orders. Then a right order P can be defined on G such that every subgroup of $\mathfrak{C}(G)$ is convex with respect to P and the prescribed right orders on the factors of the system $\mathfrak{C}(G)$ are orders induced by P.*

Proof. Let $\{A_\alpha \subset B_\alpha \mid \alpha \in I\}$ be the set of all jumps in the system $\mathfrak{C}(G)$. By assumption, each subgroup of A_α is normal in B_α and on each quotient group B_α/A_α a right order P_α is defined. On the group G we introduce an order P as follows: $x \in P$ if $x = e$ or $xA_\alpha \in P_\alpha$. The conditions O.1, O.2, and O.5 from Theorem 1.5.1 are directly verified for P. □

Corollary 3.2.1 (Zaĭtseva [25]). *If there exists a complete subnormal linearly-ordered-by-inclusion system $\mathfrak{C}(G)$ of subgroups of a group G such that all its factors are Abelian torsion-free groups, then the group G admits a right order.*

Proof. The assertion follows immediately from the fact that any Abelian torsion-free group admits a total order (cf. Propositions 1.6.6 and 3.2.1). □

3.2.2. Let $\mathcal{X}$ be an arbitrary group class.

- A group G is called *locally projectable on* $\mathcal{X}$ if any of its finitely generated subgroups has a nonunit homomorphic image that is isomorphic to some group of $\mathcal{X}$.

Theorem 3.2.1 (Burns and Hale [26]). *If a group is locally projectable on the class of right-orderable groups, then it is right-orderable.*

PROOF. We assume that there is not a right-orderable group G every finitely generated subgroup of which has a nontrivial right-orderable homomorphic image. By Theorem 3.1.2, we can assume that the group $G = \operatorname{gp}(g_1, \dots, g_k)$ has the smallest possible number of generators. Then $e \in \operatorname{sgr}(g_1^{\varepsilon_1}, \dots, g_k^{\varepsilon_k})$ for any integers $\varepsilon_1, \dots, \varepsilon_k$ ($\varepsilon_i = \pm 1$). By assumption, the group G has a proper normal subgroup K such that the group G/K is right-orderable. Consequently, at least one of the generators $g_1, \dots, g_k$ does not belong to K. Let $g_1, \dots, g_k \notin K$, $g_{r+1}, \dots, g_k \in K$, $1 \leqslant r \leqslant k$, and let $\delta_1, \dots, \delta_r$ ($\delta_j = \pm 1$) be such that $\operatorname{sgr}(g_1^{\delta_1}, \dots, g_r^{\delta_r}) \cap K = \varnothing$.

Since k is the minimal number of generators for groups that are not right-orderable but locally projectable on the class of right-orderable, there are integers $\delta_{r+1}, \dots, \delta_k$ ($\varepsilon_i = \pm 1$) such that $e \notin \operatorname{sgr}(g_{r+1}^{\delta_{r+1}}, \dots, g_k^{\delta_k})$. Nevertheless, $e \in \operatorname{sgr}(g_1^{\delta_1}, \dots, g_k^{\delta_k})$, i.e., there are positive integers $n_1, \dots, n_s$ such that

$$e = g_{i_1}^{n_1 \delta_{i_1}} \dots g_{i_s}^{n_s \delta_{i_s}}, \quad 1 \leqslant i_j \leqslant k$$

where at least one of the numbers i_j satisfies the condition $i_j \leqslant r$. Hence $K = g_{i_1}^{n_1 \delta_{i_1}} \dots g_{i_s}^{n_s \delta_{i_s}} K$, which contradicts $\operatorname{sgr}(g_1^{\delta_1}, \dots, g_r^{\delta_r}) \cap K = \varnothing$. □

- A group is called *indicable* if it is projectable on an infinite cyclic group.

Corollary 3.2.2. (Burns and Hale [26]). *Any locally indicable group is right-orderable; moreover, it admits a Conradian right order.*

3.2.3. As already mentioned, right-orderable groups are torsion-free. However, not every torsion-free group admits a total right order. The first example of such a group was constructed by Wigold [27].

- A subgroup H of a group G is called *isolated* if $x^n \in H$ implies $x \in H$.

- The *isolator of* a *subgroup* H in the group G is the least isolated subgroup of G containing H.

Theorem 3.2.2 (Rhemtulla [13]). *Let G be a nonunit finitely generated group, the quotient group G/G' be finite, and K be a subgroup of finite index in G such that every right order on it is Conradian. Then the group G is not right-orderable.*

PROOF. We assume that a group G is right-orderable with right order $\leqslant$. We denote by K_1 the isolator of the derived subgroup of K' in K. Then

$$K_1 = \{x \in K \mid x^n \in K'\}$$

for some integer n. We note that K is a finitely generated right-orderable group such that all the jumps of its convex subgroups are Archimedean with respect to every right order. Hence the quotient group K/K_1 is a free Abelian group of finite rank. Let $e = x_1 < x_2 < \ldots < x_n$ be representatives of left cosets x_iK by the normal subgroup K. We consider the mapping τ: $G \to K/K_1$ such that

$$g\tau = x_n g(\overline{x_n g})^{-1} x_{n-1} g(\overline{x_{n-1} g})^{-1} \ldots x_1 g(\overline{x_1 g})^{-1} K_1$$

where $\overline{x_i g}$ is a coset representative (i.e., one of the elements $x_1, \ldots, x_n$) of the coset $x_i gK$. It is easy to see that τ is a homomorphism from G into K/K_1. Since K/K_1 is a free Abelian group and the quotient group G/G' is finite, τ maps the group G onto the unity subgroup of the group K/K_1. Since every right order on K is Conradian and the subgroup K is finitely generated, we conclude that K has a maximal convex normal subgroup C (with respect to the induced right order on K) such that K/C is an Archimedean totally ordered group. The index of the subgroup K in G is finite. Therefore, there is a natural number p such that $x_n^p \in K$. Let $g \in K \setminus C$, $g > e$, be such that $g > x$. Then in the group G the inequalities $g > x_n >, \ldots, > x_2 > x_1 = e$ are fulfilled. Hence $gx_i^{-1} > e$ for all i, $1 \leqslant i \leqslant n$, and $x_i g x_i^{-1} > e$ in G for $1 \leqslant i \leqslant n$. Therefore,

$$x_n g x_n^{-1} x_{n-1} g x_{n-1}^{-1} \ldots x_1 g x_1^{-1} > e$$

Since $\overline{x_i g} = x_i$ and $x_n g x_n^{-1} x_{n-1} g x_{n-1}^{-1} \ldots x_1 g x_1^{-1} = \tau(g) \in K_1$, we have

$$x_n g x_n^{-1} x_{n-1} g x_{n-1}^{-1} \ldots x_1 g x_1^{-1} = g\tau \in C$$

Taking into account the inequality $x_n g x_n^{-1} x_{n-1} g x_{n-1}^{-1} \cdots x_2 g x_2^{-1} > e$, we conclude that

$$x_n g x_n^{-1} x_{n-1} g x_{n-1}^{-1} \cdots x_2 g x_2^{-1} g > g$$

However, $g \in K \setminus C$, $g > e$, and $g\tau \in C$, which is impossible. This contradiction completes the proof of Theorem 3.2.2. □

In view of Theorem 3.2.2, it is easy to construct examples of torsion-free but not right-orderable groups.

EXAMPLE 3.2.1 (Smirnov [28]). The group

$$G = \mathrm{gp}\,(a, b, c, \mid [[a, b], a] = [[a, b], b] = e, c^{-1}ac = a^{-1}$$
$$c^{-1}bc = b^{-1}, [a, b] = c^{2k}), \quad k \geqslant 2$$

is not right-orderable by Theorem 3.2.2. Indeed, as a subgroup K we can take the subgroup $\mathrm{gp}\,(a, b)$. The latter has a finite index in G and is torsion-free nilpotent of class 2. Moreover, by Theorem 2.4.2, the system of its convex subgroups under every right order is subnormal and has Archimedean factors. The derived subgroup of the group G coincides with the subgroup $\mathrm{gp}\,(a^2, b^2, [a, b])$ and has a finite index in G.

EXAMPLE 3.2.2. The group

$$G = \mathrm{gp}\,(a, b, c, \mid b_1^{-1} a_1 b_1 = a_1^{-1}, b_2^{-1} a_2 b_2 = a_2^{-1}$$
$$[a_1, a_2] = [b_1, b_2] = b_1^2 a_2^{-2} = b_2^2 a_1^{-2} = e)$$

is not right-orderable by Theorem 3.2.2. As a subgroup K we can take the subgroup $\mathrm{gp}\,(a_1, a_2)$. The latter has a finite index in G and is a torsion-free Abelian group. The derived subgroup G' of G coincides with $\mathrm{gp}\,(a_1^2, a_2^2)$ and also has a finite index in G. We note that G is a torsion-free metabelian group.

3.3. Group Conditions for Total Orderability

3.3.1. In this section, we follow Mal'cev [12], Podderyugin [29], and Rieger [30] in characterizing the orderability of groups in terms of abstract group theory.

- If there is a group of operators A, acting on a group G, a partial order on G is called an *A-order* provided that for all elements

$g \in G$ and $\alpha \in A$ the relation $g \geqslant e$ implies the inequality $g^\alpha \geqslant e$.

We denote by g^α the image of g under the action of α. For $g \in G$, $\alpha_1, \alpha_2, \ldots, \alpha_n \in A$ and $\varepsilon_1, \varepsilon_2, \ldots, \varepsilon_n$ ($\varepsilon = \pm 1$) we use the following notation:

$$g^{\varepsilon_1\alpha_1+\varepsilon_2\alpha_2+\ldots+\varepsilon_n\alpha_n} = (g^{\alpha_1})^{\varepsilon_1} \cdot (g^{\alpha_2})^{\varepsilon_2} \ldots (g^{\alpha_n})^{\varepsilon_n}$$

We denote by $S_A(M)$ the subsemigroup of G generated by the subset M in G, $S_A(M)' = S_A(M) \cup \{e\}$, which is invariant under A.

Lemma 3.3.1 (Podderyugin [29]). *Let G be an Abelian group with a commutative group of operators A. Every partial A-order on G can be extended to a total A-order if and only if for any element $g \in G$ and every operator $\alpha_1, \ldots, \alpha_n$ from A the relation $g^{\alpha_1} \ldots g^{\alpha_n} = e$ implies the equality $g = e$.*

PROOF. Let a group G admit a total order P stable under the action of operators from A and $g^{\alpha_1} \ldots g^{\alpha_n} = e$ for some element $g \in G$. Then $g^{\alpha_i} > e$ for any $\alpha_i \in A$ and $g^{\alpha_1} \ldots g^{\alpha_n} > e$, which contradicts the hypothesis of the lemma.

Conversely, let in the group G for any element g and operators $\alpha_1, \ldots, \alpha_n$ from A the equality $g^{\alpha_1} \ldots g^{\alpha_n} = e$ imply $g = e$. Let P be a partial A-order on G. We extend P to a maximal partial A-order P_1 on the group G and show that the order P_1 is total. We assume that the order P_1 is not total. Then in G there is an element g such that $g \notin P_1$ and $g \notin P_1^{-1}$. Therefore, $S_A(g) \cap P_1 \neq \varnothing$. Otherwise the semigroup $P_1 S_A(g^{-1})'$ is a pure A-invariant subsemigroup of G and determines the partial A-order on G which is greater than P_1. Indeed, it is evident that $P_1 S_A(g^{-1})'$ is an A-invariant semigroup. Let

$$p_1 g^{-\alpha_1-\ldots-\alpha_s} = (p_2 g^{-\beta_1-\ldots-\beta_r})^{-1}$$

for some elements $p_1, p_2 \in P_1$ and operators $\alpha_1, \ldots, \alpha_s, \beta_1, \ldots, \beta_r$ from A. In view of the commutativity of the group G and of operators from A, we obtain

$$g^{\beta_1+\ldots+\beta_r+\alpha_1+\ldots\alpha_s} = p_1 p_2$$

This contradicts the equality $S_A(g^{-1}) \cap P_1 = \varnothing$. Thus, $P_1 S_A(g^{-1})'$ is a partial A-order on G that is an extension of P_1. We arrive at a contradiction with the maximality of the order P_1. Hence

$$S_A(g^{-1})' \cap P_1 \neq \varnothing$$

We can obtain

$$S_A(g^{-1})' \cap P_1 \neq \varnothing$$

in a similar way. Then in A there are operators $\alpha_1, \dots, \alpha_r$ and $\beta_1, \dots, \beta_s$ such that $g^{\alpha_1+\dots+\alpha_r} \geqslant e$ and $g^{\beta_1+\dots+\beta_s} \leqslant e$. Therefore,

$$\begin{aligned} e \leqslant &(g^{\alpha_1+\dots+\alpha_r})^{\beta_1} \dots (g^{\alpha_1+\dots+\alpha_r})^{\beta_s} = g^{\alpha_1\beta_1+\dots+\alpha_i\beta_j++\dots\alpha_r\beta_s} \\ =&g^{\beta_1\alpha_1+\dots+\beta_j\alpha_i+\dots+\beta_s\alpha_r} = (g^{\beta_1+\dots+\beta_s})^{\alpha_1} \dots (g^{\beta_1+\dots+\beta_s})^{\alpha_r} \leqslant e \end{aligned}$$

Consequently,

$$g^{\alpha_1\beta_1+\dots+\alpha_r\beta_1} \dots g^{\alpha_1\beta_s+\dots+\alpha_r\beta_s} = e$$

which contradicts the above assumption. □

- A subset H of a group G is called *strongly isolated* if for elements $x, g_1, \dots, g_n$ of the group G the relation

  $$g^{1+g_1+\dots+g_n} = x \cdot g_1^{-1} x g_1 \dots g_n^{-1} x g_n \in H$$

  implies the inclusion $x \in H$.

- A group G is called a *group with strongly isolated identity element* or a *G-torsion-free group* if its identity element e is a strongly isolated subset.

We denote by $[A, B]$ the commutator subgroup of subsets A and B of a group G, i.e., the subgroup of G generated by all commutators $[a, b] = a^{-1}b^{-1}ab$, where $a \in A$ and $b \in B$. We denote by $\boldsymbol{n}(A) = N_G(A)$ the normalizer of a subset A of G, i.e., the set of all elements $g \in G$ such that $g^{-1}Ag = A$. Now, we describe a group admitting a total order in terms of a system of its subgroups.

Theorem 3.3.1 (Mal'cev [12], Podderyugin [29], and Rieger [30]). *A group G is orderable if and only if there exists a soluble infrainvariant linearly-ordered-by-inclusion system $\mathfrak{C}(G)$ of its subgroups such that:*

(a) *every subgroup of $\mathfrak{C}(G)$ is strongly isolated in G;*

(b) *for any jump $A \prec B$ in $\mathfrak{C}(G)$ the relation $[[N_G(A), N_G(A)], B] \subseteq A$ holds.*

The subgroups of $\mathfrak{C}(G)$ are convex with respect to some total order on the group G.

PROOF. Let G be a totally ordered group. Theorem 2.3.1 guarantees the existence of the required system for a totally ordered group G. Indeed, as $\mathfrak{C}(G)$ we can consider the system of convex subgroups of G. We note that the condition (b) is equivalent to the commutativity of automorphisms of the group B/A induced by the inner automorphisms of the group G generated by the elements of $N_G(A)$. The condition (a) is also fulfilled for the system $\mathfrak{C}(G)$ of convex subgroups of a totally ordered group G: if $x \cdot g_1^{-1}xg_1 \dots g_n^{-1}xg_n \in H$, where $x, g_1, \dots, g_n \in G$, $H \in \mathfrak{C}(G)$, and $x > e$, then under the assumption that $x \notin H$ we can deduce that $x > h$ for any element $h \in H$. Consequently,

$$x \cdot g_1^{-1}xg_1 \dots g_n^{-1}xg_n > e$$

which contradicts our assumption. Thus, the necessity follows from Theorem 2.3.1.

We now prove the converse assertion. Let $\mathfrak{C}(G)$ be a system of subgroups of G which is subject to the hypotheses of the theorem. We divide the set of jumps $\{A_\alpha \prec B_\alpha\}$ in $\mathfrak{C}(G)$ into disjoint classes of conjugate jumps. In each class we choose a representative $\{A_\alpha^* \prec B_\alpha^*\}$. The set of such jumps will be denoted by $\mathfrak{C}^*$. We show that in the quotient group B_α^*/A_α^* one can introduce a total order invariant under the action of automorphisms induced on B_α^*/A_α^* by the inner automorphisms of the group $N_G(A_\alpha^*)$.

The group of automorphisms $N_G(A_\alpha^*)$ acts on the group B_α^*/A_α^* as follows: $(bA_\alpha^*)^g = g^{-1}bgA_\alpha^*$ for $b \in B_\alpha^*$ and $g \in N_G(A_\alpha^*)$. The condition (b) guarantees the commutativity of operators of $N_G(A_\alpha^*)$ on B_α^*/A_α^*. Moreover, for all elements $b \in B_\alpha^*$ and $g_1, \dots, g_n \in N_G(A_\alpha^*)$ the equality $(bA_\alpha^*)^{1+g_1+\dots+g_n} = A_\alpha^*$ implies the inclusion $bg_1^{-1}bg_1 \dots g_n^{-1}bg_n \in A_\alpha^*$. Since subgroup A_α^* is strongly isolated, we obtain $b \in A_\alpha^*$, $bA_\alpha^* = A_\alpha^*$ in B_α^*/A_α^*. Applying Lemma 3.3.1, we can conclude that on B_α^*/A_α^* the trivial $N_G(A_\alpha^*)$-order can be extended to a total $N_G(A_\alpha^*)$-order $\overline{Q}_\alpha^*$. We now can define the desired total order on the group G. For an element $x \in G$, $x \neq e$, we take a jump $A_\alpha \prec B_\alpha$ in $\mathfrak{C}(G)$ such that $x \in B_\alpha \setminus A_\alpha$. Starting from this jump, we choose a representative $A_\alpha^* \prec B_\alpha^*$ in $\mathfrak{C}^*$ which is conjugated with the jump $A_\alpha \prec B_\alpha$. For this jump we find an element $g \in G$ such that $A_\alpha^g = g^{-1}A_\alpha g = A_\alpha^*$. Let $x > e$ in G if and only if $g^{-1}xgA_\alpha^* > A_\alpha^*$ in the quotient group $A_\alpha^* \prec B_\alpha^*$. We note that such a definition of $\leqslant$ is independent of the choice of $g \in G$. Indeed, if the equalities $A_\alpha^g = A_\alpha^h = A^*$ and $A_\alpha^h = A_\alpha^*$ hold, then $(A_\alpha^*)^{g^{-1}h} = A_\alpha^*$, i.e., $g^{-1}h \in N_G(A_\alpha^*)$ and $h = g \cdot u$, where $u \in N_G(A_\alpha^*)$. This means that $h^{-1}xhA_\alpha^* = u^{-1}g^{-1}xguA_\alpha^* = u^{-1}(g^{-1}xgA_\alpha^*)u$. Since $g^{-1}xgA_\alpha^* > A_\alpha^*$ in

B^*_α/A^*_α, $u \in N_G(A^*_\alpha)$, and the order $\overline{Q}_{\alpha^*}$ is invariant under the action of automorphisms of $N_G(A^*_\alpha)$, the above definition of the order $x > e$ is independent of the choice of the element g subject to the condition $A^g_\alpha = A^*_\alpha$.

It is clear that the set P of elements $x \in G$ such that $x > e$ is pure, i.e., $P \cap P^{-1} = \varnothing$. Moreover, $P \cup P^{-1} \cup \{e\} = G$. We further prove that P is a semigroup. Let $x > e$, $y > e$, and $x \in B_\alpha \setminus A_\alpha$, $y \in B_\beta \setminus A_\beta$, where $A_\alpha \prec B_\alpha$ and $A_\beta \prec B_\beta$ are jumps in $\mathfrak{C}(G)$. If $B_\beta \subseteq A_\alpha$, then $xy \in B_\alpha \setminus A_\alpha$. For an element $g \in G$ such that $A^*_\alpha = A^g_\alpha$ we have $g^{-1}xyg \in B^*_\alpha \setminus A^*_\alpha$, $g^{-1}xygA^*_\alpha = g^{-1}xgA^*_\alpha$, and $g^{-1}xgA^*_\alpha > A^*_\alpha$. Hence $xy > e$. We obtain $xy > e$ provided that $B_\alpha \subseteq A_\beta$. If $B_\alpha = B_\beta$, then for $g \in G$ such that $A^*_\alpha = A^*_\alpha$ we have $g^{-1}xgA^*_\alpha > A^*_\alpha$ and $g^{-1}ygA^*_\alpha > A^*_\alpha$ in B^*_α/A^*_α. We see that the inequality $xy > e$ is also fulfilled. Thus, P is a semigroup.

We now show that the subsemigroup P is normal in G. Let $x > e$, $h \in G$, and $x \in B_\alpha \setminus A_\alpha$ for a jump $A_\alpha \prec B_\alpha$ in $\mathfrak{C}(G)$. Then the element $h^{-1}xh$ belongs to the jump $A^h_\alpha \prec B^h_\alpha$ in $\mathfrak{C}^*$. If $g^{-1}A_\alpha g = A^*_\alpha$, then

$$g^{-1}h(A^h_\alpha h^{-1}g) = g^{-1}h \cdot h^{-1}A_\alpha h \cdot h^{-1}g = g^{-1}A_\alpha g = A^*_\alpha$$

and $g^{-1}h(h^{-1}xh)h^{-1}gA^*_\alpha = g^{-1}xgA^*_\alpha A^*_\alpha$ in B^*_α/A^*_α with respect to the order $\overline{Q}^*_\alpha$, i.e., $h^{-1}xh > e$ in G. The normality of the subsemigroup P in G is established. Therefore, the semigroup P satisfies the hypotheses of Theorems 1.5.1 and 1.5.2. Consequently, P defines a total order on G.

It is easy to see that every lower subgroup of a jump in $\mathfrak{C}(G)$ is a convex subgroup with respect to the constructed order P. Since each subgroup of the system $\mathfrak{C}(G)$ is either an intersection or union of lower subgroups of jumps, any subgroup of the system $\mathfrak{C}(G)$ is convex, which completes the proof of the theorem. □

Corollary 3.3.1 (Kokorin [31]). *A metabelian group with strongly isolated identity element is orderable.*

PROOF. We first prove that a group with strongly isolated identity element is a group with unique roots. Indeed, if for some elements $a, b \in G$ and natural number n the equality $a^n = b^n$ is fulfilled, then the following chain of equalities holds:

$$e = b^{-n}a^n = b^{-n+1}(b^{-1}a)b^{n-1} \cdot b^{-n+2}(b^{-1}a)b^{n-2} \cdot \ldots \cdot b^{-2}(b^{-1}a)b^2 \cdot b^{-1}(b^{-1}a)b \cdot (b^{-1}a)$$

This implies $a = b$. Since the derived subgroup of G is Abelian and the isolator of an Abelian subgroup of the group with unique roots is also Abelian, the isolator $J(G')$ of the derived subgroup of G is an Abelian subgroup of G. Direct verification shows that the system of subgroups $E \subset J(G) \subset G$ satisfies the hypotheses of Theorem 3.3.1. □

The following theorem is often useful. Its proof is similar to that of Theorem 3.3.1.

Theorem 3.3.2 (Kokorin [32]). *A group G is orderable if and only if it has an infrainvariant system $\mathfrak{C}(G)$ of subgroups such that for any jump $A \prec B$ in $\mathfrak{C}(G)$ the quotient group B/A admits a total order invariant under the action of those automorphisms that are induced by the inner automorphisms of the normalizer $N_G(A)$. Moreover, on G there is a total order with respect to which all subgroups of $\mathfrak{C}(G)$ are convex.*

Corollary 3.3.2. *A torsion-free locally nilpotent group is orderable.*

PROOF. Let G be a torsion-free nilpotent group. Then its upper central series $E = \zeta_0(G) \subseteq \zeta_1(G) \subseteq \dots \subseteq \zeta_n(G) = G$ satisfies all hypotheses of Theorem 3.3.2. The direct application of Corollary 3.3.1 completes the proof. □

3.3.2. Applying Theorem 3.1.1, Darnel, Glass, and Rhemtulla [33] have proved that if every total right order on a right-orderable group is two-sided, then the group is a torsion-free Abelian one. Kim and Rhemtulla [34] have obtained the following useful result: any right-orderable n-engel group (i.e., a group satisfying the identity

$$[a, \underbrace{y, \dots, y}_{t}] = e$$

for some natural number t) is nilpotent.

3.4. Groups of Automorphisms

The following result is fundamental in the right-ordered group theory. It gives the possibility of applying the sophisticated techniques of the lattice-ordered group theory to the study of right-ordered groups.

Theorem 3.4.1 (Cohn [35], Zaĭtseva [25], and Conrad [8]). *A group G is right-orderable if and only if it is isomorphically embedded into a group of automorphisms of some totally ordered set.*

PROOF. Let G be a right-ordered group. We consider the right regular representation $R\colon G \to \mathcal{A}(G)$ of the group G which for each element $g \in G$ defines the mapping $R(g)$ of the group G onto itself as follows: $(x)R(g) = xg$ for $x \in G$. As is well known, R is an isomorphic embedding of the group G into the group $\mathcal{A}(G)$ of automorphisms of the totally ordered set $\{G; \leqslant\}$. The converse assertion was proved in Example 1.4.5. □

Corollary 3.4.1. *The class of right-orderable groups coincides with the class of all groups which are isomorphically embeddable into lattice-ordered groups.*

PROOF. The assertion follows from the following facts:

— the group $\mathcal{A}(X)$ of all order automorphisms of a totally ordered set X is a lattice-ordered group with the pointwise order,

— any lattice-ordered group is isomorphic to an l-subgroup of the group $\mathcal{A}(X)$ of automorphisms of some totally ordered set X (the Holland theorem on the representation of l-groups (cf., for example, [36, Theorem 4.1.1] or [37, Theorem 4.1.1]). □

• A group G is called *divisible* (or *radicable*) if for any element $g \in G$ and natural number n the equation $x^n = g$ has at least one solution in G.

Corollary 3.4.2 (Holland [38]). *Any right-orderable group G is isomorphically embedded into a divisible (radicable) right-orderable group.*

PROOF. It suffices, using Theorem 3.4.1, to construct an embedding of the group G into the group $\mathcal{A}(G)$ of automorphisms of the totally ordered set $\{G; \leqslant\}$ of the right-ordered group G and then to construct an embedding of the group $\mathcal{A}(G)$ into a divisible group of automorphisms $A(X)$ of some totally ordered set X (cf., for example, [36, Theorem 7.7.1] or [37, Theorem 4.4.3]). □

The following interesting more precise version of Theorem 3.4.1 was obtained by Smirnov [5]. Before formulating it we introduce the necessary definitions. Let K be an associative ring with unit e. A unitary right K-module M is called *free* if there is a subset X such that each element $a \in M$ can be uniquely represented as a finite sum $a = x_1k_1 + \ldots + x_nk_n$ of some elements $x_1, \ldots, x_n$ of the set X with coefficients $k_1, \ldots, k_n$ from K. In this case, the set X is called a *basis* in M over K. If $\mathbb{Z}[G]$ is the integral group ring of the group G, then each right $\mathbb{Z}[G]$-module will be called a *G-module*. Let M be a module over the ring K. We suppose that on M the order relation $\leqslant$ is defined with respect to which M is a totally ordered group. Moreover, we assume that this order is preserved under multiplication by elements of some subset Γ of the ring K. Then M is called a *Γ-ordered K-module.* If $K = \mathbb{Z}[G]$ and $\Gamma = G$, then M is called a *G-ordered module.*

Theorem 3.4.2 (Smirnov [5]). *Any right-ordered group G is isomorphic to a subgroup of the group of all order automorphisms of a totally ordered free Abelian group A.*

PROOF. Let G be a right-ordered group. We denote by A the additive group of the integral group ring $\mathbb{Z}[G]$ of the group G. It is clear that A is a free Abelian group with free basis G. We first prove that A is a G-ordered $\mathbb{Z}[G]$-module. Indeed, any element $a \in A$ is uniquely represented in the form $a = n_1g_1 + \ldots n_sg_s$, where $g_1, \ldots, g_s \in G$ and $g_1 < \ldots < g_s$ with respect to a given right order on G and $n_1, \ldots, n_s$ are nonzero integers. We assume that $a \in P$ for $n_s > 0$. Direct verification shows that the set $P \cup \{0\}$ defines on A a total order which is preserved under multiplication from the right by elements of the group G. It is also evident that, for the mapping R corresponding to each element $g \in G$, the right translation $R(g)$ of the group G defines an isomorphism of the group G into the group of order automorphisms of the totally ordered group A. □

3.5. Connection with Totally Ordered Groups

Let G be a group on which a group Γ of endomorphisms acts.

- The group G is called *Γ-orderable* if it is possible to introduce a total order $\leqslant$ on G such that G becomes a totally ordered group and the order is preserved under the action of endomorphisms of Γ.

Lemma 3.5.1 (Smirnov [5]). *The quotient group F/A of a non-commutative free group F by a normal subgroup A is right-orderable if and only if the quotient $A/[A,A]$ of the group A by its derived subgroup $[A,A]$ is $F/[A,A]$-orderable.*

PROOF. We assume that the group $G = F/A$ is right-orderable. We set $M = A/[A,A]$ and define the multiplication of elements of the group M from the right by elements of the group G as follows: $a \cdot g = f^{-1}af$, where $a \in M$, $f \in F$, and $g = Af$ in F/A. Then M becomes a right G-module. By the Magnus theorem (cf., for example, [3]) M is a submodule of a free G-module S. However, any free G-module is a direct sum of free cyclic G-modules, i.e., modules that are isomorphic to the additive group of the integral group ring $\mathbb{Z}[G]$ of the group G over itself. Since the group G is right-orderable, the module $\mathbb{Z}[G]$ regarded as a G-module is G-orderable. Thus, the modules S and M are G-orderable modules. This means that the group $A/[A,A]$ admits a total order which is preserved under the action of any inner automorphism of the group $F/[A,A]$.

Conversely, let the group $A/[A,A]$ admit a total order P which is preserved under the action of any automorphism of the group $A/[A,A]$ induced by the inner automorphisms of the group F. We show that the group F/A is right-orderable. Let C be a centralizer of the quotient group $A/[A,A]$ in F. It is easy to see that $C \supseteq A$ and the quotient group F/C is isomorphic to the group of automorphisms of the group $A/[A,A]$ induced by the inner automorphisms of the group F. Since the group $A/[A,A]$ is $F/[A,A]$-orderable, the group F/C admits a right order. To complete the proof it remains to show that $C = A$. To prove the last equality we consider an element $c \in C$ and the quotient group $H/[A,A]$, where H is the subgroup of the group F generated by A and c. It is clear that $H/[A,A]$ is an Abelian group. Therefore, $[H,H] = [A,A]$. In noncommutative free groups this means that $H = A$ (cf. the Auslander–Lyndon theorem in [39, Theorem 43.21]), i.e., $c \in A$ and $C = A$. □

Theorem 3.5.1 (Smirnov [5]). *If the quotient group F/A of a free group F by a normal subgroup A has a soluble infrainvariant system of subgroups, all the factors of which are torsion-free Abelian groups, then the group $F/[A,A]$ admits a total order.*

PROOF. Assume that F is a free group of rank at least 1 and $A \neq E$. By assumption, in F there exists an infrainvariant system

$\mathfrak{C}(F)$ of subgroups that connects F with A so that

$$\mathfrak{C}(F)\colon F = F_1 \supset \ldots \supset F_\alpha \supset F_{\alpha+1} \supset \ldots \supset F_\xi = A$$

and all the factors $F_\alpha/F_{\alpha+1}$ of jumps are torsion-free Abelian groups. We set $F_0' = F$ and $F_\alpha' = [F_\alpha, F_\alpha]$, where $\alpha \neq 0$. Then the system of subgroups

$$\mathfrak{C}'(F) : F = F_0' \supset F_1' \supset \ldots F_\alpha' \supset F_{\alpha+1}' \supset \ldots \supset F_\xi' = [A, A]$$

is complete under taking unions and intersections (cf. [40]). Since $F_\alpha \supset F_{\alpha+1} \supseteq F_\alpha' \supseteq F_{\alpha+1}'$ and $F_{\alpha+1}/F_{\alpha+1}'$ is a free Abelian group, we conclude that $F_\alpha'/F_{\alpha+1}'$ is a free Abelian group too. Let N_α be the normalizer of the subgroup $F_{\alpha+1}'$ of F. It is easy to check that N_α is the normalizer of F_α'. By Theorem 3.3.2, it suffices to show that for any α the quotient group $F_\alpha'/F_{\alpha+1}'$ admits a total order which is preserved under the action of any inner automorphism of the group N_α. If $\alpha = 0$, then $F_\alpha' = F \supset F_1' = F'$, and this assertion is evident for the quotient group F/F'. If $\alpha \neq 0$, then

$$F \supseteq N_\alpha \supseteq F_\alpha \supset F_{\alpha+1} \supseteq F_\alpha' \supseteq F_{\alpha+1}'$$

Let $N(F_\alpha)$ be the normalizer of the subgroup F_α of F. We show that $N(F_\alpha) = N_\alpha$. Indeed, since the subgroup F_α' is characteristic in F_α, it follows that $N(F_\alpha) \supseteq N_\alpha$. Conversely, let $g \in N_\alpha$, and let $F_\beta = g^{-1}F_\alpha g$. The infrainvariance of the system of subgroups $\mathcal{C}(F)$ in F implies that the subgroup F_β is contained in $\mathcal{C}(F)$, i.e., $g \in N(F_\alpha)$.

Thus, N_α is the normalizer of both subgroups F_α and $F_{\alpha+1}$ of F. By assumption, for F there is an infrainvariant system of subgroups passing through $F_{\alpha+1}$ and possessing Abelian factors. Thus, the quotient group $N_\alpha/F_{\alpha+1}$ has a soluble subnormal system of subgroups with torsion-free factors. By Corollary 3.2.1, the group $N_\alpha/F_{\alpha+1}$ admits a right order. By Lemma 3.3.1, the quotient group $F_{\alpha+1}/F_{\alpha+1}'$ admits a total order which is invariant under the action of inner automorphisms of the group N_α. Therefore, its subgroup $F_\alpha'/F_{\alpha+1}'$ has the same property. □

Corollary 3.5.1 (Smirnov [5]). *If the quotient group F/A of a free group F by some normal subgroup A has a soluble infrainvariant system of subgroups, then the quotient group F/A' is right-orderable and the quotient group F/A'' admits a total order (here $A' = [A, A]$, $A'' = [A', A']$).*

PROOF. If

$$\mathfrak{C}(F) : F = F_1 \supset \ldots \supset F_\alpha \supset F_{\alpha+1} \supset \ldots \supset F_\xi = A$$

is a soluble infrainvariant system of subgroups in F and

$$\mathfrak{C}'(F) : F_0' = F \supset F_1' \supset \ldots \supset F_\alpha' \supset F_{\alpha+1}' \supset F_\xi' = A'$$

then, as in the proof of Theorem 3.5.1, the system $\mathfrak{C}'(F)$ is soluble and all its factors are torsion-free Abelian groups. By Corollary 3.2.1, the quotient group F/A' is right-orderable. The orderability of the quotient group F/A'' follows from Theorem 3.5.1. □

3.6. Fully Orderable Groups

3.6.1. We introduce the following definition.

- A group G is called *fully right-orderable* if each of its partial right orders can be extended to a total right order.

We indicate the necessary and sufficient conditions for fully right-orderability of groups in terms of semigroups.

RO* An element $a \neq e$ of a group G has the following properties:

RO*.1 the element a has an infinite order,

RO*.2 for any subset $S \neq \{e\}$ of G, a nonidentity element $b \in \operatorname{sgr}(S^{-1} \setminus \{e\}, a)$, and an element $c \in \operatorname{sgr}(S^{-1} \setminus \{e\}, a^{-1})$, the subsemigroup $\operatorname{sgr}(S \setminus \{e\}, b, c)$ contains the identity element e.

Theorem 3.6.1 (Todorinov [41]). *A group G is fully right-orderable if and only if each of its nonidentity elements satisfies the condition* RO*.

PROOF. Let G be a fully right-orderable group and an element $a \in G$, $a \neq e$, satisfy the condition RO*. Then there is a subset $S \neq \{e\}$ and elements $b \in \operatorname{sgr}(S^{-1} \setminus \{e\}, a)$, $b \neq e$, and $c \in \operatorname{sgr}(S^{-1} \setminus \{e\}, a^{-1})$ such that the semigroup $\operatorname{sgr}(S \setminus \{e\}, b, c)$ does not contain the identity element e of the group G. The partial right order $P = \operatorname{sgr}(S \setminus \{e\}, b, c) \cup \{e\}$ on the group G can be extended to a total right order $\overline{P}$. Since $\overline{P}$ is a total right order, we have either $a > e$ or $a < e$ with respect to the order $\overline{P}$. We assume that $a > e$. Since $S \setminus \{e\} \subseteq \overline{P}$, we have $c < e$ with respect to the order $\overline{P}$. This contradicts the relation $c > e$ with respect to the order $\overline{P}$. We consider the case corresponding to $a < e$ with respect to the order P in a similar way.

Now, let an element a satisfy the condition RO* and P be a maximal partial right order on the group G. We assume that one of the semigroups $\operatorname{sgr}(P^{-1} \setminus \{e\}, a)$ and $\operatorname{sgr}(P^{-1} \setminus \{e\}, a^{-1})$ does not contain e. In view of the maximality of the partial right order P^{-1}, we have $P^{-1} = \operatorname{sgr}(P^{-1} \setminus \{e\}, a^{-1})$ or $P = \operatorname{sgr}(P^{-1} \setminus \{e\}, a)$. Therefore, the elements a and e are comparable with respect to the maximal order P.

Let $e \in \operatorname{sgr}(P^{-1} \setminus \{e\}, a), \operatorname{sgr}(P^{-1} \setminus \{e\}, a^{-1})$. Then for some elements $p_1, \ldots, p_{n+1} \in P$ the following equality holds:

$$e = (p_1^{-1})^{\varepsilon_1} a (p_2^{-1})^{\varepsilon_2} a \ldots (p_n^{-1})^{\varepsilon_n} a (p_{n+1}^{-1})^{\varepsilon_{n+1}}$$

where $\varepsilon_i = 0, 1$ and $(p_i^{-1})^0 = e$. If $\varepsilon_1 = \ldots = \varepsilon^{n+1} = 0$, then $a^n = e$, which is impossible. Let $\varepsilon_1 \neq 0$. Then

$$p_1 = a(p_2^{-1})^{\varepsilon_2} a \ldots (p_n^{-1})^{\varepsilon_n} a (p_{n+1}^{-1})^{\varepsilon_{n+1}} = b \in \operatorname{sgr}(P^{-1} \setminus \{e\}, a) \cap P$$

We now assume that $\varepsilon_1 = \varepsilon_{n+1} = 0$. The equality

$$e = a(p_2^{-1})^{\varepsilon_2} a \ldots (p_{n-m}^{-1})^{\varepsilon_{n-m}} a (p_{n+1-m}^{-1})^{\varepsilon_{n+1-m}} a^m$$

where $(p_{n+1-m}^{-1})^{\varepsilon_{n+1-m}} a^m \neq e$, $1 \leqslant m \leqslant n-1$, implies

$$p_{n+1-m} = a^{m+1} (p_2^{-1})^{\varepsilon_2} a \ldots (p_{n-m}^{-1})^{\varepsilon_{n-m}} a = b \in \operatorname{sgr}(P^{-1} \setminus \{e\}, a) \cap P$$

By analogy, we find $c \in \operatorname{sgr}(P^{-1} \setminus \{e\}, a^{-1}) \cap P$, $c \neq e$. Then $\operatorname{sgr}(P \setminus \{e\}, b, c) \subseteq P \setminus \{e\}$. Thus, the element a does not satisfy the condition RO*. This contradiction completes the proof. □

Corollary 3.6.1 (Todorinov [41]). *The quotient group G/H of a fully right-orderable group G by a normal isolated subgroup H is fully right-orderable.*

- The group class $\mathcal{X}$ is called *locally closed* if any group such that each of its finitely generated subgroups is contained in $\mathcal{X}$, is contained in $\mathcal{X}$ itself.

Theorem 3.6.2 (Rhemtulla [13]). *The class of all fully right-orderable groups is locally closed.*

PROOF. We assume that every finitely generated subgroup of a group G is fully right-orderable, but the group G itself is not fully right-orderable. Then there exists a maximal partial right order P which is not a total right order. This means that there is an element $x \in G$ such that $x \notin P \cup P^{-1}$. It follows that $e \in \mathrm{sgr}\,(P, x) \cup \mathrm{sgr}\,(P, x^{-1})$. Therefore, there are elements $p_1, \dots, p_n, q_1, \dots, q_m \in P$ such that

$$e = x^{r_1} p_1 \dots x^{r_n} p_n = x^{-s_1} q_1 \dots x^{-s_m} q_m$$

for some nonnegative exponents $r_1, \dots, r_n, s_1, \dots, s_m$. Thus, in the subgroup $H = \mathrm{gp}\,(p_1, \dots, p_n, q_1, \dots, q_m, x)$, the partial right order $P \cap H$ cannot be extended to a total right order, which contradicts the assumption. □

It is not hard to note that the condition RO* is satisfied for any torsion-free Abelian group. We now consider the question on the extension of partial right orders on torsion-free nilpotent groups.

Theorem 3.6.3 (Rhemtulla [13]). *Any torsion-free nilpotent group is fully right-orderable.*

PROOF. Let, on the contrary, G be a torsion-free nilpotent group with a partial maximal but not total right order P. Then there is an element $x \notin P \cup P^{-1} \cup \{e\}$ such that the semigroups $\mathrm{sgr}\,(P, x) \cup \{e\}$ and $\mathrm{sgr}\,(P, x^{-1}) \cup \{e\}$ do not define the right order, i.e.,

$$e = x^{s_1} p_1 \dots x^{s_m} p_m = x^{-r_1} q_1 \dots x^{-r_n} q_n$$

for some nonnegative exponents $r_1, \dots, r_n, s_1, \dots, s_m$ and elements $p_i, q_j \in P$, $1 \leqslant i \leqslant m$, $1 \leqslant j \leqslant n$. As usual, $\zeta_i(G)$ denotes the ith member of the upper central series of the group G.

We note that $x \notin \zeta_1(G)$. Otherwise the last equality takes the form $x^s p = x^{-r} q = e$ with some integers $s, r \in \mathbb{Z}$, $s, r > 0$, and elements $p, q \in P$. This means that $x^{sr} \in P$ and $x^{-sr} \in P$, i.e., $x^{sr} \in P \cap P^{-1}$, which contradicts the fact that the semigroup P is pure.

Let W be the set of all words of signature $w = x^{s_1} p_1 \dots x^{s_m} p_m$, where $p_1, \dots, p_m \in P$, $s_1, \dots, s_m > 0$, such that $m \geqslant 1$ and $w = e$ in G. We define a function μ on P with values in the set of nonnegative integers so that

$$\mu(p) = \begin{cases} 0 & \text{if } [p, x] = e \\ j & \text{if } [p, x] \in \zeta_j(G) \setminus \zeta_{j-1}(G) \end{cases}$$

For a word $w = x^{s_1}p_1 \dots x^{s_m}p_m \in W$ we put $l(w) = \max\{\mu(p_i) \mid 1 \leqslant i \leqslant m\}$ and note that if $l(w) = 0$, then $w = x^s p$, where $s = \sum_{i=1}^{m} s_i > 0$ and $p = p_i \dots p_m \in P$. Consequently, $x^{-s} \in P$.

We now prove that there exists an element $w \in W$ such that $l(w) = 0$. We assume that this is not so. Then we take a subset $W_1 \subseteq W$ of elements $w \in W$ such that $l(w)$ is minimal. For the element $w = x^{s_1}p_1 \dots x^{s_m}p_m$ we refer to p_i as a *dominant component* of the word w provided that $\mu(p_i) = l(w)$. Since $w = e$ in G for any word $w \in W$, it is easy to see that every element $w \in W$ has at least two dominant components. We denote by W_2 the set of those elements $w \in W_1$ the number of dominant components of which is minimal and by W_3 the set of those words of W_2 for which $\mu(p_1) = l(w)$. If $w \in W_2$, then

$$w' = p_1^{-1}x^{-s_1}wx^{s_1}p_1 = x^{s_2}p_2 \dots x^{s_m}p_m$$

Hence the set W_3 is not empty. Let $j > 1$, $j > i$, and

$$w = x^{s_1}p_1 \dots x^{s_m}p_m \in W_3$$

Then $\mu(p_i) < \mu(p_1)$ and $\mu(p_j) = \mu(p_1)$. Let W_5 be the set of those words $w = x^{s_1}p_1 \dots x^{s_m}p_m \in W_4$ for which m is minimal. By the assumption that there is no element $w \in W$ satisfying the equality $l(w) = 0$, we have $m > 1$ for each word $w = x^{s_1}p_1 \dots x^{s_m}p_m \in W_5$. Since $x^{s_j}p_j = p_j(x[x, p_j])^{s_j}$, we can assume that $[x, p_j] < e$ in G; otherwise the following equality holds:

$$x^{s_1}p_1 \dots x^{s_m}p_m = x^{s_1}p_1 \dots x^{s_{j-1}}p_{j-1}p_j(x[x, p_j])^{s_j}x^{s_{j+1}} \dots x^{s_m}p_m = w'$$

and $l(w') = l(w)$, where the words w' and w have the same number of dominant components, but the dominant component of w' has the $(j-1)$th position, which contradicts the choice of j.

Thus, $[x, p_j] < e$ in G. If $j < m$, we consider an element $w_1 \in W_5$ of the form

$$w_1 = x^{s_1}p_1 \dots x^{s_{j-1}}p_{j-1}x^{s_j+1}p_{j1}x^{s_{j+1}-1} \dots x^{s_m}p_m$$

where $p_{j1} = x^{-1}p_jx = p_j[p_j, x] \in P$ and $\mu(p_{j1}) = \mu(p_1)$. Repeating the above procedure with the element w_1, we build the words

$$w_i = x^{s_1}p_1 \dots x^{s_{j-1}}p_{j-1}x^{s_j+i}p_{ji}x^{s_{j+1}-i} \dots x^{s_m}p_m, \quad 1 \leqslant i \leqslant s_{j+1}$$

where $p_{j,i-1} = x^{-1}p_{ij}x = p_{ij}[p_{ij}, x]$ and $\mu(p_{ij}) = \mu(p_1)$. Then $w_i \in W_5$ for $i < s_{j+1}$. It is easy to see that

$$w_{s_{j+1}} = x^{s_1}p_1 \dots p_{j-1}x^{s_j+s_{j+1}}p'x^{s_{j+2}} \dots p_m$$

where $p' = p_{js_{j+1}} \cdot p_{j+1} \in P$. It is also obvious that the inequality $\mu(p') < \mu(p_1)$ fails; otherwise this contradicts the definition of W_2. But the equality $\mu(p') = \mu(p_1)$ is also impossible because it contradicts the choice of W_5.

If $j = m$, we conjugate the element $w = x^{s_1}p_1 \dots x^{s_m}p_m \in W_5$ by $x^{s_m}p_m$ and obtain $w' = x^{s_m}p_m x^{s_1}p_1 \dots x^{s_{m-1}}p_{m-1} \in W_5$. Arguing as in the case $j < m$, we conclude that in W there is an element

$$w'' = x^{s_m+s_1}(p_m[p_m, x])^{s_1}p_1 \dots x^{m-1}p_{m-1}$$

which either has a unique dominant component $(p_m[p_m, x])^{s_1}p_1$ (if $\mu((p_m[p_m, x])^{s_1})p_1 = \mu(p_1))$ or satisfies the inequality $l(w'') < l(w)$. But both cases are impossible.

Thus, we have proved that there is an element $w \in G$ such that $l(w) = 0$, i.e., $w = x^s p = e$ for some s, p, where $s \in \mathbb{N}$ and $p \in P$. In a similar way, we can prove that there is a positive integer r and an element $q \in P$ such that $x^{-r}q = e$. Hence $x^{-s} \in P$, $x^r \in P$, and $x^{rs} \in P \cap P^{-1} = \{e\}$. But this is impossible because $x \neq e$ and G is a torsion-free group. □

Corollary 3.6.2. *Any torsion-free locally nilpotent group is fully right-orderable.*

From Theorems 3.6.2 and 3.6.3, it follows that the class of all fully right-orderable groups is sufficiently wide. Nevertheless, there are right-orderable but not fully right-orderable groups.

EXAMPLE 3.6.1 (Mura and Rhemtulla [42]). Let $G = \mathrm{gp}\,(a, b \mid b^{-1}ab = a^{-1})$. It is clear that G admits the following total right order: $x > e$ for $x = b^k a^s$ if $k > e$ or $k = 0$ and $s > 0$. The group G is not fully right-orderable. Indeed, the semigroup $P = \mathrm{sgr}\,(b^2, a, ab^{-2})$ determines a partial right order on G. We assume that the order can be extended to a total right order P'. Since $b^2 \in P'$, we have $b \in P'$, $ab^{-1} = ab^{-2}b \in P'$, and $bab^{-1} = a^{-1} \in P'$, which contradicts the inclusion $a \in P'$.

EXAMPLE 3.6.2 (Rhemtulla [13]). Let $G = \mathrm{gp}\,(a, b \mid b^{-1}ab = a^{-2})$. It is easy to show that the normal subgroup A generated by a is an Abelian group isomorphic to the subgroup $\mathbb{Q}_2 = \{m2^n \mid m, n \in \mathbb{Z}\}$

of the group $\mathbb{Q}$ and $G = A \cdot (b)$ is a split extension of the normal subgroup A by the infinite cyclic group (b). Every element of the group G can be uniquely represented as $g = b^k a^q$, where $k \in \mathbb{Z}$ and $q \in \mathbb{Q}_2$. It is clear that G admits a total right order: $g = b^k a^q > e$ if $k > 0$ or $k = 0$ and $q > 0$. We consider the subset $P_1 = \{a^q b^{2m} \mid q > 0 \text{ or } q = 0, n \geqslant 0\}$ of the group G. Direct verification shows that P_1 is a pure semigroup and determines a partial right order on G. We assume that P_1 can be extended to a total right order P. Since $b^2 \in P_1 \subseteq P$, we have $b \in P$. Therefore, $ba^q \in P$ for any $q \in \mathbb{Q}_2$. But $ba^q \cdot ba^q = a^{-q/2}a^{q/4}b^2 \in P_1^{-1}$ because $-q2^{-1} + q2^{-2} < 0$. Thus, the partial right order P_1 cannot be extended to a total right order on the group G.

We note that there are nonnilpotent fully right-orderable groups. As an example, we indicate the group $G = \mathrm{gp}\,(a, b \mid b^{-1}ab = a^2)$ (cf. [43]). The proof of this fact is left to the reader as an exercise.

3.6.2. We introduce a definition.

- A group on which each partial order can be extended to a total order is called *fully orderable.*

We give necessary and sufficient conditions for the full orderability of groups in terms of semigroups.

We say that an element $a \in G$ satisfies the property (O^*) if for any elements $b, c \in S(a)$ the following relation holds:

O^* $$S(b) \cap S(c) \neq \varnothing$$

As usual, $S(x)$ denotes a normal subsemigroup of the group G generated by the element x.

Lemma 3.6.1. *Let G be a G-torsion-free group. An element a and the identity element e of the group G are comparable with respect to any maximal partial order on G if and only if a satisfies O^*.*

Proof. We assume that an element a and the identity element e are comparable with respect to any maximal partial order on the group G and there are elements $b, c \in S(a)$ such that $S(b) \cap S(c) = \varnothing$. By Lemma 3.1.2, the semigroup $P = S(b, e) \cdot S^{-1}(c, e)$ defines a partial order on the group G. Let a maximal partial order Q be the extension of an order P on the group G. Then $b \in Q$ and $c \in Q$, which is impossible because a and e are comparable with respect to the order Q.

Conversely, let P be a partial order on the group G and an element a satisfy the property O*. We note that $P \cap S(a) \neq \varnothing$ implies $P \cap S^{-1}(a) \subseteq \{e\}$. Indeed, we assume the contrary: $x \in \cap S(a)$ and $y \in P \cap S^{-1}(a)$. Then the relation $x^{-1}, y \in S^{-1}(a) = S(a^{-1})$ and the property O* imply $S(x^{-1}) \cap S(y) \neq \varnothing$, which contradicts the inclusions $S(x^{-1}) \subseteq P^{-1}$, $S(y) \subseteq P$ and the property $P^{-1} \cap P = \{e\}$ of the partial order P.

If P is a maximal partial order, then, as above, either $a \in P$ or $a^{-1} \in P$. □

A subgroup H of a group G is called *G-fully orderable* if any maximal partial order on the group G induces a total order on H.

Corollary 3.6.3. *A subgroup H of a G-torsion-free group G is G-fully orderable if and only if all its elements satisfy the property* O*.

Theorem 3.6.4 (Ohnishi [44]). *A group G is fully orderable if and only if it is a G-torsion-free group and all its elements satisfy the property* O*.

Corollary 3.6.4. *A group is fully orderable if and only if each of its finitely generated subgroups is fully orderable.*

We note that analogs of Lemma 3.6.1 and Theorem 3.6.4 are true for operator groups with the domain of operators Ω. Moreover, the minimal Ω-invariant subsemigroup $S_\Omega(a, \ldots)$ generated by the set $\{a, \ldots\}$ plays the role of the invariant semigroup $S(a, \ldots)$.

Corollary 3.6.5 (Podderyugin [29]). *An Abelian group with the commutative domain of operators Ω is Ω-fully orderable if and only if it is a Ω-torsion-free group.*

Proof. Let $b, c \in S_\Omega(a)$. Then $b = a^\omega$ and $c = a^\nu$, where $\omega = \omega_1 + \ldots + \omega_n$, $\nu = \nu_1 + \ldots + \nu_m$, and $\omega_i, \nu_j \in \Omega$. Hence

$$b^\nu = c^\omega = a^{\omega\nu} = a^{\nu\omega} \in S_\Omega(b) \cap S_\Omega(c)$$

From the above it follows that the group G is Ω-fully orderable. The converse assertion is obvious. □

Lemma 3.6.2. *Let an element s of a totally ordered group G be represented in the form $s = g^{x_1 + \cdots + x_n}$, where $g, x_i \in G$. Then there are elements $y, z \in G$ such that $s^y \geqslant g^n$ and $s^z \leqslant g^n$.*

PROOF. Indeed, for elements x_k, x_r such that $g^{x_k} \leqslant g^{x_i}$, $1 \leqslant i \leqslant n$, and $g^{x_r} \geqslant g^{x_i}$, $1 \leqslant i \leqslant n$, we set $y = x_k^{-1}$ and $z = x_r^{-1}$. □

Lemma 3.6.3. *Let H be a normal G-fully orderable subgroup of an orderable group G such that the quotient group G/H is orderable. If an element $a \in G$ is such that for any finite set of elements $g_1, \dots, g_n$ of the group G the relation $a^{g_1 + \cdots g_n} a^{-n} \in H$ holds, then the elements a and e are comparable.*

PROOF. We assume the contrary: the elements a and the identity element e are incomparable with respect to some maximal partial order P on the group G. By Lemma 3.1.2, $S(a^{-1}) \cap P \neq \varnothing$ and $S(a) \cap P \neq \varnothing$. Therefore, there are $s_1, s_2 \in P$ such that $s_1 = a^{g_1 + \cdots + g_n}$, $s_2 = (a^{-1})^{h_1 + \cdots + h_n}$. On the group G we introduce a total order Q such that the relation $H \cap Q = P_1 = H \cap P$ holds and the subgroup H is convex. This is possible in view of the orderability of the quotient group G/H and Theorem 1.6.2. By Lemma 3.6.2, there are elements y and x such that $s_1^y \leqslant a^n$ and $a^n \leqslant (s_2^{-1})^x$.

Multiplying the last inequalities, we obtain $s_1^y \leqslant (s_2^{-1})^x$. Hence $z = s_1^y \cdot s_2^x \leqslant e$, which contradicts the conditions $z \in P_1 \subseteq Q$ and $s_1, s_2 \in P$. □

We recall that a group is called *hypercentral* if it coincides with some term of its upper central series.

Theorem 3.6.5 (Kokorin [32]). *If the quotient group G/H of a group G by a G-fully orderable normal subgroup H is a torsion-free hypercentral group, then the group G is fully orderable.*

PROOF. The assertion follows from the fact that the hypercentral torsion-free group is orderable, since it is a locally nilpotent group (cf., for example, [45]). □

Corollary 3.6.6. *A torsion-free locally nilpotent group is fully orderable.*

Theorem 3.6.6 (Kokorin [32]). *A metabelian G-torsion-free group G is fully orderable.*

PROOF. As was shown in Corollary 3.3.1, a metabelian G-torsion-free group G is orderable. The isolator J of its derived subgroup G', as an isolator of an Abelian subgroup of a G-torsion-free group, is an Abelian subgroup. By Corollary 3.6.4, J is a G-fully orderable subgroup. The application of Theorem 3.6.5 completes the proof because the quotient group G/J is a torsion-free Abelian group. □

Theorem 3.6.7 (Kargapolov [46] and Kokorin [47]). *The direct product of fully orderable groups is fully orderable.*

PROOF. Let $G = A \times B$, where A and B are fully orderable groups and P is a maximal order on G. Since all the elements of subgroups A and B satisfy the property O^*, the subgroups A and B are totally ordered with respect to any maximal order P on the group G. We proceed to show that $s = ab$ for any element $s \in G$, where $a \in A$, $b \in B$, and either $s \in P$ or $s^{-1} \in P$. To the contrary, let $S(s) \cap P \neq \varnothing$ and $S(s^{-1}) \cap P \neq \varnothing$. Then there are elements $s_1, s_2^{-1} \in P$ such that

$$s_1 = (ab)^{x_1+\dots+x_n} = a^{x_1+\dots+x_n} b^{x_1+\dots+x_n}$$
$$s_2^{-1} = ((ab)^{-1})^{y_1+\dots+y_n} = (a^{-1})^{y_1+\dots+y_n} (b^{-1})^{y_1+\dots+y_n}$$

In view of Lemma 3.6.2 and the total orderability of the groups A and B with respect to the order P, we conclude that there are elements $a_1, a_2 \in A$ and $b_1, b_2 \in B$ such that

$$(a^{x_1+\dots+x_n})^{a_1} \leqslant (a^{y_1+\dots+y_n})^{a_2}, \quad (a^{x_1+\dots+x_n})^{b_1} \leqslant (b^{y_1+\dots+y_n})^{b_2}$$

Therefore, $s_3 = s_1^{b_1 a_1}(s_2^{-1})^{b_2 a_2} \leqslant e$, which contradicts the inclusions $s_1, s_2^{-1} \in P$. □

We note that the class of fully orderable groups is not closed under taking the Cartesian product. This result belongs to Kargapolov and can be found in [48]. Since the class of fully orderable groups is closed under taking direct products and is not closed under taking the Cartesian product, it follows that the class of fully orderable groups is not axiomatizable.

We now show that there are orderable groups which are not fully orderable.

EXAMPLE 3.6.3. We consider the group $G = (\mathbb{R}\oplus\mathbb{R})\cdot A$ which is the split extension of the direct sum $K = \mathbb{R}\oplus\mathbb{R}$ of two isomorphic copies of the group $\mathbb{R}$, written in the additive form by means of the group A of automorphisms of the group K, generated by the automorphisms α, β acting on K by the following rule:

$$(x, y)\alpha = (y, x), \quad (x, y)\beta = (x, 2^{-1}y)$$

for any $x, y \in \mathbb{R}$. Elements of the group G will be denoted by (φ, v), where $\varphi \in A$ and $v \in K$. Then the group operation in G is defined as follows:

$$(\varphi, v) \cdot (\varphi', v') = (\varphi\varphi', v\varphi' + v')$$

for $\varphi, \varphi' \in A$ and $v, v' \in K$.

We now consider the elements $h_1 = (\alpha, (0,0))$, $h_2 = (\beta, (0,0))$, and $h_3 = (\varepsilon, (-3, 4))$ of the group G. Here $(0,0)$ and $(-3,4)$ belong to K and ε is the identity element of A. Let

$$P = \{(\varepsilon, v) \mid v = (x, y) \in K, \quad x, y > 0 \text{ in } \mathbb{R}\}$$

It is clear that P is a normal semigroup of G and $e \notin P$. Further,

$$h = h_3 h_1^{-1} h_3 h_1 = h_3(\alpha^{-1}, (0,0)) h_3(\alpha, (0,0)) = (\varepsilon, (1,1)) \in P$$

$$h_1^{-1} h_2^{-1} h_3^{-1} h_2 h_1 h_2^{-1} h_3^{-1} h_2 = (\varepsilon, (-2, 3)) + (\varepsilon, (3, -2)) = (\varepsilon, (1,1)) = h$$

Since $h \in P$, the invariant semigroup G generated by the element h does not contain the identity element e of the group G.

Theorem 3.6.8 (Holland [49]). *If a group X has a quotient group X/N which is isomorphic to the group G from Example* 3.6.3, *then it is not fully orderable.*

Proof. We assume that the group X is fully orderable. Then there is a homomorphism θ: $X \to G$ from X onto G. Let $a, b, c \in X$ be such that $\theta(a) = h_1$, $\theta(b) = h_2$, and $\theta(c) = h_3$. We consider the elements $f_1 = ca^{-1}ca$, $f_2 = a^{-1}b^{-1}c^{-1}bab^{-1}c^{-1}b$ of the group and denote by S the invariant semigroup of X generated by the elements f_1 and f_2. Since

$$\theta(f_1) = \theta(ca^{-1}ca) = h_3 h_1^{-1} h_3 h_1 = h = \theta(f_2) = h_1^{-1} h_2^{-1} h_3^{-1} h_2 h_1 h_2^{-1} h_3^{-1} h_2$$

we have $\theta(S) \subseteq P$. Since P is a pure semigroup, S is a semigroup too. We now show that the partial order S cannot be extended to a total order on X. Indeed, if the partial order S can be extended to the total order $\hat{S}$, then f_2 is negative and f_1 is positive. If c is negative, then f_1 is negative and f_2 is positive. Both cases are impossible. □

Corollary 3.6.7. *A free group, the rank of which is at least* 3, *is not fully orderable.*

Corollary 3.6.8. *For $n \geqslant 3$ and $k \geqslant 3$ a free n-soluble group with k generators is not fully orderable.*

Other examples of not fully orderable groups were obtained by Kargapolov [46], Fuchs and Sasiada [50], and Kopytov [51]. For detailed information on fully orderable groups, refer to [37, 48].

Chapter 4

Groups of Order Automorphisms

For simplicity of notation, we omit parentheses when no confusion can arise. For example, for Ω_1, Ω_2, $\alpha \in \Omega_1$, $g \in \mathcal{A}(\Omega_1)$, $\varphi: \Omega_1 \to \Omega_2$, $\psi: \mathcal{A}(\Omega_1) \to \mathcal{A}(\Omega_2)$, we write an element $((\alpha)\varphi)((g)\psi)$ of Ω_2 briefly as $\alpha\varphi \cdot g\psi$.

By Theorem 3.4.1, any right-orderable group G is isomorphic to a subgroup of the group $\mathcal{A}(\Omega)$ constituted by all order automorphisms of a suitable totally ordered set Ω. In this chapter, we study the properties of groups of automorphisms of totally ordered sets.

4.1. Preliminaries

4.1.1. Let Ω be a totally ordered set and $\mathcal{A}(\Omega)$ be the group of all order automorphisms of the set Ω.

- Any subgroup of $\mathcal{A}(\Omega)$ regarded as a group acting on Ω is called a *group of order automorphisms* of Ω and is denoted by (G, Ω).
- Any homomorphism $\psi: G \to \mathcal{A}(\Omega)$ is called a *representation* of G.
- A representation is called *faithful* if its kernel is trivial.

The groups (G, Ω) and (H, Λ) of order automorphisms of totally ordered sets Ω and Λ are said to be *isomorphic as groups of order automorphisms* provided that there exists an isomorphism π between the groups G and H and an order isomorphism τ between the totally ordered sets Ω and Λ such that $(\alpha g) = (\alpha\tau)g\pi$ for any $g \in G$ and $\alpha \in \Omega$, denoted by (π, τ).

If $\varphi\colon \Omega \to \Lambda$ is an order mapping and $\psi\colon G \to H$ is a homomorphism of G into H such that the equality $(\alpha g)\varphi = (\alpha\varphi)(g\psi)$ is fulfilled for all $\alpha \in \Omega$ and $g \in G$, the pair (ψ, φ) defines a homomorphism of the group (G, Ω) into the group (H, Λ).

Let (G, Ω) be the group of order automorphisms; moreover, on Ω there is an equivalence relation θ such that

— $\alpha \equiv \beta \pmod{\theta}$ for $\alpha, \beta, \gamma \in \Omega$ such that $\alpha \leqslant \beta \leqslant \gamma$ and $\alpha \equiv \gamma \pmod{\theta}$,

— $\alpha g \equiv \beta g \pmod{\theta}$ for $\alpha, \beta \in \Omega$ such that $\alpha \equiv \beta \pmod{\theta}$ and $g \in G$

(we refer to θ as a *convex G-congruence*).

From the definition it directly follows that the convex G-congruence θ divides Ω into disjoint sets $\alpha\theta = \{\beta \in \Omega \mid \alpha g \equiv \beta g \pmod{\theta}\}$, each of which is a convex subset of Ω. If $\alpha\theta \neq \beta\theta$ and $\alpha_1 < \alpha_2$, then $\beta_1 < \beta_2$ for any elements $\beta_1, \beta_2 \in \Omega$ such that $\beta_1 \in \alpha_1\theta$ and $\beta_2 \in \alpha_2\theta$. This means that on the quotient set Ω/θ there is a natural total order induced by the total order on the set Ω. Furthermore, if $g \in G$ and $\alpha \in \Omega$, then the set $(\alpha\theta)g = \{(\beta)g \mid \beta \in \alpha\theta\}$ is a class of congruent elements. Moreover, if $\alpha_1\theta < \alpha_2\theta$ in Ω/θ, then $(\alpha_1\theta)g < (\alpha_2\theta)g$, i.e., every element $g \in G$ defines an order automorphism of Ω/θ.

- A subset $\Delta \neq \varnothing$ of a set Ω is called a *G-block* if Δ is a convex subset of Ω and $\Delta g = \Delta$ or $\Delta g \cap \Delta = \varnothing$ for every $g \in G$.

Proposition 4.1.1. *Let (G, Ω) be a group of order automorphisms of a totally ordered set Ω. Then an equivalence class of an arbitrary convex G-congruence is a G-block. Conversely: each G-block is an equivalence class of some convex G-congruence on Ω.*

PROOF. Let Δ be a G-block. We set $\alpha \equiv \beta \pmod{\zeta}$ if $\alpha = \beta$ or $\alpha, \beta \in \Delta g$ for some element $g \in G$. It is easy to verify that ζ is an equivalence relation. Further, if $\alpha \leqslant \beta \leqslant \gamma$ and $\alpha \equiv \gamma \pmod{\zeta}$, then $\alpha, \gamma \in \Delta g$ for some $g \in G$. However, Δg is a convex subset of Ω. Hence $\beta \in \Delta g$. Consequently, $\alpha \equiv \beta \pmod{\zeta}$. If $\alpha = \beta$, then $\alpha f = \beta f$ for every element $f \in G$ and $\alpha g, \beta g \in \Delta$ if $\alpha, \beta \in \Delta f$ for some $f \in G$. Thereby, we have proved that the equivalence relation ζ is a convex G-congruence.

The converse assertion follows from the definition of a convex G-congruence. □

Let L_θ be a set of elements $g \in G$ such that $(\alpha\theta)g = \alpha\theta$ for every $\alpha \in \Omega$. For any element $g \in L_\theta$ the following equalities are obvious:

$$(\alpha\theta)x^{-1}gx = (\alpha x^{-1}\theta)gx = (\alpha x^{-1})\theta x = \alpha\theta$$

If $g \in G$, then elements of the coset $\overline{g} = gL_\theta$ induce the same order automorphism of the totally ordered quotient set Ω/θ, i.e., the quotient group G/L_θ can be regarded as a subgroup of the l-group $\mathcal{A}(\Omega/\theta)$ (cf. Example 1.4.5). Let ψ: $G \to G/L_\theta$ be the natural homomorphism of G onto G/L_θ, and let φ: $\Omega \to \Omega/\theta$ be the natural order mapping of Ω onto Ω/θ. It is easy to check that the pair (ψ, φ) defines the homomorphism of (G, Ω) onto $(G/L_\theta, \Omega/\theta)$. In this case the group $(G/L_\theta, \Omega/\theta)$ is called the *quotient group* of (G, Ω) by the convex G-congruence θ, and the pair (ψ, φ) is called the *natural homomorphism* of (G, Ω) onto $(G/L_\theta, \Omega/\theta)$.

- The normal subgroup L_θ is called the *kernel of congruence* θ.

Theorem 4.1.1 (on homomorphisms). *For any homomorphism (π, τ) of the group (G, Ω) into the group (H, Λ) the mapping τ defines a convex G-congruence θ on Ω such that*

$$\alpha \equiv \beta \pmod{\theta} \Leftrightarrow \alpha\tau = \beta\tau$$

The proof of the theorem is standard and is left to the reader.

Let Δ be a subset of Ω such that $\Delta g = \Delta$ for any $g \in G$. In this case the set Δ is called *invariant under the action of* G (G*-invariant*).

Every element $g \in G$ induces an order automorphism of Δ. In this case we obtain the representation of the group G by the automorphisms of Δ. The representation is not necessarily faithful.

- If $\Delta \subseteq \Omega$, then the set of $g \in G$ such that $(\Delta)g = \Delta$ is called the *stabilizer* (denoted by $G_{(\Delta)}$) of Δ in G.
- The set of elements $g \in G$ such that $\alpha g = \alpha$ for any $\alpha \in \Delta$ is called the *point-wise stabilizer* (denoted by G_Δ) of $\Delta \subseteq \Omega$ in G.

If $\Delta = \{\alpha\}$ is a single-element subset, then $G_\alpha = G_{(\alpha)}$.

It is easy to see that for any subset $\Delta \subseteq \Omega$ the groups G_Δ and $G_{(\Delta)}$ are subgroups of the group $\mathcal{A}(\Omega)$. Moreover, if the subset Δ is invariant under the action of G, then G_Δ is a normal subgroup of G and any two elements g_1 and g_2 of G such that $G_\Delta g_1 = G_\Delta g_2$ induce on Δ the same order automorphism. Therefore, the pair $(G/G_\Delta, \Delta)$ can be regarded as a group of order automorphisms of the totally ordered set Δ. This group is called *restriction* of G on Δ, and the mapping φ, where $g\varphi = gG_\Delta$, is a representation of the group G by order automorphisms of Δ.

If Δ is a convex G-invariant subset of the set Ω, then a convex G-congruence θ: $\alpha \equiv \beta \pmod{\theta}$ can be introduced on Ω if and only if α, β is either less than any element $\gamma \in \Delta$ or greater than any element $\gamma \in \Delta$. The quotient set Ω/θ is order-isomorphic to one of the four sets

$$\{u\} \cup \Delta \cup \{v\}, \quad \{u\} \cup \Delta, \quad \Delta \cup \{v\}, \quad \Delta$$

and $u < \delta < v$ for any $\delta \in \Delta$, depending on the availability or unavailability of lower and upper bounds for Δ in $\Omega \setminus \Delta$.

The mapping (π, ε), where $(G_\Delta g)\pi = gL_\theta$ and $(y)\varepsilon = y$, sets up an isomorphism of the subgroup G/G_Δ of $\mathcal{A}(\Delta)$ onto the group G/L_θ of automorphisms of the totally ordered set Ω/θ.

It is evident that the mapping π, $g\pi = g$, is an isomorphism of the group $G_{(\Delta)}$ into the group $\mathcal{A}(\Omega)$ of order automorphisms of the totally ordered set Ω.

4.1.2. We consider the question of the extension of order automorphisms of a totally ordered set Ω to a larger totally ordered set containing Ω.

- A bounded-from-above subset I of a totally ordered set Ω is called an *ideal* of Ω if $x \in I$ for any $\alpha \in I$ and $x \in \Omega$ such that $x \leqslant \alpha$.

The set of all ideals of a totally ordered set Ω is denoted by $\overline{\Omega}$. On the set $\overline{\Omega}$, the natural total order can be introduced by means of inclusion. The mapping

$$\varepsilon\colon \Omega \to \overline{\Omega}, \quad x\varepsilon = I_x = \{y \in \Omega \mid y \leqslant x\}$$

is an order isomorphism of Ω into $\overline{\Omega}$.

Having identified $\Omega\varepsilon$ with $\overline{\Omega}$, the set $\overline{\Omega}$ will be referred to as the *Dedekind completion* of Ω and its elements will sometimes be denoted by $\overline{x}$.

We note that for any $\overline{x} \in \overline{\Omega}$ the relation

$$\overline{x} = \sup\{x \in \Omega \mid x \leqslant \overline{x}\}$$

is fulfilled and any bounded-from-above subset Y of the set Ω has the least upper bound in $\overline{\Omega}$.

Proposition 4.1.2 (Holland [52]). *An order automorphism g of a totally ordered set Ω can be uniquely extended to an automorphism $\overline{g}$ of the Dedekind completion $\overline{\Omega}$.*

PROOF. For elements $\overline{x} \in \overline{\Omega}$ and $g \in \mathcal{A}(\Omega)$ we set

$$(\overline{x})\overline{g} = \sup\{(x)g \mid x \leqslant \overline{x},\ x \in \Omega\}$$

It is easy to show that $\overline{g} \in \mathcal{A}(\overline{\Omega})$ and $(x)\overline{g} = (x)g$ for any $x \in \Omega$. □

Theorem 4.1.2 (Holland [38]). *For any totally ordered set Ω there exists an isomorphic embedding (π, ε) of the group $(\mathcal{A}(\Omega), \Omega)$ into the group $(\mathcal{A}(\overline{\Omega}), \overline{\Omega})$.*

PROOF. Let $(g)\pi = \overline{g}$ for $g \in \mathcal{A}(\Omega)$ and $\varepsilon\colon \Omega \to \overline{\Omega}$. By Proposition 4.1.2, $g \in \mathcal{A}(\overline{\Omega})$. If $\alpha \in \Omega$ and $f, g \in \mathcal{A}(\Omega)$, then

$$(\alpha)((g\pi)(f\pi)) = ((\alpha)\overline{g})\overline{f} = (\alpha g)f = (\alpha)fg = \overline{gf}$$

We note that the automorphisms $(g\pi)(f\pi) = \overline{g} \cdot \overline{f}$ and $\overline{gf}$ coincide on Ω. Consequently, they are equal as order automorphisms of the set $\overline{\Omega}$. Thereby, we have proved that π is an isomorphism of $\mathcal{A}(\Omega)$ into $\mathcal{A}(\overline{\Omega})$. Since $((\alpha)g)\varepsilon = (\alpha)g\pi$ for any $\alpha \in \Omega$ and $g \in \mathcal{A}(\Omega)$, we conclude that (π, ε) defines an isomorphic embedding of $(\mathcal{A}(\Omega), \Omega)$ into $(\mathcal{A}(\overline{\Omega}), \overline{\Omega})$. □

Corollary 4.1.1. *For any totally ordered set Ω and a group (G, Ω) there exists an isomorphic embedding (π, ε) of (G, Ω) into $(\mathcal{A}(\overline{\Omega}), \overline{\Omega})$ such that $(\alpha)(g\pi) = \alpha g$ and $\alpha\varepsilon = \alpha$ for any $\alpha \in \Omega$.*

In the sequel, we identify G with $G\pi$.

EXAMPLE 4.1.1. The group $\mathcal{A}(\mathbb{Q})$ of order automorphisms of the totally ordered set $\mathbb{Q}$ of rational numbers is a subgroup (moreover, l-subgroup) of the group $\mathcal{A}(\mathbb{R})$ of order automorphisms of the set $\mathbb{R}$ of real numbers.

4.1.3. We introduce some definitions.

- A group (G, Ω) of automorphisms of a totally ordered set Ω is called *transitive* on Ω if for any $\alpha, \beta \in \Omega$ there exists an automorphism $g \in G$ such that $\alpha g = \beta$.

In the case of transitive group (G, Ω), we say that *G acts transitively on* Ω. A representation φ of a group G by order automorphisms of a totally ordered set Ω is called *transitive* if $(G\varphi, \Omega)$ is a transitive group.

- A subset $\alpha G = \{\alpha g \mid g \in G\}$ of the set Ω is called the *G-orbit* of $\alpha \in \Omega$ or the *transitivity* set of G.

Proposition 4.1.3. *A subset $\Delta \subseteq \Omega$ is the G-orbit of some element $\alpha \in \Omega$ if and only if Δ is a G-invariant set and G acts transitively on Δ.*

PROOF. Let $\Delta = \alpha G$ for some $\alpha \in \Omega$. Then for any $\beta \in \Delta$ there is an element $h \in G$ such that $\beta = \alpha h$. Hence $\beta g = \alpha h g \in \Delta$ for any $g \in G$, i.e., Δ is a G-invariant subset of the set Ω. Further, if $\beta, \gamma \in \Delta$, then $\beta = \alpha h$ and $\gamma = \alpha f$ for some elements $h, f \in G$. Consequently,

$$\beta h^{-1} f = \alpha h h^{-1} f = \alpha f = \gamma$$

which proves the fact that the group G acts transitively on Δ.

We now pass to proving the converse assertion. Let ΔG be a G-invariant set on which the group G acts transitively. We choose an arbitrary element $\alpha \in \Delta$ and consider the G-orbit of α. It is obvious that $\alpha G \subseteq \Delta$. Moreover, in view of the transitivity of the action of G on Δ, for any $\beta \in \Delta$ there is an element $g \in G$ such that $\beta = \alpha g$. This means that $\Delta = \alpha G$. □

- We say that a group G *is approximated* by the group class $\mathfrak{M}$ if for any nonidentity element $g \in G$ there is a group H which is isomorphic to a group of $\mathfrak{M}$, and a homomorphism φ of the group G onto the group H such that $g\varphi \neq e$.

As is well known (cf. [53]), a group G is approximated by a group class $\mathfrak{M}$ if and only if there is a family $\{N_\alpha \mid \alpha \in I\}$ of normal subgroups N_α of the group G such that for any $\alpha \in I$ the quotient group G/N_α is isomorphic to some group of $\mathfrak{M}$ and $\bigcap_{\alpha \in I} N_\alpha = E$.

Proposition 4.1.4. *A totally ordered set Ω can be represented as the set-theoretical union of the mutually disjoint G-orbits $\Omega_\alpha = \alpha G$. Moreover, the group G is approximated by the class of groups $\overline{G_\alpha} = G/G_{\alpha G}$, $\alpha \in \Omega$.*

PROOF. It is easy to check that $\alpha G = \beta G$ or $\alpha G \cap \beta G = \varnothing$ for any $\alpha, \beta \in \Omega$. The stabilizer $G_{\alpha G}$ of the set αG is a normal subgroup of G for any G-orbit αG. Moreover, $\bigcap_{\alpha\in\Omega} G_{\alpha G} = E$ because $\alpha g = \alpha$ for any $g \in \bigcap_{\alpha\in\Omega} G_{\alpha G}$ and $\alpha \in \Omega$. □

Proposition 4.1.5. *The equality $G_\beta = g^{-1}G_\alpha g$ is fulfilled for any $\alpha, \beta \in \Omega$ such that $\beta = \alpha g$, $g \in G$. In particular, in the transitive group (G, Ω) the stabilizers of points are conjugate.*

PROOF. Let $\alpha g = \beta$ for some $\alpha, \beta \in \Omega$ and $g \in G$. Then $\beta(g^{-1}hg) = \alpha hg = \alpha g = \beta$ for any $h \in G_\alpha$. Hence

$$g^{-1}G_\alpha g \subseteq G_\beta$$

The inverse inclusion is established in a similar way. □

We note that any right-ordered group has "sufficiently many" transitive representations. Let G be a right-ordered group, H be its convex subgroup, and $R(G : H)$ be the totally ordered set of right cosets of the group G by H. For any $g \in G$ we define the mapping $R_H(g)$: $R(G : H) \to R(G : H)$ as follows:

$$(Hx)R_H(g) = Hxg$$

It is evident that R_H is a representation of the group G by order automorphisms of the totally ordered set $R(G : H)$. The representation is faithful provided that H contains no nontrivial normal subgroups.

Representations of such a type form the set of all faithful transitive representations of groups of automorphisms of a totally ordered set. Namely, in the above notation the following assertion is true.

Theorem 4.1.3. *Any faithful transitive representation (G, Ω) of a right-ordered group G by automorphisms of a totally ordered set Ω is isomorphic to the representation $(G, R(G : H))$, where H is a convex subgroup of the group G with some right order P, and the set of right cosets $R(G : H)$ is totally ordered with respect to the induced total order.*

PROOF. Let (G,Ω) be a faithful transitive representation of a group G and G_α be the stabilizer of some $\alpha \in \Omega$. On the set Ω we introduce a well-order $\prec$ so that α is the least element in Ω. By using the relation $\prec$ and the action of G on Ω, we construct a right order P on G in a standard way (cf. Example 1.4.5). It is not hard to show that the subgroup G_α is convex with respect to such a right order. We set $H = G_\alpha$ and denote by τ a mapping of $R(G : H)$ onto Ω such that

$$(Hx)\tau = \alpha g$$

Then the pair of mappings (R_H, τ) is an isomorphism of the group $(R_H(G), R(G : H))$ onto the group (G,Ω). □

4.1.4. In this section, we consider properties of convex congruences on transitive groups (G,Ω). We recall that for convex G-congruences ζ_1 and ζ_2 we have $\zeta_1 \leqslant \zeta_2$ if and only if $\alpha\zeta_1 \subseteq \alpha\zeta_2$ for any $\alpha \in \Omega$.

Proposition 4.1.6 (Holland and McCleary [54]). *Let α be an arbitrary element of a totally ordered set Ω. Then a mapping $\varphi\colon \zeta \to \alpha\zeta$ from the set $\mathcal{K}$ of convex G-congruences onto the set of G-blocks containing α has the following properties:*

(a) *φ is a one-to-one mapping,*

(b) *$\zeta_1 \leqslant \zeta_2$ if and only if $\alpha\zeta_1 \subseteq \alpha\zeta_2$.*

PROOF. In accordance with Proposition 4.1.1 we can assume that φ maps $\mathcal{K}$ onto the set of G-blocks. We also assume that there exist convex G-congruences ζ_1 and ζ_2 such that $\zeta_1 \neq \zeta_2$ and $\alpha\zeta_1 = \Delta = \alpha\zeta_2$. By assumption, there exist $\alpha, \beta \in \Omega$ such that $\alpha \equiv \beta \pmod{\zeta_1}$, but the relation $\alpha \equiv \beta \pmod{\zeta_2}$ fails. In view of the transitivity of (G,Ω), there exists an element $g \in G$ such that $\alpha g \in \Delta$. Hence $\alpha g \equiv \beta g \pmod{\zeta_1}$ and $\alpha g, \beta g \in \Delta = \alpha\zeta_2$. Consequently, $\alpha g \equiv \beta g (\mathrm{mod}\, \zeta_2)$. Since ζ_2 is a convex G-congruence, we have $\alpha \equiv \beta \pmod{\zeta_2}$, which contradicts the assumption. Property (a) is established. Property (b) is obvious. □

For subsets A and B of a totally ordered set Ω, the notation $A < B$ means that $\alpha < \beta$ for any $\alpha \in A$ and $\beta \in B$.

Proposition 4.1.7 (Holland and McCleary [54]). *Let α be an arbitrary element of a totally ordered set Ω. Then the set of G-blocks of Ω which contain α is totally ordered by inclusion.*

PROOF. Let Δ_1 and Δ_2 be G-blocks and $\alpha \in \Delta_1 \cap \Delta_2$. If $\Delta_1 \not\subseteq \Delta_2$ and $\Delta_1 \not\supseteq \Delta_2$, then there exist elements $\delta_1, \delta_2 \in \Omega$ such that $\delta_1 \in \Delta_1 \backslash \Delta_2$ and $\delta_2 \in \Delta_2 \setminus \Delta_1$. Without loss of generality, we can assume that $\delta_1 < \beta$ if $\beta \in \Delta_2$ and $\delta_2 > \beta$ if $\beta \in \Delta_1$. In view of the transitivity of (G, Ω), there exists an element $g \in G$ such that $\alpha g = \delta_2$. Therefore, $\Delta_2 g \cap \Delta_2 \neq \varnothing$. Consequently, $\Delta_2 = \Delta_2 g$ and $\Delta_2 g^{-1} = \Delta_2$. If $\Delta_1 g \cap \Delta_1 = \varnothing$, then $\Delta_1 g < \Delta_1$ or $\Delta_1 g > \Delta_1$. But the first inequality is impossible because $\Delta_1 g$ contains the element αg which is equal to δ_2 and the relation $\delta_2 > \beta$ holds for every element $\beta \in \Delta_1$. Hence $\Delta_1 g > \Delta_1$. In particular, it follows that all elements of the set $\Delta_1 g$ are greater than α. On the other hand, $\delta_1 < \Delta_2 g = \Delta_2 = \Delta_2 g^{-1}$. Consequently, $\delta_1 g < \Delta_2$ and $\delta_1 g < \alpha$, which is impossible. Hence $\Delta_1 g = \Delta_1$, $\alpha \in \Delta_1$, and $\alpha g = \delta_2 \notin \Delta_1$ in accordance with the choice of $g \in G$, which contradicts our assumption. □

Corollary 4.1.2. *The set $\mathcal{K}$ of convex G-congruences on Ω is totally ordered with respect to the original partial order on $\mathcal{K}$.*

The assertion directly follows from Propositions 4.1.6 and 4.1.7.

It is clear that the totally ordered set $\mathcal{K}$ of convex G-congruences contains the greatest congruence and the least one. In the sequel, they will be denoted by ε and χ respectively and will be referred to as *trivial G-congruences.*

Corollary 4.1.3. *Let $\{\zeta_i \mid i \in I\}$ be an arbitrary family of convex G-congruences. Then $\bigcap_{i\in I} \zeta_i$ and $\bigcup_{i \in I} \zeta_i$ are convex G-congruences.*

The assertion follows from the fact that the union and intersection of any family of G-blocks is a G-block itself.

4.1.5. We consider a group (G, Ω) of order automorphisms of a totally ordered set Ω and introduce the following definitions.

- A group (G, Ω) is called *o-primitive* if there exist no nontrivial convex G-congruences on Ω.

- A group (G, Ω) has the *separation property* if for every nonempty convex set $\Delta \subset \Omega$ and any two elements $\alpha, \beta \in \Omega$ there exists an element $g \in G$ such that only one of the elements αg and βg belongs to the subset Δ (i.e., α and β are separated by the interval Δ).

Proposition 4.1.8 (Holland and McCleary [54]). *A transitive group (G, Ω) of order automorphisms is o-primitive if and only if it has the separation property.*

PROOF. We assume that (G, Ω) is not o-primitive. Then there exists a bounded G-block Δ such that $\Delta \neq \Omega$ and $|\Omega| > 1$. We choose two arbitrary elements $\alpha, \beta \in \Delta$ such that $\alpha \neq \beta$. It is obvious that $\alpha g \in \Delta g$ only if $\beta g \in \Delta g$ for an arbitrary element $g \in G$. Thus, the separation property for (G, Ω) fails.

To prove the converse assertion we assume that elements α and β, $\alpha \neq \beta$, are contained in Ω and cannot be separated by a nonempty convex set $\Delta \neq \Omega$. We introduce the relation ζ by setting $\sigma \equiv \tau \pmod{\zeta}$ if $\sigma g, \tau g \in \Delta$ or $\sigma g, \tau g \notin \Delta$ for $g \in G$. Then it is easy to check that ζ is a convex G-congruence and $\alpha \equiv \beta \pmod{\zeta}$. Taking elements $\delta \in \Delta$ and $\delta' \in \Omega \setminus \Delta$, we obtain $\delta' \notin \delta\zeta$, which shows that the congruence ζ is nontrivial. Consequently, (G, Ω) is not o-primitive. □

- A group (G, Ω) is called *o-2-transitive* if for any elements $\alpha_1, \alpha_2, \beta_1, \beta_2 \in \Omega$ such that $\alpha_1 < \alpha_2$ and $\beta_1 < \beta_2$ there exists an element $g \in G$ such that $\alpha_1 g = \beta_1$ and $\alpha_2 g = \beta_2$.

It is easy to check that if a group (G, Ω) is o-2-transitive, then it is o-primitive.

EXAMPLE 4.1.2. We consider the naturally ordered set $\mathbb{R}$ of real numbers and the group $\mathcal{A}(\mathbb{R})$ of all order automorphisms of the set $\mathbb{R}$. The group $\mathcal{A}(\mathbb{R})$ is o-2-transitive and, consequently, o-primitive.

4.1.6. Let a group (G, Ω) be transitive, ζ and θ, $\zeta \leqslant \theta$, be convex G-congruences, and Δ be an arbitrary congruence class of θ. We set $\overline{G}_\Delta = G_{(\Delta)}/G_\Delta$. Then $(\overline{G}_\Delta, \Delta)$ is a transitive group of order automorphisms of the totally ordered set Δ. The restriction $\zeta\,|_{\Delta\times\Delta} = \zeta'$ is a convex $\overline{G}_\Delta$-congruence, hence the group $(\widehat{G}_\Delta, \Delta/\zeta')$, where $\widehat{G}_\Delta \cong \overline{G}_\Delta / L_{\zeta'}$, is transitive. It is called the *(ζ, θ)-component* of the group (G, Ω) on Δ.

We show that the notion of (ζ, θ)-component is independent of the choice of the equivalence class Δ.

Proposition 4.1.9 (Holland and McCleary [54]). *Let ζ and θ, $\zeta \leqslant \theta$, be convex G-congruences on a totally ordered set Ω, and let Δ_1 and Δ_2 be arbitrary congruence classes of θ. Then (ζ, θ)-components of (G, Ω) are isomorphic as groups of order automorphisms.*

PROOF. Since the group (G,Ω) is transitive, there exists an element $g \in G$ such that $\Delta_1 g = \Delta_2$ and

$$g^{-1}G_{(\Delta_1)}g = G_{(\Delta_2)}$$
$$g^{-1}G_{\Delta_1}g = G_{\Delta_2}$$

Let $\zeta_i' = \zeta\,|_{\Delta_i\times\Delta_i}$, $i = 1,2$. It is easy to check that $g^{-1}L_{\zeta_1'}g = L_{\zeta_2'}$. For an arbitrary element $\delta_1 \in \Delta_1$ we put $(\delta_1\zeta_1')\varphi = (\delta_1 g\zeta_2')$ and $(L'_{\zeta_1}f)\psi = L_{\zeta_2'}g^{-1}fg$. Then the mapping $\psi\colon \widehat{G}_{(\Delta_1)} \to \widehat{G}_{(\Delta_2)}$ is an isomorphism of the groups $\widehat{G}_{(\Delta_1)}$ and $\widehat{G}_{(\Delta_2)}$. Moreover, $\varphi\colon \Delta_1/\zeta_1' \to \Delta_2/\zeta_2'$ is an order isomorphism between totally ordered sets Δ_1/ζ_1' and Δ_2/ζ_2'. Further,

$$((\delta_1\zeta_1')\varphi)(L_{\zeta_1'}f)\psi = (\delta_1 g\zeta_2')(L_{\zeta_2'}g^{-1}fg) = (\delta_1' f)\varphi$$

Consequently, the pair (ψ,φ) sets up an isomorphism of the groups $(\widehat{G}_{(\Delta_1)}, \Delta_1/\zeta_1')$ and $(\widehat{G}_{(\Delta_2)}, \Delta_2/\zeta_2')$ of order automorphisms. □

- If (G,Ω) is a transitive group of order automorphisms and ζ_1 and ζ_2 are convex G-congruences such that $\zeta_1 \leqslant \zeta_2$ and $\zeta_1 \neq \zeta_2$, then we say that ζ_2 *covers* ζ_1 provided that there exists no convex G-congruence θ such that $\zeta_1 \leqslant \theta \leqslant \zeta_2$ and $\theta \neq \zeta_1, \zeta_2$.

Proposition 4.1.10 (Holland and McCleary [54]). *Let ζ and θ be convex G-congruences on a totally ordered set Ω such that $\zeta \leqslant \theta$ and $\zeta \neq \theta$. Then the (ζ,θ)-component of the group (G,Ω) is primitive if and only if θ covers ζ.*

PROOF. Let Δ be an arbitrary equivalence class of θ. It is evident that any $(\widehat{G}_{(\Delta)}, \Delta/\zeta)$-block has the form T/ζ, where $T \subseteq \Delta$, and is a $(G_{(\Delta)},\Omega)$-block. For any element $g \in G$ the inequality $Tg \cap T \neq \varnothing$ implies the inequality $\Delta g \cap \Delta \neq \varnothing$. Since Δ is a (G,Ω)-block, the equality $\Delta g = \Delta$ holds. Hence $g \in G_{(\Delta)}$ and $Tg = T$. Consequently, T is a (G,Ω)-block. The application of Proposition 4.1.1 completes the proof. □

EXAMPLE 4.1.3. Let $\Omega = \mathbb{R} \overleftarrow{\times} \mathbb{Q}$ be the lexicographic product of totally ordered sets of real and rational numbers. We define the equivalence relation ζ on Ω by setting $(\alpha,\beta) \equiv (\alpha',\beta') \pmod{\zeta}$ for $\beta = \beta'$ $(\alpha,\alpha' \in \mathbb{R},\ \beta,\beta' \in \mathbb{Q})$. We also set

$$G = \{g \in \mathcal{A}(\Omega) \mid \sigma \equiv \tau \pmod{\zeta} \Leftrightarrow \sigma g \equiv \tau g \pmod{\zeta}\}$$

It is easy to show that ζ is a convex G-congruence and (G,Ω) is a transitive group of order automorphisms of the totally ordered set Ω.

Therefore, $\varepsilon \leqslant \chi$, where ε and χ are the least convex G-congruence and the greatest one respectively.

We now describe the (ε, ζ)-component of the group (G,Ω). To this end, we consider an arbitrary ζ-class Δ. For example, we can take $\Delta = \{(\alpha, \beta_0 \mid \alpha \in \mathbb{R})$, where β_0 is a fixed rational number. It is evident that Δ is order-isomorphic to the totally ordered set $\mathbb{R}$. Therefore, $\mathcal{A}(\Delta) \cong \mathcal{A}(\mathbb{R})$. For an arbitrary $s \in \mathcal{A}(\mathbb{R})$ we consider an element $g_s \in G_{(\Delta)}$ such that

$$(\alpha,\beta)g_s = \begin{cases} (\alpha_s, \beta_0) & \text{if } \beta = \beta_0 \\ (\alpha, \beta) & \text{if } \beta \neq \beta_0 \end{cases}$$

The mappings $s \to L_\varepsilon g_s$ and $\alpha \to (\alpha, \beta_0)$ define an isomorphism between $(\mathcal{A}(\mathbb{R}), \mathbb{R})$ and $(\widehat{G}_{(\Delta)}, \Delta)$. Consequently, the ($\varepsilon$, ζ)-component of the group (G,Ω) is o-primitive (cf. Example 4.1.2).

We now describe the (ζ,χ)-component of the group (G,Ω). It is obvious that Ω is a single equivalence class of χ and

$$L_\zeta = \{g \in G \mid Tg = T \ \text{ for all equivalence classes of } \ \zeta\}$$

For an arbitrary element $s \in \mathcal{A}(\mathbb{Q})$ we take an element $\overline{g}_s \in G$ such that $(\alpha,\beta)\overline{g}_s = (\alpha, \beta s)$. It is easy to see that the mappings $s \to L_\zeta \overline{g}_s$ and $\beta \to (\alpha,\beta)\zeta$ define an isomorphism of the groups $(\mathcal{A}(\mathbb{Q}), \mathbb{Q})$ and $(G/L_\zeta, \Omega/\zeta)$ of order automorphisms. Since $(\mathcal{A}(\mathbb{Q}), \mathbb{Q})$ is o-2-transitive, the (ζ,χ)-component of (G,Ω) is an o-primitive group. □

Let ζ and θ be convex G-congruences such that $\zeta \leqslant \theta$, and let ζ cover θ. In this case (ζ,θ) is called a *covering pair* of convex G-congruences. We denote by $k(G,\Omega)$ the set of all covering pairs of convex G-congruences. We write $(\zeta_K, \zeta^K) = K$ in $k(G,\Omega)$ if ζ^K covers ζ_K. By Corollary 4.1.2, the set $\mathcal{K}$ of convex G-congruences is totally ordered. Setting $K < K'$, for $\zeta^K \leqslant \zeta_{K'}$, we find that the set $k(G,\Omega)$ is also totally ordered. For $\alpha, \beta \in \Omega$, $\alpha \neq \beta$, we denote by ζ_K and ζ^K the union and the intersection of all convex G-congruences ζ such that $\beta \notin \alpha\zeta$ and $\beta \in \alpha\zeta$. By Corollary 4.1.3, ζ_K and ζ^K are convex G-congruences and, obviously, $K = (\zeta_K, \zeta^K)$ is a covering pair described by elements α and β. In this case K is called the *value* of the pair α, β and is denoted by $\mathrm{Val}(\alpha,\beta)$.

Proposition 4.1.11 (Holland and McCleary [54]). *For every pair of elements α and β, $\alpha \neq \beta$, of a totally ordered set Ω the inclusion* $\mathrm{Val}(\alpha,\beta) \in k(G,\Omega)$ *is fulfilled. Conversely, for every $K \in k(G,\Omega)$ in Ω there exist elements α and β such that $K = \mathrm{Val}(\alpha,\beta)$.*

PROOF. The first assertion of the proposition has been proved earlier. To prove the second one we assume that $K \in k(G,\Omega)$. Since $\zeta_K \neq \zeta^K$, in Ω there exist elements α and β such that $\alpha \in \beta\zeta^K$, but $\alpha \notin \beta\zeta_K$. Taking into account the relation $\alpha \in \beta\zeta^K$ and the definition of total order on $k(G,\Omega)$, we obtain $K \leqslant \mathrm{Val}(\alpha,\beta)$. Since $\beta \notin \alpha\zeta_K$, by the definition of linear order on $k(G,\Omega)$, we obtain the inequality $K \geqslant \mathrm{Val}(\alpha,\beta)$. Consequently, $K = \mathrm{Val}(\alpha,\beta)$. □

Proposition 4.1.12 (Holland and McCleary [54]). *Let ζ be a G-congruence. Then*

$$\cap\{\zeta^K \mid K \in k(G,\Omega),\ \zeta \leqslant \zeta_K\} = \zeta = \cup\{\zeta^K \mid K \in k(G,\Omega),\ \zeta^K \leqslant \zeta\}$$

PROOF. It is obvious that

$$\zeta' = \cup\{\zeta^K \mid K \in k(G,\Omega), \zeta^K \leqslant \zeta\} \leqslant \zeta$$

In the case $\zeta' \neq \zeta$, we take elements $\alpha,\beta \in \Omega$ such that $\alpha \in \beta\zeta$, but $\alpha \notin \beta\zeta'$. We set $K_0 = \mathrm{Val}(\alpha,\beta)$. It is evident that $\zeta' \leqslant \zeta_{K_0} \leqslant \zeta^{K_0} \leqslant \zeta$ and $\zeta_{K_0} \neq \zeta^{K_0}$. Since $\zeta^{K_0} \leqslant \zeta$, by the definition of ζ' we obtain the relation $\zeta^{K_0} \leqslant \zeta'$. But then $\zeta^{K_0} \leqslant \zeta' \leqslant \zeta^{K_0}$ and $\zeta' \neq \zeta^{K_0}$, which is impossible. The equality

$$\cap\{\zeta^K \mid K \in k(G,\Omega),\ \zeta \leqslant \zeta_K\} = \zeta$$

can be proved in a similar way. □

4.2. Wreath Products

4.2.1. Following Holland and McCleary [54], we define the wreath product of groups of automorphisms of a totally ordered set. Let (G,Ω) and (H,T) be transitive groups of order automorphisms. For every element $\tau \in T$ we fix a group G_τ which is identically isomorphic to the group G. An element g of the Cartesian product $\overline{\prod\limits_{\tau\in T}} G_\tau$ will be regarded as a function with domain of definition T and range of values G and will be denoted by $\{g_\tau : \tau \in T\} = \{g_\tau\}$. This means that $g_\tau \in G_\tau$ for any $g \in \overline{\prod\limits_{\tau\in T}} G_\tau$ and $\tau \in T$.

Let

$$\Lambda = \Omega \overleftarrow{\times} T$$

On the set

$$W = \{(\{g_\tau\}, h) \mid h \in H, g_\tau \in G_\tau, \quad \tau \in T\} \tag{4.2.1}$$

we introduce the binary operation of multiplication $\cdot$ and the unary operation $^{-1}$ in accordance with the following rule:

$$(\{f_\tau\}, h_1) \cdot (\{g_\tau\}, h_2) = (\{a_\tau\}, h_1 h_2) \tag{4.2.2}$$

where $a_\tau = f_\tau g_{\tau h_1}$, $\tau \in T$, and

$$(\{g_\tau\}, h)^{-1} = (\{f_\tau\}, h^{-1}) \tag{4.2.3}$$

where $f_\tau = g_{\tau h^{-1}}$, $\tau \in T$. For $(\alpha, \sigma) \in \Lambda$ we set

$$(\alpha, \sigma)(\{g_\tau\}, h) = (\alpha g_\sigma, \sigma h)$$

where $\alpha \in \Omega$ and $\sigma \in T$. It is easy to see that the element $(\{g_\tau\}, h) \in W$ is an order automorphism of the totally ordered set $\Lambda = \Omega \overleftarrow{\times} T$.

Lemma 4.2.1. *If (G, Ω) and (H, T) are transitive groups of order automorphisms, then (W, Λ) is a transitive group of order automorphisms of a totally ordered set Λ with the identity element $(\{e_\tau\}, e)$, where $e_\tau = e$ for any $\tau \in T$.*

PROOF. It is easy to check that elements of the group (W, Λ) keep the order introduced on Λ. We show that the group (W, Λ) is transitive. Let (α, σ) and (α', σ') be elements of the set Λ. By the transitivity of (G, Ω) and (H, T), there exist elements $g \in G$ and $h \in H$ such that $\alpha g = \alpha'$ and $\sigma h = \sigma'$. We take an element $w = (\{g_\tau\}, h) \in W$, where $g_\sigma = h$. By definition, $(\alpha, \sigma)(\{g_\tau\}, h) = (\alpha g_\sigma, \sigma h) = (\alpha', \sigma')$, which proves the transitivity of (W, Λ). □

- The above-mentioned group (W, Λ) of order automorphisms is called the *Cartesian wreath product* of the groups (G, Ω) and (H, T) (denoted by $(G, \Omega)\mathrm{Wr}\,(H, T)$).
- An element $h \in H$ is called a *global component*, and $\{g_\tau \mid \tau \in T\}$ is called the *set of local components* of the element $w = (\{g_\tau\}, h) \in (G, \Omega)\mathrm{Wr}\,(H, T)$,

We denote by $(G, \Omega)\mathrm{wr}\,(H, T)$ the subgroup $(G, \Omega)\mathrm{Wr}\,(H, T)$ generated by elements $w = (\{g_\tau\}, h) \in (G, \Omega)\mathrm{Wr}\,(H, T)$ such that the equality $g_\tau = e$ is valid for all $\tau \in T$ (except for a finite number of points).

- The subgroup $(G, \Omega)\mathrm{wr}\,(H, T)$ is called the *direct wreath product* of the groups (G, Ω) and (H, T).

It is easy to see that if (G, Ω) and (H, T) are transitive groups of order automorphisms, then $(G, \Omega)\mathrm{wr}\,(H, T)$ is also a transitive group of order automorphisms of the totally ordered set Λ.

Let (G, Ω) be a transitive group of order automorphisms and ζ be an arbitrary convex G-congruence, $\varepsilon \leqslant \zeta \leqslant \chi$, where ε and χ are the least and the greatest G-congruences respectively. Then the (ε, ζ)-component of the group (G, Ω) is equal to $(\widehat{G}_{(\Delta)}, \Delta/\varepsilon) = (\widehat{G}_{(\Delta)}, \Delta)$, where Δ denotes an arbitrary equivalence class of ζ and the (ζ,χ)-component of the group (G, Ω) is equal to $(G/L_\zeta, \Omega/\zeta)$ (cf. Proposition 4.1.9).

Theorem 4.2.1 (Holland and McCleary [54]). *Let (G, Ω) be a transitive group of order automorphisms, ζ be an arbitrary convex G-congruence, and Δ be the equivalence class of ζ. Then (G, Ω) is isomorphic to a subgroup of the wreath product*

$$(W, \Lambda) = (\widehat{G}_{(\Delta)}, \Delta)\mathrm{Wr}\,(G/L_\zeta, \Omega/\zeta)$$

PROOF. For every class A of the congruence ζ we choose an element $c_A \in G$ such that $\Delta c_A = A$ and $c_\Delta = e$. The element c_A is called the *coordinate mapping.* We define the mapping $\varphi : \Omega \to \Lambda$, where $\Lambda = \Delta \overleftarrow{\times} \Omega/\zeta$, by setting $\alpha\varphi = (\alpha c_{\alpha\zeta}^{-1}, \alpha\zeta)$, and show that φ is an order isomorphism between Ω and Λ.

It is obvious that φ is a mapping "onto." We assume that $\alpha \neq \beta$ and $\alpha, \beta \in \Omega$, but $\alpha\varphi = \beta\varphi$. Then $\alpha\zeta = \beta\zeta$ and $\alpha c_{\alpha\zeta}^{-1} = \beta c_{\alpha\zeta}^{-1}$. Consequently, $\alpha = \beta$, which contradicts our assumption. Let $\alpha \leqslant \beta$. Then $\alpha\zeta \leqslant \beta\zeta$ implies $\alpha\varphi < \beta\varphi$. If $\alpha\zeta = \beta\zeta$, then

$$\alpha\varphi = (\alpha c_{\alpha\zeta}^{-1}, \alpha\zeta) = (\alpha c_{\beta\zeta}^{-1}, \beta\zeta) < (\beta c_{\beta\zeta}^{-1}, \beta\zeta) = \beta\varphi$$

Let $\overline{g}$ designate the image of an element g under the action of the natural homomorphism of G onto G/L_ζ. For an arbitrary class A of the congruence ζ we set

$$\widehat{g}_A = c_A g c_{Ag}^{-1} \mid_\Delta \in \widehat{G}_{(\Delta)}$$

Let $\psi : g \to (\{\widehat{g}_A\}, \overline{g})$ and $h \in G$. Then

$$\widehat{fh}_A = c_A fh c_{Afh}^{-1} = c_A f c_{Af}^{-1} c_{Af} h c_{Afh}^{-1} = \widehat{f}_A \widehat{h}_A$$

Consequently,

$$(fh)\psi = (\{\widehat{fh}_A\}, \overline{fh}) = (\{\widehat{f}_A\}, \overline{f}) \cdot (\{\widehat{h}_A, \}\overline{h})$$

Thus, ψ is an isomorphism of the group G into the group W. If $\alpha \in \Omega, g \in G$, then

$$(\alpha g)\varphi = (\alpha g c_{\alpha g \zeta}^{-1}, \alpha g \zeta) = (\alpha c_{\alpha \zeta}^{-1}, \alpha \zeta)(\{\widehat{g}_A\}, \overline{g}) = (\alpha\varphi)((g)\psi)$$

and the pair (ψ, φ) sets up an isomorphism between the groups (G, Ω) and (W, Λ). □

Lemma 4.2.2 (associativity of wreath products). *Let (G, Ω), (H, T), and (L, B) be groups of order automorphisms. Then the groups $((G, \Omega)\mathrm{Wr}\,(H, T))\mathrm{Wr}\,(L, B)$ and $(G, \Omega)\mathrm{Wr}\,((H, T)\mathrm{Wr}\,(L, B))$ regarded as groups of order automorphisms are isomorphic.*

The assertion follows from the definition of wreath product and can be verified immediately.

We now introduce the notion of the wreath product for an arbitrary collection of groups of order automorphisms. Let K be an arbitrary totally ordered set of indices. With each index $k \in K$ we associate the group (G_k, Ω_k) of order automorphisms of a totally ordered set Ω_k. We consider the set $\Omega^+ = \overline{\prod_{k \in K}} \Omega_k$, and fix an arbitrary element $o \in \Omega^+$. With every element $\alpha \in \Omega^+$ we associate its support $\mathrm{supp}\,(\alpha) = \{k \in K \mid \alpha_k \neq o_k\}$. Let $\widehat{\Omega}$ denote the subset of Ω^+, which is formed by those $\alpha \in \Omega^+$ whose support is a dually well-ordered set. We note that if $\alpha, \beta \in \widehat{\Omega}$ and $\alpha \neq \beta$, then

$$\varnothing \neq K(\alpha, \beta) = \{k \in K \mid \alpha_k \neq \beta_k\} \subseteq \mathrm{supp}\,(\alpha) \cup \mathrm{supp}\,(\beta)$$

Consequently, $K(\alpha, \beta)$ is also dually well-ordered. Thus, in $K(\alpha, \beta)$ there exists the greatest element k_0 such that $\alpha_{k_0} \neq \beta_{k_0}$. We introduce a total order on $\widehat{\Omega}$ by setting $\alpha < \beta$ if $\alpha_{k_0} < \beta_{k_0}$. On the totally ordered set $\widehat{\Omega}$ we consider the natural equivalence relations. For $k \in K$ we set

$$\alpha \equiv^k \beta$$

if $\alpha_{k'} = \beta_{k'}$ for any $k' \in K(\alpha, \beta)$, $k' > k$, and

$$\alpha \equiv_k \beta$$

if $\alpha_{k'} = \beta_{k'}$ for each $k' \in K(\alpha, \beta)$, $k' > k$.

It is obvious that $\equiv^k$ and $\equiv_k$ are the equivalence relations on $\widehat{\Omega}$. Moreover, the equivalence classes are convex subsets of $\widehat{\Omega}$. If $\alpha, \beta \in \widehat{\Omega}$, $\alpha \neq \beta$, and k_0 is the greatest element of the set $K(\alpha, \beta)$, then $\alpha \equiv^k \beta$

for $k > k_0$ and $\alpha \equiv_k \beta$ for $k \geqslant k_0$. In $\mathcal{A}(\widehat{\Omega})$ we consider the subset

$$W_1 = \{g \mid (\forall k \in K)(\forall \alpha, \beta \in \widehat{\Omega})((\alpha \equiv^k \beta \\ \Leftrightarrow \alpha g \equiv^k \beta g) \& (\alpha \equiv_k \beta \Leftrightarrow \alpha g \equiv_k \beta g))$$

It is obvious that the quotient set $\alpha \equiv^k / \equiv_k$ coincides with Ω_k for every $\alpha \in \widehat{\Omega}$ and $k \in K$; moreover, $\equiv^k$ and $\equiv_k$ are convex W_1-congruences. Further, for every $k \in K$ and $\alpha \in \widehat{\Omega}$, we denote by α^k an element of $\overline{\prod_{k'>k}} \Omega_{k'}$ such that

$$(\alpha^k)_{k'} = \alpha_{k'}$$

for any k', $k' > k$.

We note that the element α^k is unique and the equality $\alpha^k = \beta^k$ is fulfilled if and only if $\alpha \equiv^k \beta$. For $\alpha \in \widehat{\Omega}$, $k \in K$, an element $g \in W_1$ induces an order automorphism g_{k,α^k} of Ω_k as follows:

$$\sigma g_{k,\alpha^k} = (\alpha' g)_k, \quad \alpha' \equiv \alpha, \quad \alpha'_k = \sigma$$

Lemma 4.2.3. *The mapping g_{k,α^k} is an order automorphism of the totally ordered set Ω_k.*

PROOF. If $\alpha \equiv^k \beta$ and $\beta_k = \sigma$, then $\alpha' \equiv_k \beta$. Since $g \in W_1$, we have $\beta g \equiv_k \alpha' g$. Consequently, $(\beta g)_k = (\alpha' g)_k$ and the definition of g_{k,α^k} is correct.

If $\tau \in \Omega_k$, $\tau < \sigma$, $\gamma \equiv^k \alpha$, and $\gamma_k = \tau$, then $\gamma \equiv^k \alpha'$ and $\alpha'_k < \gamma_k$. Hence $\alpha' < \gamma$ and $(\alpha' g)_k < (\gamma g)_k$. Consequently, g_{k,α^k} is an order automorphism of the totally ordered set Ω_k. □

We set

$$W = \{g \in W_1 \mid g_{k,\alpha^k} \in G_k \text{ for all } k \in K, \alpha \in \widehat{\Omega}\}$$

It is obvious that $(W, \widehat{\Omega})$ is a group of order automorphisms of the totally ordered set $\widehat{\Omega}$. The group is called the *generalized wreath product* of groups (G_k, Ω_k), $k \in K$, and is denoted by $\mathrm{Wr}\{(G_k, \Omega_k) \mid k \in K\}$. Elements of the generalized wreath product can be represented as "$K \times \widehat{\Omega}$-matrices" $(g_{k,\alpha})$, where $g_{k,\alpha} = g_{k,\beta}$ for $\alpha^k = \beta^k$. The multiplication of such "matrices" is defined by the following rule:

$$(f_{k,\alpha}) \cdot (h_{k,\alpha}) = (h_{k,\alpha})$$

where $h_{k,\alpha} = f_{k,\alpha} \cdot g_{k,\alpha f}$ $(k \in K, \alpha \in \widehat{\Omega})$ and $(\delta(g_{k,\alpha}))_k = \delta g_{k,\delta}$ $(\delta \in \widehat{\Omega},\ k \in K)$.

Lemma 4.2.4. *Let (G_k, Ω_k), $k \in K$, be groups of order automorphisms. Then the generalized wreath product* $\mathrm{Wr}\,\{(G_k, \Omega_k) \mid k \in K\}$ *is also a group of order automorphisms.*

PROOF. The assertion immediately follows from the definition of generalized wreath product. □

Generally speaking, the definition of generalized wreath product depends on the choice of an element $o \in \Omega^+$. We show that in the case in which all the groups (G_k, Ω_k), $k \in K$, are transitive, such a dependence disappears.

Lemma 4.2.5. *Let (G_k, Ω_k), $k \in K$, be transitive groups of order automorphisms, $(W, \widehat{\Omega})$ be the generalized wreath product of the groups (G_k, Ω_k), $k \in K$, with a fixed element $o \in \Omega^+$, and $(W', \widehat{\Omega}')$ be the generalized wreath product of the groups (G_k, Ω_k), $k \in K$, with a fixed element o'. Then $(W, \widehat{\Omega})$ and $(W', \widehat{\Omega}')$ are transitive groups of order automorphisms of $\widehat{\Omega}$ and $\widehat{\Omega}'$ respectively. Moreover, the groups regarded as groups of order automorphisms are isomorphic.*

PROOF. Let $\alpha, \beta \in \widehat{\Omega}$ be arbitrary elements. For every $k \in K$ there exists an element $g_k \in G_k$ such that $\alpha_k g_k = \beta_k$; moreover, $g_k = e$ for $k \in K(\alpha, \beta)$. We define a mapping $g \in \widehat{\Omega}$ by setting $(\gamma g)_k = \gamma_k g_k$ $(\gamma \in \widehat{\Omega},\ k \in K)$. It is obvious that $\alpha g = \beta$. Since $\gamma \in \widehat{\Omega}$, we have $\mathrm{supp}\,(\gamma g) \subseteq \mathrm{supp}\,(\gamma) \cup K(\alpha, \beta)$. Consequently, $\mathrm{supp}\,(\gamma g)$ is a dually well-ordered set. It is easy to check that g is an order automorphism of $\widehat{\Omega}$ and $g_{k,\gamma} = g_k \in G_k$. Hence $g \in W$, and $(W, \widehat{\Omega})$ is a transitive group.

For an arbitrary $k \in K$ there exists an element $f_k \in G_k$ such that $o_k f_k = o'_k$. We define the mapping $\varphi : \Omega^+ \to \Omega^+$ by setting $(\alpha\varphi)_k = \alpha_k f_k$ $(\alpha \in \widehat{\Omega})$. It is evident that φ is an order automorphism of Ω^+. Moreover, if $\alpha \in \widehat{\Omega}$, then $\alpha\varphi \in \widehat{\Omega}'$. We consider the mapping $\psi : W \to W'$, where $(g)\psi = \varphi^{-1} g \varphi$. The simplest computation yields $(\alpha\varphi)(g\psi) = \alpha g \psi$. Therefore, the pair (ψ, φ) defines an isomorphism between the groups $(W, \widehat{\Omega})$ and $(W', \widehat{\Omega}')$. □

Let H be a group and L be its subgroup.

- The set $T(L)$ of elements of a group H is called a *transversal* of L if for each coset L the set $T(L)$ contains exactly one element $(h \in H)$ of Lh.

Lemma 4.2.6. *In every group H there exists a family of transversals $\{T(L \mid L\}$, where L "runs over" the set of all subgroups of H such that*

(a) $T(L_1) \subseteq T(L_2)$ for $L_2 \subseteq L_1$,

(b) $T(L) \cap L = \{e\}$.

PROOF. We introduce a dual well-order for the set of elements of the group H, starting with the element e. Let L be a subgroup of the group H and $h \in H$. As elements of the transversal $T(L)$ (i.e., representatives of cosets Lh) we choose the least elements of the sets Lh. It is obvious that the constructed family of transversals $\{T(L) \mid L$ is a subgroup of $H\}$ satisfies the required properties. □

Theorem 4.2.2 (Holland and McCleary [54]). *Let (G, Ω) be a transitive group of order automorphisms, $K = K(G, \Omega)$, $\{(G_k, \Omega_k) \mid k \in K\}$ be the set of primitive components of the group (G, Ω), and $(W, \widehat{\Omega}) = \mathrm{Wr}\,\{(G_k, \Omega_k) \mid k \in K\}$. Then there exists an isomorphic embedding (ψ, φ) of (G, Ω) into $(W, \widehat{\Omega})$.*

PROOF. We fix an element $\sigma_0 \in \Omega$ and set $\Omega_k = \sigma_0\zeta^k/\zeta_k$ for every $k \in K$. Further, we designate $T = \{T(G_{(\sigma_0\zeta^k)}) \mid k \in K\}$ and set $c(\sigma, k)$ equal to a single element g of $T(G_{(\sigma_0\zeta^k)})$ such that

$$\sigma_0 c(\sigma, k) \equiv \sigma g(\pmod{\zeta)^k})$$

By Lemma 4.2.5, $c(\sigma_0, k) = e$ for every $k \in K$. We now introduce a mapping $\varphi : \Omega \to \Omega^+ = \prod_{k \in K} \Omega_k$ such that

$$(\sigma_0\varphi)_k = \sigma c(\sigma, k)^{-1}\zeta_k$$

It is clear that the following inclusion holds:

$$(\sigma_0\varphi)_k \in \Omega_k$$

for $\sigma \in \Omega$, $k \in K$. Let $o = \sigma_0\varphi$. Then $o_k = (\sigma_0\zeta_k)$ and $\widehat{\Omega}$ is a subset of Ω^+ constituted by those elements of Ω^+ whose supports (with respect to o) are well-ordered. We now show that

$$\Omega\varphi \subseteq \widehat{\Omega}$$

where φ stands for a one-to-one order mapping. We consider $\sigma \in \Omega$ and K' such that $\varnothing \neq K' \subseteq \operatorname{supp}(\sigma\varphi)$. We choose $f \in G$ so as to satisfy the condition $\sigma_0 f = \sigma$ and set $H = \bigcup_{k\in K'} G_{(\sigma_0\zeta^k)}$. Since the family $\{G_{(\sigma_0\zeta^k)} \mid k \in K\}$ of subgroups is totally ordered by inclusion, we conclude that H is a subgroup of G. Let $T(H) \cap Hf = g$. Then $fg^{-1} \in H$. Consequently, $fg^{-1} \in G_{(\sigma_0\zeta^{k_1})}$ for some element $k_1 \in K'$. We show that k_1 is the greatest element of K'. Indeed, if there exists a $k \in K'$ such that $k > k_1$, then

$$G_{(\sigma_0\zeta^{k_1})} \subseteq G_{(\sigma_0\zeta^k)} \subseteq H$$

In view of the choice of T, we have $g \in T(H)$, i.e., $g \in T(G_{(\sigma_0\zeta^k)})$. However,

$$fg^{-1} \in G_{(\sigma_0\zeta^{k_1})} \subseteq G_{(\sigma_0\zeta^k)}$$

$$\sigma_0 fg^{-1} \equiv \sigma_0 \pmod{\zeta_k}$$

Consequently, $\sigma g^{-1} \equiv \sigma_0 \pmod{\zeta_k}$, $\sigma \equiv \sigma_0 g \pmod{\zeta_k}$, and $g = c(\sigma, k)$. Therefore,

$$(\sigma\varphi)_k = \sigma c(\sigma,k)^{-1}\zeta_k = \sigma g^{-1}\zeta_k = o_k$$

and $k \notin \operatorname{supp}(\sigma\varphi)$. Hence k_1 is the greatest element of K' and $(\sigma\varphi)$ is a well-ordered subset of K. Thus, $\Omega\varphi \subseteq \widehat{\Omega}$. In the case of $\sigma, \tau \in \Omega$, $\sigma < \tau$, we take an element $k \in \operatorname{Val}(\sigma,\tau)$ and

$$\sigma \equiv \tau \pmod{\zeta)^k}, \; c(\sigma,k) = c(\tau,k)$$

$$(\sigma\varphi)_k = \sigma c(\sigma,k)^{-1}\zeta_k < \tau c(\sigma,k)^{-1}\zeta_k = (\tau\varphi)_k$$

For $k' > k$ the following inclusion holds:

$$\zeta_{k'} \supseteq \zeta^k$$

Consequently,

$$(\sigma\varphi)_{k'} = \sigma c(\sigma,k')^{-1}\zeta k' = \tau c(\sigma,k')^{-1}\zeta_{K'} = \tau c(\tau,k')^{-1}\zeta_{k'} = (\tau\varphi)_{k'}$$

Hence $\sigma\varphi < \tau\varphi$. We now show that the relations $\sigma \equiv \tau \pmod{\zeta^k}$ and $\sigma \equiv \tau \pmod{\zeta_k}$ are fulfilled if and only if $\sigma\varphi \equiv^k \tau\varphi$ and $\sigma\varphi \equiv_k \tau\varphi$ respectively. We assume that $\sigma \equiv \tau \pmod{\zeta^k}$. Then $\sigma \equiv \tau \pmod{\zeta_{k'}}$

for all $k' > k$. Therefore, $c(\sigma, k') = c(\tau, k')$. Consequently,

$$(\sigma\varphi)_{k'} = \sigma c(\sigma, k')^{-1}\zeta_{k'} = (\tau\varphi)_{k'}$$

and $\sigma\varphi \equiv^k \tau\varphi$. Conversely, for $\tau \notin \sigma\zeta^k$ we consider $k' = \mathrm{Val}\,(\sigma, \tau)$. Then $k' > k$, $c(\sigma, k') = c(\tau, k')$, and

$$(\sigma\varphi)_{k'} = \sigma c(\sigma, k')^{-1}\zeta_{k'} = \sigma c(\tau, k')^{-1}\zeta_{k'} \neq (\tau\varphi)_{k'}$$

Consequently, $\sigma\varphi \equiv^k \tau\varphi$. In the case $\sigma \equiv \tau \pmod{\zeta_k}$, the proof is similar.

We now take elements $\sigma \in \Omega$ and $\widehat{\alpha} \in \widehat{\Omega}$ such that

$$\sigma\varphi \equiv^k \widehat{\alpha}$$

$$\widehat{\alpha}_k \in \Omega_k = \sigma_0\zeta^k/\zeta_k$$

We have $\widehat{\alpha}_k = \tau\zeta^k$ for some $\tau \in \sigma_0\zeta^k$. We choose an element τ' so as to satisfy the equality $\tau' = \tau c(\sigma, k) \in \Omega$. Then $\tau' = \sigma_0 c(\sigma, k)^{-1}\zeta^k = \sigma\zeta^k$. Therefore, $\tau'\varphi \equiv^k \sigma\varphi \equiv^k \widehat{\alpha}$ and

$$(\tau'\varphi)_k = \tau' c(\tau', k)^{-1}\zeta_k = \tau\zeta_k = \widehat{\alpha}_k$$

Consequently, $\tau'\varphi \equiv_k \widehat{\alpha}$. Thus, we have shown that any equivalence class $\equiv_k$ is contained in $(\sigma\varphi) \equiv^k \cap\Omega\varphi$.

Let $\widehat{\alpha} \in \widehat{\Omega}$, $k \in K$, $g \in G$. We introduce a mapping ψ: $G \to W$ by the following rule:

$$(\widehat{\alpha} \cdot g\psi)_k = \begin{cases} ((\sigma g)\varphi)_k & \text{if } \sigma\varphi \equiv_k \widehat{\alpha} \text{ for some } \sigma \in \Omega \\ \widehat{\alpha}_k & \text{otherwise} \end{cases}$$

The definition of ψ is correct since $\sigma\varphi \equiv_k \tau\varphi$ implies $\sigma \equiv \tau \pmod{\zeta_k}$ and, consequently, $\sigma g \equiv \tau g \pmod{\zeta_k}$. Moreover, from the definition it follows that

$$(\sigma\varphi)(g\psi) = (\sigma g)\varphi$$

We now show that $(g)\psi \in W$ and that the mapping ψ is an isomorphism. We consider an arbitrary element $\widehat{\alpha} \in \widehat{\Omega}$. Let

$$\varnothing = K' \subseteq \mathrm{supp}\,(\widehat{\alpha} \cdot g\psi)$$

If $K' \not\subseteq \mathrm{supp}\,(\widehat{\alpha})$, then there exist elements $k \in K' \setminus \mathrm{supp}\,(\widehat{\alpha})$ and $\sigma \in \Omega$ such that $\sigma\varphi \equiv_k \widehat{\alpha}$. For any element k' such that $k' > k$, the following equality holds:

$$(\widehat{\alpha} g\psi)_{k'} = ((\sigma g)\varphi)_{k'}$$

Consequently,

$$\begin{aligned}\varnothing &\neq \{k' \in K \mid k' > k\}\\ &\subseteq \{k' \in K' \mid k' \geqslant k,\ k' \in \operatorname{supp}((\sigma g)\varphi)\}\\ &\subseteq \operatorname{supp}((\sigma g)\varphi)\end{aligned}$$

Since the support $\operatorname{supp}((\sigma g)\varphi)$ is a well-ordered set, K' has the greatest element. Therefore, the support $\operatorname{supp}((\widehat{\alpha})(g)\psi)$ is also well-ordered and $\widehat{\alpha}((g)\psi) \in \widehat{\Omega}$. Further, $(g)\psi \in \mathcal{A}(\widehat{\Omega})$, $\equiv^k$ and $\equiv_k$ are convex G-congruences, and $(g\psi)_{k,\widehat{\alpha}} \in G_k$ for all $k \in K$, $\widehat{\alpha} \in \widehat{\Omega}$. Thus, $(g)\psi \in W$. Further,

$$(\widehat{\alpha}((f)\psi(g)\psi))_k = \begin{cases} (((\sigma f)g)\varphi)_k & \text{if } \sigma\varphi \equiv_k \widehat{\alpha} \text{ for some } \sigma \in \Omega \\ \widehat{\alpha}_k & \text{otherwise} \end{cases}$$

$$(\widehat{\alpha}((f)\psi(g)\psi))_k = (\widehat{\alpha}((fg)\psi))_k$$

Consequently, ψ is an isomorphism. It is obvious that the pair (ψ, φ) defines an isomorphic embedding of (G, Ω) into $(W, \widehat{\Omega})$. □

4.2.2. Theorem 4.2.2 shows the significance of *o*-primitive groups of order automorphisms. Such groups are those "small bricks" out of which all groups of order automorphisms of totally ordered sets can be built. Alternatively, *o*-primitive groups regarded as groups of order automorphisms do not have nontrivial homomorphisms, so they are close to simple groups. McCleary [55, 56] has proved a theorem describing the *o*-primitive groups of order automorphisms. We formulate this theorem without proofs.

As was proved in Chapter 1, the group $\mathcal{A}(\Omega)$ of all order automorphisms of a totally ordered set Ω is an l-group with composition and lattice operations $\vee$ and $\wedge$ which are defined by the following equalities:

$$\begin{aligned}(\alpha)(f \vee g) &= \max\{\alpha f, \alpha g\}\\ (\alpha)(f \wedge g) &= \min\{\alpha f, \alpha g\}\end{aligned}$$

where $f, g \in \mathcal{A}(\Omega)$ and $\alpha \in \Omega$. Therefore, any group (G, Ω) of order automorphisms is contained in the l-group $\mathcal{A}(\Omega)$. We denote by (G_l, Ω) the l-subgroup of the group $\mathcal{A}(\Omega)$ generated by (G, Ω).

- A group (G, Ω) is called a *periodic* group of automorphisms of a totally ordered set Ω if there exists an automorphism t of the Dedekind completion $\overline{\Omega}$ of the set Ω, $t \in \mathcal{A}(\overline{\Omega})$, such that
  — if $g \in G$ and $x \in X$, then $(\alpha)tg = (\alpha)gt$,
  — t has no fixed points in $\overline{\Omega}$ and for every element $\alpha \in \Omega$ the set $\{\alpha t^n \mid n \in \mathbb{Z}\}$ is unbounded in Ω,
  — for any element $\alpha \in \Omega$ an open interval $((\alpha, \alpha t))$ in Ω coincides with the orbit $\beta G_\alpha = \Omega(\alpha, \beta)$ of the stabilizer G_α for some $\beta \in \Omega$.
- An o-2-transitive group (G, Ω) is called *pathological* if it does not contain elements with bounded supports.

Theorem 4.2.3 (McCleary [56]). *A transitive o-primitive lattice-ordered group (G, Ω) of order automorphisms of a totally ordered set Ω is a group of one of the following types:*

— *an Archimedean totally ordered group given by its right regular representation,*

— *an o-2-transitive group on Ω,*

— *a periodic group of order automorphisms; moreover, the stabilizer G_α in G of every point α of the set Ω acts on each set $\Omega(\alpha, \beta)$ and is an o-2-transitive and nonpathological group of order automorphisms of the totally ordered set $\Omega(\alpha, \beta)$.*

The proof can be found in [37] and [36].

Corollary 4.2.1. *A primitive group (G, Ω) of order automorphisms is a subgroup of the l-group (G_l, Ω) of one of the types indicated in Theorem* 4.2.3.

4.2.3. We use the structure of the wreath product of groups of automorphisms of a totally ordered set to establish a connection between Conradian right-ordered groups and normal-valued lattice-ordered groups. We recall the definition of a normal-valued lattice-ordered group with subnormal jumps. Beforehand, we note that the system $\mathfrak{C}(G)$ of convex l-subgroups of a lattice-ordered group is a complete lattice.

- If, for any jump $A \prec B$, $A, B \in \mathfrak{C}(G)$, in $\mathfrak{C}(G)$, A is an ideal of B, then the l-group G is called an *l-group with subnormal jumps* or a *normal-valued l-group*.

For the properties of l-groups with subnormal jumps and their significance in the theory of lattice-ordered groups, we refer the reader to [36, 37]. It suffices for our purposes to know the fact that the direct product and the Cartesian product of l-groups with subnormal jumps, as well as the lexicographic extension of such a group by a totally ordered group, are again normal-valued l-groups.

The class of l-groups with subnormal jumps is closed under taking l-subgroups. Therefore, from the above-mentioned properties it is not hard to deduce that the wreath product W of any collection of groups $(\mathbb{R}, \mathbb{R})$ of automorphisms, which are right regular representatives of the additive group $\mathbb{R}$ with the natural order, is an l-group with subnormal jumps.

Let G be a right-ordered group. Then for every element $g \in G$ a mapping $R : G \to G$ such that $(x)R(g) = xg$ is an order automorphism of the totally ordered set G. Thus, the mapping R of the group G onto the group $R(G) = (G, G)$ of order automorphisms is a faithful transitive representative of the right-ordered group G into the group $\mathcal{A}(G)$ of order automorphisms of the totally ordered set G.

Let $\mathfrak{C}(G) = \{G_\alpha \mid \alpha \in I\}$ be the system of convex subgroups of a right-ordered group G. It is easy to show that for any convex subgroup G_α from $\mathfrak{C}(G)$ the relation ζ_α introduced by the rule

$$x \equiv y \pmod{\zeta_\alpha} \Leftrightarrow xy^{-1} \in G_\alpha$$

is a convex G-congruence on the totally ordered set G. Conversely, if ζ is a convex G-congruence on G, then the congruence class of ζ, containing the identity element e of the group G, will be a convex subgroup of the group G. Thus, totally ordered sets of convex G-congruences and those of convex subgroups of G are order-isomorphic.

We suppose that the system $\mathfrak{C}(\sigma) = \{G_\alpha \mid \alpha \in I\}$ of convex subgroups of a right-ordered group G is subnormal (i.e., the order on G is Conradian). Let $K = K(G, \Omega)$ be the set of all covering pairs of convex G-congruences. As already mentioned, with an arbitrary covering pair $k = (\zeta_k, \zeta^k)$ we can associate the jump $G_{\alpha(k)} \subseteq \overline{G}_{\alpha(k)}$ in the system $\mathfrak{C}(G)$.

In the notation of Theorem 4.2.2 we have

$$\Omega_k = \overline{G}_{\alpha(k)}/G_{\alpha(k)}$$

$$G_k = \overline{G}_{\alpha(k)}/G_{\alpha(k)}$$

By Theorem 4.2.2, the right-ordered group (G, G) regarded as a group of order automorphisms is isomorphic to a subgroup of the group $(W, \widehat{\Omega}) = \mathrm{Wr}\{(G_k, \Omega_k) \mid k \in K\}$. Since the order on the group G is Conradian, the group (G_k, Ω_k) is order-isomorphic to a subgroup of $(\mathbb{R}, \mathbb{R})$ of the group $\mathbb{R}$ with its right regular representation.

Thus, the group (G, G), as a group of order automorphisms, is isomorphic to a subgroup of the group $(W, T) = \mathrm{Wr}\{(\mathbb{R}_k, \mathbb{R}_k) \mid k \in K\}$. Therefore, the l-subgroup (G^*, G) of the group $\mathcal{A}(G)$ generated by the right-ordered group (G, G), as a group of order automorphisms, is also isomorphic to a subgroup of the group (W, T). As already mentioned, (W, T) is an l-group with subnormal jumps. Thus, we have proved the following theorem.

Theorem 4.2.4 (Kopytov [57] and Varaksin [58]). *If G is a Conradian right-ordered group, then the l-subgroup G^* of the l-group $\mathcal{A}(G)$, which is generated by the right regular representation $R(G) = (G, G)$, is a lattice-ordered group with subnormal jumps.*

If the system

$$E \subseteq G_1 \subseteq \ldots \subseteq G_n = G$$

of convex subgroups of G is finite and contains $n + 1$ elements, then the l-subgroup G^* of the l-group $\mathcal{A}(G)$ generated by the right regular representation $R(G)$ of the group G is a solvable group whose soluble length is at most n.

We note that similar questions for nilpotent right-ordered groups are discussed in detail by Varaksin [58] and Darnel [59, 60].

4.3. Order Types of Right-Ordered Groups

In this section, the problem of the transformation of a totally ordered set into a right-ordered or totally ordered group is studied.

Let η and σ be the order types of the sets $\mathbb{Z}$ and $\mathbb{Q}$ with the natural orders. If ξ and τ are order types of totally ordered sets X and Y, we write $\xi\tau$ for the order type of the lexicographic product $Y \vec{\times} X$. In other words, $\xi\tau$ is the order type of the totally ordered set of the pairs (y, x), $y \in Y$, $x \in X$, with the following order: $(y_1, x_1) \leqslant (y_2, x_2)$ if $y_1 \leqslant y_2$ or $y_1 = y_2$, $x_1 \leqslant x_2$.

If α is an ordinal, then ξ^α is the order type of the set X^α defined by induction. If all X^β have been already defined for $\beta < \alpha$ and there exists $\alpha - 1$, then $X^\alpha = X^{\alpha-1}\vec{\times}X$. If α is a limit ordinal, then we assume that X^α is equal to the union of increasing sequences X^β $(\beta < \alpha)$ with the natural embedding. It is clear that $\xi^1 = \xi$. We denote by ξ^0 the order type of a single-element set.

We introduce the following notation:

G is a transitive group of order automorphisms of a totally ordered set X,

θ is a convex G-congruence on the totally ordered set X,

ξ is the order type of the totally ordered set X,

$\overline{\eta}$ is the order type of some class $Y = x\eta$ of the congruence η,

ζ is the order type of the totally ordered quotient set X/θ of the totally ordered set X by the congruence θ.

Lemma 4.3.1. *In the above notation the following equality holds*: $\zeta = \overline{\eta}\zeta$.

The proof is trivial.

Theorem 4.3.1 (Zaĭtseva [9]). *If the group $\mathcal{A}(X)$ of order automorphisms of a countable totally ordered set X is transitive, then the order type of X is equal to $\eta^\alpha\sigma^\varepsilon$ for some ordinal α and number ε equal to* 0 *or* 1.

PROOF. Let X be a nondiscrete totally ordered set. In view of the countability of the set X and the transitivity of the group $\mathcal{A}(X)$, the totally ordered set X is order-isomorphic to the set of rational numbers with the natural order, i.e., $\xi = \eta^0\sigma$.

In the case of a discrete totally ordered set X we assume that for an ordinal β we have already defined the $\mathcal{A}(X)$-congruence θ_β on X which has order type η^β. Then there exists a representation φ_β of the group $\mathcal{A}(X)$ by order automorphisms of the quotient set X/θ_β; moreover, the group $(\mathcal{A}(X))\varphi_\beta$ acts on the totally ordered set $X/\theta_\beta = X_\beta$ transitively. By Lemma 4.3.1, the order type ξ of the totally ordered set X is equal to $\eta^\beta\xi_\beta$, where ξ_β is the order type of the totally ordered set X_β. If X_β is not discrete, then $\xi_\beta = \sigma$ and $\xi = \eta^\beta\sigma$. For discrete X_β we fix an element $x_\beta \in X_\beta$. It is evident that there is an element $x'_\beta \in X_\beta$ such that $x_\beta < x'_\beta$; moreover, between x_β and x'_β there are no elements of X_β. Consequently, there exists an order automorphism g in $\mathcal{A}(X_\beta)$ such that $(x_\beta)g = x'_\beta$. We consider the subset

$$C_\beta = \{y_\beta = (x_\beta)g^n | n \in \mathbb{Z}\}$$

of the set X_β. It is clear that $(x_\beta)g^k < (x_\beta)g^{k+1}$ for any $k \in \mathbb{Z}$. If $(x_\beta)g^k \leqslant z_\beta < (x_\beta)g^{k+1}$, then $x_\beta \leqslant (z_\beta)g^k < (x_\beta)g = x'_\beta$ and, due to the choice of the order automorphism g, we have the equality $(z_\beta)g^{-k} = x_\beta$, which implies the convexity of the set C_β. Let $h \in \mathcal{A}(X)$, $z_\beta = (x_\beta)g^k \in C_\beta$, and $(x_\beta)g^n h = (\alpha_\beta)g^{k+r} \in C_\beta$. Then $(x_\beta)g^n h = (x_\beta)g^{n+r} \in C_\beta$ for any integer n. By Proposition 4.1.1, the set C_β defines a convex $\mathcal{A}(X_\beta)$-congruence $\overline{\theta}_{\beta+1}$ on the totally ordered set $X_\beta : x_\beta \equiv y_\beta \pmod{\overline{\theta}_{\beta+1}}$ if and only if $x_\beta, y_\beta \in (C)h$ for some order automorphism $h \in \mathcal{A}(X_\beta)$. It is obvious that the totally ordered set C_β has order type η. Therefore, $\xi_\beta = \eta\xi_{\beta+1}$ where $\xi_{\beta+1}$ is the order type of the totally ordered set $X_\beta/\overline{\theta}_{\beta+1}$. Then $\xi = \eta^{\beta+1}\xi_{\beta+1}$ and, starting with the relation $\overline{\theta}_{\beta+1}$, we can define a convex congruence $\theta_{\beta+1}$ on the totally ordered set X by setting $x \equiv y \pmod{\theta_{\beta+1}}$ if $x \equiv y \pmod{\overline{\theta}_{\beta+1}}$, where x_β and y_β are classes of θ_β-congruent elements of X containing x and y respectively. It is clear that $\theta_{\beta+1}$ is a convex $\mathcal{A}(X)$-congruence on the totally ordered set X.

Thus, we have built the following increasing sequence of convex $\mathcal{A}(X)$-congruences:

$$\theta_1 < \theta_2 \ldots \theta_\gamma < \theta_{\gamma+1} < \ldots$$

(on the limit places we put the union of congruences). Moreover, the classes of θ_γ-congruences have order type η^γ and the group $\mathcal{A}(X/\theta_\gamma)$ is transitive on the totally ordered set X/θ_γ. For some countable ordinal α the process must terminate: either the totally ordered set X coincides with the class of θ_α-congruent elements or the totally ordered set X/θ_α is not discrete. In the first case $\xi = \eta^\alpha = \eta^\alpha\sigma^0$; in the second case $\xi = \eta^\alpha\sigma$. □

Corollary 4.3.1 (Mal'cev [12]). *A countable totally ordered group has order type $\eta^\alpha\sigma^\varepsilon$, where α is a countable ordinal and ε is equal to 0 or 1.*

PROOF. If a totally ordered group G has order type $\eta^\alpha\sigma^\varepsilon$, there is the upper central series of its convex subgroups

$$E \subseteq Z_1 \subseteq Z_2 \subseteq \ldots Z_\alpha$$

with infinite cyclic factors such that the totally ordered quotient group G/Z_α is not discrete for $\varepsilon = +1$ and $G = Z_\alpha$ for $\varepsilon = 0$. □

Corollary 4.3.2 (Zaĭtseva [9]). *A countable right-ordered group G has order type $\eta^\alpha\sigma^\varepsilon$, where α is some countable ordinal and ε is equal to 0 or 1.*

Corollary 4.3.3 (Zaĭtseva [9] and Mal'cev [12]). *For any countable right-ordered group G there is an Abelian totally ordered group A whose order type is the same as that of the right-ordered group G.*

PROOF. Let the order type of a right-ordered group G be equal to $\eta^\alpha\sigma$. As the required totally ordered group, we can take the direct product of the totally ordered groups $\mathbb{Z}_\gamma$, $\gamma \leqslant \alpha$, and $\mathbb{Q}$ with the following lexicographic order: for

$$x = x_{\alpha(1)} \cdots x_{\alpha(k)} q$$

where $\alpha(1) < \ldots \alpha(k)$, $x_{\alpha(i)} \in \mathbb{Z}_{\alpha(i)}$, $q \in \mathbb{Q}$, we set $x > e$ if $q > 0$ in $\mathbb{Q}$ or $q = 0$, $x_{\alpha(k)} > 0$ in $\mathbb{Z}_{\alpha(k)}$. The case $\varepsilon = 0$ is considered in a similar way. □

An analogous result for uncountable groups is false. Recently, Holland, Mekler, and Shelah [61] have established the existence of uncountable totally ordered groups such that there are no Abelian totally ordered groups having the same order type.

4.4. The Chehata Groups and the Dlab Groups

We begin by proving Higman's result that a group of permutations of an infinite set is simple. We also give some examples of simple orderable groups of order automorphisms.

4.4.1. Let G be a group of permutations of an infinite set Ω. As above, for $g \in G$ we denote by $\operatorname{supp} g$ the set of elements $\alpha \in \Omega$ such that $\alpha g \neq \alpha$.

- A subset T of a set Ω is called *unstable* with respect to an element $g \in G$ if $T \cap (T)g = \varnothing$.

Theorem 4.4.1 (Higman [62]). *If for any three elements f, g, h, $f \neq e$, of a group G of permutations of Ω there is an element $r \in G$ such that the set $(\operatorname{supp} g \cup \operatorname{supp} f)r$ is unstable with respect to f, then the derived subgroup G' of the group G is simple.*

PROOF. We first show that a nontrivial normal subgroup N of a group G contains the derived subgroup G' of the group G. We take a nonidentity element f of the subgroup N and arbitrary elements g and h of G. By assumption, there is an element $r \in G$ such that the set $(\operatorname{supp} g \cup \operatorname{supp} h)r$ is unstable with respect to f. We denote $d = rfr^{-1}$. It is clear that the intersection of $((\operatorname{supp} g \cup \operatorname{supp} h)r)f$ and $(\operatorname{supp} g \cup \operatorname{supp} h)r$ is empty. Consequently,

$$(\operatorname{supp} g \cup \operatorname{supp} h)d \cap (\operatorname{supp} g \cup \operatorname{supp} h) = \varnothing$$

This means that

$$(\operatorname{supp} g)d \cap \operatorname{supp} h = \varnothing$$
$$\operatorname{supp} d^{-1}gd \cap \operatorname{supp} h = \varnothing$$

Hence the elements $d^{-1}gd$ and h commute and

$$\begin{aligned} g^{-1}h^{-1}gh &= g^{-1} \cdot d^{-1}gd \cdot h^{-1} \cdot d^{-1}g^{-1}d \cdot gh \\ &= g^{-1}d^{-1}g \cdot d \cdot h^{-1}d^{-1} \cdot h^{-1}g^{-1} \cdot d \cdot gh \end{aligned}$$

Since d and f are conjugate, the element $g^{-1}h^{-1}gh$ belongs to N. Consequently, $G' \subseteq N$. We show that the second derived group G'' of the group G coincides with G'. Indeed, we can assume that $G' \neq E$. It suffices to prove that $G'' \neq E$, since in this case we have $G'' \supseteq G'$. We choose elements $g, f, h \in G'$ such that $g = h = f \neq e$. As above, we can show that the sets $\operatorname{supp} f$ and $\operatorname{supp} d^{-1}fd$ are mutually disjoint for a suitable $d \in G$. It is evident that $f \neq d^{-1}fd$. Therefore, $f^{-1}d^{-1}fd \neq e$.

We now show that the group G' has the required properties. For given elements $g, h, f \in G'$, $f \neq e$, we choose an element $r \in G$ such that

$$(\operatorname{supp} g \cup \operatorname{supp} h)rf \cap (\operatorname{supp} g \cup \operatorname{supp} h)r = \varnothing$$

and then an element s such that the set $(\operatorname{supp} f \cup \operatorname{supp} r)s$ is unstable with respect to f. As above, we can conclude that for $k = sfs^{-1}$ we have $\operatorname{supp} f \cap \operatorname{supp} k^{-1}rk = \varnothing$. In particular, $k^{-1}rk$ leaves fixed the elements of the set $(\operatorname{supp} g \cup \operatorname{supp} h)r \subseteq \operatorname{supp} f$. But

$$(\operatorname{supp} g \cup \operatorname{supp} h)(rk^{-1}r^{-1}r) = (\operatorname{supp} g \cup \operatorname{supp} h)r$$

and the set $(\operatorname{supp} g \cup \operatorname{supp} h)(rk^{-1}r^{-1}k)$ is unstable with respect to f. Thus, for any elements $g, h, f \in G'$, $f \neq e$, there is an element $r' \in G'$, $r' = rk^{-1}r^{-1}k$, such that the set $(\operatorname{supp} g \cup \operatorname{supp} h)(r')$ is unstable with respect to f. Thus, we have proved that the derived subgroup G' has the required properties.

We note that $G'' \subseteq K$ for a normal subgroup K of the group G'. Since $G' = G''$, we have $K = G'$. The simplicity of the group G' is established. □

Corollary 4.4.1. *The derived subgroup G' of an o-2-transitive group (G, Ω) of bounded order automorphisms of a totally ordered set Ω is a simple group.*

Proof. Indeed, for any elements g, h, f, $f > e$, of the group G there is an element $\alpha \in \Omega$ such that $(\alpha)f \neq \alpha$. For definiteness, we set $\beta = (\alpha)f > \alpha$ and $\alpha_1 < \operatorname{supp} g \cup \operatorname{supp} h < \beta_1$. In view of the o-2-transitivity of (G, Ω), there is an element $r \in G$ such that $(\alpha_1)r = \alpha$ and $(\beta_1)r = \beta$. Therefore,

$$(\operatorname{supp} g \cup \operatorname{supp} h)rf \cap (\operatorname{supp} g \cup \operatorname{supp} h)r = \varnothing$$

i.e., the group (G, Ω) satisfies the hypotheses of Theorem 4.4.1. Consequently, the derived subgroup G' is a simple group. □

4.4.2. We give an example by Chehata [63] of a simple orderable group of automorphisms of a totally ordered set. Describing the example, we use the traditional denotations: at a point $\xi \in \mathbb{R}$ the values of piecewise linear functions f and g defined on $\mathbb{R}$ will be denoted by $f(\xi)$ and $g(\xi)$ and the composition of f and g will be written as $fg(\xi) = g(f(\xi))$. We also denote by $D^0 = D^0(\mathbb{R})$ the set of monotonically increasing functions $f \in \mathcal{A}(\mathbb{R})$ having the following properties: there exist numbers $\lambda = \lambda(f)$ and $\mu = \mu(f)$ such that $f(\xi) = \xi$ if ξ lies outside the closed interval $[\![\lambda, \mu]\!] = \{\xi \mid \xi \in \mathbb{R},\ \lambda \leqslant \xi \leqslant \mu\}$; moreover, f is continuous and piecewise linear on the interval $[\![\lambda, \mu]\!]$. For the identical function $e(\xi) = \xi$, $\xi \in \mathbb{R}$, the quantities $\lambda(e)$ and $\mu(e)$ are not fixed. For a convex subset $C \subseteq \mathbb{R}$ we introduce the group $D^0(C)$ consisting of e and all functions $f \in D^0$ such that $\lambda(f), \mu(f) \in C$. We note that if $I = ((\alpha, \beta))$ denotes an open interval in $\mathbb{R}$ with the endpoints α, β and $\overline{I} = [\![\alpha, \beta]\!]$ is its closure in $\mathbb{R}$, then functions $f \in D^0(\mathbb{R})$ such that $\alpha \leqslant \lambda(f) \leqslant \mu(f) \leqslant \beta$ form a subgroup $D^0(I)$ of the group $D^0(\overline{I})$. Moreover, $D^0(\overline{I}) \subseteq D^0(\overline{I}')$ if $\overline{I} \subseteq \overline{I}'$, i.e., $\overline{I}' = [\![\alpha', \beta']\!]$ and $\alpha' \leqslant \alpha < \beta \leqslant \beta'$.

Let $f \in D^0$, $\lambda(f) = \xi_1 = \lambda < \xi_2 < \ldots < \xi_{n-1} < \mu(f) = \xi_n = \mu$, and let the function f be linear on each of the intervals $[\![\xi_1, \xi_2]\!]$, $[\![\xi_2, \xi_3]\!]$, $\ldots$, $[\![\xi_{n-1}, \xi_n]\!]$. If the values of f at $\xi_1, \xi_2, \ldots, \xi_{n-1}, \xi_n$ are equal to $\xi_1 = \eta_1, \eta_2, \ldots, \eta_{n-1}, \eta_n = \xi_n$ respectively, we use the following notation for f:

$$f = \left\{ \begin{array}{ccccccccccc} & \eta_1 & & \eta_2 & \cdots & & \eta_{n-1} & & \eta_n & \\ 1 & & \theta_1 & & \theta_2 \;\cdots & \theta_{n-2} & & \theta_{n-1} & & 1 \\ & \xi_1 & & \xi_2 & \cdots & & \xi_{n-1} & & \xi_n & \end{array} \right\} \tag{4.4.1}$$

where $\theta_0 = 1$, $\theta_i = (\eta_{i+1} - \eta_i)/(\xi_{i+1} - \xi_i) > 0$ for $i = 1, 2, \ldots, n-1$ and

$$\theta_1(\xi_2 - \xi_1) + \theta_2(\xi_3 - \xi_2) + \cdots + \theta_{n-1}(\xi_n - \xi_{n-1}) = \eta_n - \eta_1 = \mu - \lambda$$

The values of the function f at an arbitrary point x can be found from the equality

$$f(\xi) = \eta_k x + \sum_{k=0}^{k-1} (\theta_i - \theta_{i+1})\xi_{i+1}$$

or from the equality

$$f(\xi) = \eta_k(\xi - \xi_k)$$

provided that $\xi_k \leqslant \xi \leqslant \xi_{k+1}$, $k = 1, 2, \ldots, n-1$, where $\theta_0 = 1$.

In the sequel, we assume that θ_1 and θ_{n-1} differ from 1 for $f \neq e$. If any two neighboring slopes θ_i, θ_{i+1} are different, then we say that the *function f has n fractures.* The quantity θ_i is referred to as the ith *slope* and the elements $\xi_1, \xi_2, \ldots, \xi_n$ are called *fracture points* of the function f. The set of fracture points for a function f is denoted by Σ_f.

The function

$$f^{-1} = \left\{ \begin{array}{ccccccccccc} & \xi_1 & & \xi_2 & \cdots & & \xi_{n-1} & & \xi_n & \\ 1 & & \theta_1^{-1} & & \theta_2^{-1} \;\cdots & \theta_{n-2}^{-1} & & \theta_{n-1}^{-1} & & 1 \\ & \eta_1 & & \eta_2 & \cdots & & \eta_{n-1} & & \eta_n & \end{array} \right\}$$

is the inverse one to a function f. The graph of the function f^{-1} on the plane with the Cartesian coordinates ξ, η can be presented as the line that is symmetric to the graph of f with respect to the line $\xi = \eta$. We emphasize that the introduced notation for elements of the group D^0 is convenient for computation.

Lemma 4.4.1. *For any elements $f, g \in D^0$ the following inclusion holds:*

$$\Sigma_{fg} \subseteq \Sigma_f \cup f^{-1}(\Sigma_g) \tag{4.4.2}$$

PROOF. It is clear that fracture points of any function $h(\xi)$ of D^0 coincide with those points of $\mathbb{R}$ at which the derivative $h'(\xi)$ does not exist. Therefore, the set of fracture points for the function $h(\xi) = fg(\xi) = g(f(\xi))$ is contained in the set on which the function $g'(f(\xi)) \cdot f'(\xi)$ is not defined. It is obvious that the set of points at which the function f' is not defined contains the set Σ_f and the set of points at which the function $g'(f(\xi))$ is not defined contains the set of those points ξ for which $f(\xi) \in \Sigma_g$. Consequently, the inclusion $\Sigma_{fg} \subseteq \Sigma_f \cup f^{-1}(\Sigma_g)$ holds. □

Proposition 4.4.1. *The group D^0 is an o-2-transitive group of order automorphisms of $\mathbb{R}$.*

PROOF. Let $\alpha_1, \alpha_2, \beta_1, \beta_2 \in \mathbb{R}$, $\alpha_1 < \alpha_2$, and $\beta_1 < \beta_2$. Then $f(\xi) = (\beta_2 - \beta_1)(\alpha_2 - \alpha_1)^{-1}(\xi - \alpha_1) + \beta_1$ is an order mapping from $\mathbb{R}$ onto $\mathbb{R}$ such that $f(\alpha_1) = \beta_1$ and $f(\alpha_2) = \beta_2$, i.e., $\mathcal{A}(\mathbb{R})$ is an o-2-transitive group. It is easy to construct a piecewise monotonically increasing linear function $g \in D^0$ such that g coincides with f on the interval $[\![\alpha_1, \alpha_2]\!]$. □

Proposition 4.4.2. *For any open interval*

$$I = (\!(\alpha, \beta)\!) = \{x \mid x \in \mathbb{R},\ \alpha < x < \beta\}$$

the group $D^0(I)$ is an o-2-transitive group of order automorphisms of the totally ordered set I.

The proof is similar to that of Proposition 4.4.1.

For $I = (\!(\alpha, \beta)\!)$ we denote by $\Phi(I)$ the subset of those elements of the group $D^0(I)$ which have only three fracture points and by Φ the set of all elements with three fractures.

Lemma 4.4.2. *The group $D^0(I)$ is generated by the set $\Phi(I)$, and the group D^0 is generated by the set Φ.*

PROOF. It suffices to prove only the first assertion of the lemma. We show that any element of the group $D^0(I)$ can be represented as a product of elements of the set $\Phi(I)$. To this end, we use induction on the number of fracture points of f. By definition, every element with three fracture points is contained in $\Phi(I)$. We assume that any element of $D^0(I)$ having at most $n > 3$ fracture points is a product of elements of $\Phi(I)$. Let f have n fracture points and look like (4.4.1) with $\alpha < \lambda(f) = \xi_1 < \xi_n = \mu(f)\beta$. We take an element $g \in \Phi(I)$ of the form

$$g = \left\{ \begin{matrix} & \xi_1 & & \delta & & \xi_n & \\ 1 & & \theta & & \theta' & & 1 \\ & \xi_1 & & \eta_2 & & \xi_n & \end{matrix} \right\}$$

where $\eta_2 = f(\xi_2)$, and a parameter δ satisfying the inequalities $\xi_1 < \delta < \xi_n$, which will be specified later. The element g will be thought of as a function in variables ξ and δ, i.e., $g = g(\xi, \delta)$. By assumption, we have $\xi_2 < \xi_3 < \xi_n$ and $\eta_2 < \eta_3 < \eta_n$. The lemma will be proved if we are able to choose the value of $\delta = \overline{\delta}$ such that for $\overline{g}(\xi) = g(\xi, \overline{\delta})$ the element $f\overline{g}$ has at most n fracture points. From Lemma 4.4.1, it follows that

$$\Sigma_{fg} \subseteq \Sigma_f \cup f^{-1}(\Sigma_g) = \{\xi_1, \xi_2, \ldots, \xi_n\} \cup \{f^{-1}(\eta_2)\} = \Sigma_f$$

since $f^{-1}(\eta_2) = \xi_2$ by the definition of f and by the choice of g. Therefore, the function fg has at least n fracture points. Consequently, fg assumes the form

$$fg = \left\{ \begin{matrix} & \xi_1 & & \delta_2 & & \cdots & & \delta_{n-1} & & \xi_n & \\ 1 & & \theta_1^* & & \theta_2^* & \cdots & \theta_{n-2}^* & & \theta_{n-1}^* & & 1 \\ & \xi_1 & & \xi_2 & & \cdots & & \xi_{n-1} & & \xi_n & \end{matrix} \right\}$$

Moreover, for a given function f the numbers $\delta_2, \ldots, \delta_{n-1}$ and slopes $\theta_1^*, \ldots, \theta_{n-1}^*$ are uniquely determined by the value of $\delta \in ((\xi_1, \xi_n))$.

We choose δ so that the slopes θ_1^* and θ_2^* are equal. It easy to see that at each point ξ the right slope of the function fg is equal to the value of the right derivative $(fg)'(\xi + 0)$ at the same point ξ. Hence

$$\theta_1^* = \theta_1\theta = \frac{\eta_2 - \xi_1}{\xi_2 - \xi_1} \cdot \frac{\delta - \xi_1}{\eta_2 - \xi_1} = \frac{\delta - \xi_1}{\xi_2 - \xi_1}$$

For θ_2^* we have

$$\theta_2^* = g(f(\xi_2 + 0))' = g'(\eta_2 + 0)f'(\xi_2 + 0) = \theta'\theta_2 = \frac{\xi_n - \delta}{\xi_n - \eta_2}\frac{\eta_3 - \eta_2}{\xi_3 - \xi_2}$$

It is clear that for any δ, $\xi_1 < \delta < \xi_n$, the inequalities $\theta_1^* > 0$ and $\theta_2^* > 0$ are fulfilled. We consider the difference $\Delta(\delta)$ at the endpoints of the interval $[\![\xi_1, \xi_n]\!]$. By (4.4.2), we have

$$\Delta(\xi_1) = \frac{\xi_n - \xi_1}{\xi_n - \eta_2} \frac{\eta_3 - \eta_2}{\xi_3 - \xi_2} > 0$$
$$\Delta(\xi_1) = -\frac{\xi_n - \xi_1}{\xi_2 - \xi_1} < 0$$

Since the function $\Delta(\delta)$ is linear in δ and takes values of different signs at the endpoints of the interval $[\![\xi_1, \xi_2]\!]$, there is a unique point $\overline{\delta} \in ((\xi_1, \xi_2))$ such that $\Delta(\overline{\delta}) = 0$, i.e., there is a function $\overline{g}$ corresponding to the parameter $\overline{\delta}$ such that the number of fracture points of $f\overline{g}$ is strictly less than n. The proof of the inductive step and the lemma is complete. □

We now consider a mapping $\iota\colon D^0 \to D^0$ defined by the equality $\iota(f)(\xi) = -f^{-1}(-\xi)$. For any element $f \in D^0$ of the form (4.4.1) the following equality holds:

$$\iota(f) = \left\{ \begin{matrix} & -\xi_n & & -\xi_{n-1} & & \cdots & & -\xi_2 & & -\xi_1 & \\ 1 & & \theta_n^{-1} & & \theta_{n-1}^{-1} & \cdots & \theta_2^{-1} & & \theta_1^{-1} & & 1 \\ & -\eta_n & & -\eta_{n-1} & & \cdots & & -\eta_2 & & -\eta_1 & \end{matrix} \right\}$$

The graph of the function $\iota(f)$ is the line which is symmetric to the graph of the function f with respect to the line $\xi + \eta = 0$.

Lemma 4.4.3. *The mapping ι is an antiautomorphism of order 2 of the group D^0.*

PROOF. It is obvious that ι is a one-to-one mapping from D^0 onto itself. We consider $\iota(fg)$, where $f, g \in D^0$. For any $\xi \in \mathbb{R}$ we have the relation

$$\iota(fg)(\xi) = -(fg)^{-1}(-\xi) = -g^{-1}f^{-1}(-\xi) = -f^{-1}(g^{-1}(-\xi))$$
$$= -f^{-1}(-(-g^{-1}(\xi))) = -f^{-1}(-\iota(g)(\xi)) = \iota(f)(\iota(g)(\xi)) = (\iota(g)\iota(f))(\xi)$$

which implies that ι is an antiautomorphism of D^0. The equality $\iota^2(g) = g$ for $g \in D^0$ is obvious. □

Lemma 4.4.4. *Every element of the set $\Phi(I)$ is contained in the derived subgroup $D^0(I)'$ of the group $D^0(I)$.*

PROOF. Let h be an arbitrary element of $\Phi(I)$,

$$h = \left\{ \begin{matrix} & \zeta_1 & & \tau & & \zeta_3 & \\ 1 & & \rho & & \rho' & & 1 \\ & \zeta_1 & & \zeta_2 & & \zeta_3 & \end{matrix} \right\}$$

where ρ and ρ', $0 < \rho' < 1 < \rho$, are prescribed slopes and $\zeta_1, \zeta_2, \zeta_3, \tau$ are real numbers such that $\zeta_1 < \zeta_2 < \zeta_3$ and $\tau = \zeta_1 + \rho(\zeta_2 - \zeta_1) = \zeta_3 + \rho'(\zeta_2 - \zeta_3)$. According to the slopes ρ and ρ', three cases are possible.

CASE 1: $\rho\rho' > 1$. We consider the element h subject to the following additional restriction: $0 < \zeta_1$. Starting from the element h, we choose elements $f, g \in \Phi$ so as to satisfy the equality

$$h = gfg^{-1}f^{-1} = [g^{-1}, f^{-1}] \tag{4.4.3}$$

Moreover, we seek elements f and g in the form

$$f = \left\{\begin{matrix} & 0 & & \eta & & \mu & \\ 1 & & \eta & & \theta & & 1 \\ & 0 & & 1 & & \mu & \end{matrix}\right\}$$

$$g = \left\{\begin{matrix} & 0 & & \eta^{-1} & & \mu & \\ 1 & & \gamma^{-1}\eta^{-1} & & \theta^* & & 1 \\ & 0 & & \gamma & & \mu & \end{matrix}\right\}$$

where

$$0 < \gamma < 1 < \mu, \quad 1 < \eta, \quad \gamma\eta \neq 1$$

$$\theta = (\mu - \eta)/(\mu - 1), \quad \theta^* = (\eta - 1)/[\eta(1 - \gamma)]$$

We denote by Σ_{h^*} the set of fracture points of the function

$$h^* = [g^{-1}, f^{-1}](\xi) = gfg^{-1}f^{-1}(\xi) = f^{-1}(g^{-1}(f(g(\xi))))$$

By Lemma 4.4.1, for Σ_{h^*} we have

$$\begin{aligned} \Sigma_{h^*} &\subseteq \Sigma_g \cup g^{-1}(\Sigma_f) \cup g^{-1}(f^{-1}(\Sigma_{g^{-1}})) \cup g^{-1}(f^{-1}(g(\Sigma_{f^{-1}}))) \\ &= \{0, \gamma, 1\} \cup \{0, 1, \mu\} \cup \{0, g^{-1}(f^{-1}(\eta^{-1})), g^{-1}(f^{-1}(1))\} \cup \{0, \eta, \mu\} \\ &= \{0, \gamma, 1, \mu, g^{-1}(f^{-1}(\eta^{-1})), g^{-1}(f^{-1}(1))\} \end{aligned}$$

Since $f^{-1}(1) = g(\gamma) < 1$, we have

$$g^{-1}(f^{-1}(\eta^{-1})) = \gamma\eta^{-1} < g^{-1}(f^{-1}(1)) = \gamma$$

Consequently,

$$\Sigma_{h^*} \subseteq \{0, \gamma\eta^{-1}, \gamma, 1, \mu\}$$

Thus,

$$\Sigma_{h^*} \subseteq \{0, \eta\gamma^{-1}, \gamma, 1, \mu\}$$

We note that the function $h^*(\xi)$ depends on the parameters η, γ, μ that are subject to the following restrictions:

$$0 < \gamma < 1, \quad 0 < \eta < \mu, \quad \gamma\eta \neq 1 \tag{4.4.4}$$

We now vary these parameters so that the element h^* assumes the same form as the element h, i.e., it has only three fracture points, its first slope is equal to ρ, and its last slope is equal to ρ'.

We now study the behavior of the function h^* on the intervals $[\![0, \gamma\eta^{-1}]\!]$, $[\![1, \mu]\!]$. On the interval $[\![0, \gamma\eta^{-1}]\!]$ the function h^* is linear and can be represented as a composition of the linear functions $\tilde{f}^{-1}(\xi)$, $\tilde{g}^{-1}(\xi)$, $\tilde{f}(\xi)$, and $\tilde{g}(\xi)$, where

$$\tilde{f}(\xi) = \eta \cdot \xi, \quad \tilde{f}^{-1}(\xi) = \eta^{-1} \cdot \xi$$
$$\tilde{g}(\xi) = (\gamma\eta)^{-1} \cdot \xi, \quad \tilde{g}^{-1}(\xi) = \gamma\eta \cdot \xi$$

Therefore, on the interval $[\![0, \gamma\eta^{-1}]\!]$ the slope of the function $h^*(\xi)$ is equal to 1. This means that $\gamma\eta^{-1}$ is the first fracture point of the function h^*.

On the other hand, for any real number $\xi > 1$ we have $g(\xi) = \xi$ and $g^{-1}(\xi) = \xi$. Thus,

$$h^*(\xi) = f^{-1}(g^{-1}(f(g(\xi)))) = f^{-1}(g^{-1}f(\xi))$$

Since $\xi > 1$, we have $f(\xi) > f(1) = \eta > 1$ and $g^{-1}(f(\xi)) = f(\xi)$. This means that the last fracture point of the function h^* coincides with 1. We set $\zeta_1 = \gamma\eta^{-1}$ and $\zeta_2 = 1$. Thus, the function h^* assumes the form

$$g = \left\{ \begin{matrix} & \gamma\eta^{-1} & & \eta^{-1} & & \mu & \\ 1 & & \frac{1-\gamma}{\eta(\eta-1)} & & \frac{\eta-1}{\eta(1-\gamma)} & & 1 \\ & \gamma\eta^{-1} & & \gamma & & \mu & \end{matrix} \right\}$$

Therefore, we conclude that the function h^* is independent of the parameter μ, the single condition $\eta < 1$ is posed on η, and any real number greater than η can be chosen as μ. In order that the element h^* have the first slope ρ and the second slope ρ', the parameters γ and η must be chosen so as to satisfy the relation (4.4.4) and the following system of equations:

$$\frac{1-\gamma}{\eta(\eta-1)} = \rho$$
$$\frac{\eta-1}{\eta(1-\gamma)} = \rho'$$

Solving each equation of the system with respect to γ, we find

$$\gamma = \frac{1}{1+\rho(\eta-1)}$$
$$\gamma = \frac{1+\eta(\rho'-1)}{\rho'\eta}$$

Equating the last expressions for γ and carrying out some equivalent transformations, we obtain a quadratic equation with respect to η:

$$\rho(1-\rho')\eta^2 + (1+\rho\rho'-2\rho)\eta + (\rho-1) = 0$$

whose roots are $\eta_{(1)} = 1$ and $\eta_2 = (\rho-1)/[\rho(1-\rho')]$. By assumption, $0 < \rho' < 1 < \rho$ and $\rho\rho' > 1$. Therefore, $\eta_{(2)} > 1$ and $0 < \gamma < 1$, i.e., there are elements $f, g \in \Phi$ satisfying (4.4.3) and (4.4.4). The value of h^* corresponding to the parameters $\eta_{(2)}$ and $\gamma_{(2)} = 1/[1+\rho(\eta_{(2)}-1)]$ is assumed to be equal to h.

Thus, the element $h \in \Phi$ with slopes ρ and ρ', which are subject to the inequalities $0 < \rho' < 1 < \rho$ and $\rho\rho' > 1$ and have fracture points $\zeta_1, \zeta_2, \zeta_3$, belongs to the derived subgroup $D^{0\prime}$ of the group D^0. From the proof of Propositions 4.4.1 and 4.4.2, it follows that the same assertion is true for the group $D^0(I)$.

CASE 2: $0 < \rho' < 1 < \rho$ and $\rho \cdot \rho' < 1$. This case is reduced to Case 1 by means of the antiautomorphism ι introduced before Lemma 4.4.3. Indeed, if h satisfies the conditions of Case 2, then $\iota(h)$ satisfies those of Case 1. Hence the element $\iota(h)$ (thereby, the element h) is contained in the derived subgroup of the group D^0.

CASE 3: $h \in \Phi$ is such that the product of its first and second slopes is equal to 1. As in Cases 1 and 2, it suffices to prove that for any number $\rho > 1$ the element

$$h = \left\{ \begin{matrix} & -1 & & \zeta & & 1 & \\ 1 & & \rho & & \rho^{-1} & & 1 \\ & -1 & & -\zeta & & 1 & \end{matrix} \right\}$$

where $0 < \zeta < 1$, $\zeta = (\rho-1)/(\rho+1)$, is contained in the derived subgroup of the group D^0. We represent h as a product of three elements of the set Φ so that every factor is a commutator of elements of Φ.

We consider the elements $g, t \in \Phi$ of the form

$$g = \left\{ \begin{matrix} & -1 & & \delta & & \zeta & \\ 1 & & \theta & & \eta & & 1 \\ & -1 & & -\gamma & & \zeta & \end{matrix} \right\}$$

$$t = \left\{ 1 \begin{matrix} -1 \\ \\ -1 \end{matrix} \varphi \begin{matrix} \sigma \\ \\ -\delta \end{matrix} \psi \begin{matrix} 1 \\ \\ 1 \end{matrix} 1 \right\} \tag{4.4.5}$$

where $\zeta < \gamma$, and choose the parameters $\theta, \eta, \varphi, \psi$ so that the product hgt has exactly three fracture points and each of the elements g, t, and hgt satisfies the conditions of Case 1 or 2.

It is easy to see that for any ρ, θ, η, ζ subject to the restrictions $1 < \rho$, $0 < \eta < 1 < \theta$, $0 < \zeta < 1$, $\zeta = (\rho - 1)/(\rho + 1)$, the product hg has four fracture points:

$$\begin{aligned} \Sigma_{hg} \subseteq \Sigma_h \cup h^{-1}(\Sigma_g) &= \{-1, -\zeta, 1\} \cup \{h^{-1}(-\gamma), h^{-1}(\zeta)\} \\ &= \{-1, h^{-1}(-\gamma), -\zeta, 1\} \end{aligned}$$

where $-1 < h^{-1}(-\gamma) < -\zeta < 1$, since, in view of the above assumption, $\zeta < \gamma$. Consequently, $-\gamma < -\zeta < \zeta$ and $h^{-1}(-\gamma) < h^{-1}(\zeta) = -\zeta$; moreover, $\iota(h) = h$ and $h^{-1}(-\gamma) = -h(\gamma)$. Thus,

$$hg = \left\{ 1 \begin{matrix} -1 \\ \\ -1 \end{matrix} \varphi^* \begin{matrix} -\delta \\ \\ h^{-1}(-\gamma) \end{matrix} \psi^* \begin{matrix} \zeta \\ \\ -\zeta \end{matrix} \rho^{-1} \begin{matrix} 1 \\ \\ 1 \end{matrix} 1 \right\}$$

We now use the procedure from the proof of Lemma 4.4.2 to find an element t of the form (4.4.5) such that hgt has exactly three fracture points. In the above notation, we have

$$\begin{aligned} \Sigma_{hgt} &= \Sigma_{hg} \cup (hg)^{-1}(\Sigma_t) \\ &= \{-1, h^{-1}(-\gamma), -\zeta, 1\} \cup \{h^1(g^{-1}(-\delta))\} \\ &= \{-1, h^{-1}(-\gamma), -\zeta, 1\} \end{aligned}$$

$$h^* = hgt = \left\{ 1 \begin{matrix} -1 \\ \\ -1 \end{matrix} \varphi^{**} \begin{matrix} \sigma \\ \\ h^{-1}(-\gamma) \end{matrix} \psi^{**} \begin{matrix} t(\zeta) \\ \\ -\zeta \end{matrix} \rho^* \begin{matrix} 1 \\ \\ 1 \end{matrix} 1 \right\}$$

We can carry out no computations of the values σ and $t(\zeta)$, but restrict ourselves to the computation of the slopes $\varphi^{**}, \psi^{**}, \rho^*$ considering the latter as the values of the right (or, if necessary, left) derivatives of the function h^* at its fracture points. Thus,

$$\begin{aligned}
\varphi^{**} &= h^{*\prime}(-1+0) = (t(g(h(\xi))))'_{|\xi=-1+0} \\
&= t'(g(h(-1+0))) \cdot g'(h(-1+0)) = \varphi\theta\rho \\
\psi^{**} &= h^{*\prime}(h^{-1}(-\gamma+0)) = (t(g(h(\xi))))'_{|\xi=h^{-1}(-\gamma)+0)} \\
&= t'(g(h(h^{-1}(-\gamma)+0)))g'(h(h^{-1}(-\gamma+0)))h'(h^{-1}(-\gamma)+0) \\
&= t'(g(-\gamma+0))g'(-\gamma+0)h'(h^{-1}(-\gamma)+0) = \rho\eta\psi \\
\rho^* &= h^{*\prime}(-\zeta+0) = h^{*\prime}(1-0) \\
&= t'(1-0)g'(1-0)h'(1-0) = \psi\eta\rho^{-1}
\end{aligned}$$

Calculating the slopes, we take into account the inequality $-\gamma < -\zeta$.

We now choose the values of the parameters φ and ψ (in other words, the value of σ) so as to satisfy the equality $\varphi^{**} = \psi^{**}$ and relations

$$\theta\eta = k > 1, \quad \varphi\psi = r > 1 \tag{4.4.6}$$

These requirements can be written as follows:

$$\varphi\theta\rho = \rho\eta\psi, \quad \theta\eta = k, \quad \varphi\psi = r \quad k > 1, \quad r > 1$$

Reducing the first equation by ρ and taking the logarithm of both sides of these equations, for the sake of convenience, we conclude that the validity of (4.4.6) is equivalent to the existence of a solution to the system of equations

$$\begin{aligned}
&\ln\theta - \ln\eta + \ln\varphi - \ln\psi = 0 \\
&\ln\theta + \ln\eta = \ln k \\
&\ln\varphi + \ln\psi = \ln r
\end{aligned}$$

with unknown functions $\ln\theta$, $\ln\eta$, $\ln\varphi$, $\ln\psi$ for some values of the parameters k and r subject to the conditions $k > 1$ and $r > 1$. The rank of the system is 3, so the system is solvable. Therefore, a solution exists. We choose an arbitrary solution $\overline{\theta}, \overline{\eta}, \overline{\varphi}, \overline{\psi}$ to the system, fix functions $\overline{g}$, $\overline{t}$, and compute the product of the slopes of the function $h\overline{g}\overline{t}$ corresponding to the solution:

$$\varphi^{**}\rho^* = \rho\overline{\theta}\overline{\varphi}\overline{\psi}\overline{\eta}\rho^{-1} = kr > 1$$

Thus, starting from the function h with slopes ρ and ρ^{-1}, we have found the functions $\overline{g}$, $\overline{t}$, and $\overline{h}$ such that $h\overline{g}\overline{t} = \overline{h}$ and $h = \overline{h}\overline{t}^{-1}\overline{g}^{-1}$. Moreover, each of the functions $\overline{g}$, $\overline{t}$, $\overline{h}$ has exactly three slopes, and for each of these functions the product of its slopes differs from 1, i.e., each of these functions satisfies the conditions of Case 1 and, consequently, is contained in the derived subgroup of the group D^0. Hence h is also contained in the derived subgroup of the group D^0. As in Case 1, we note that $h \in \Phi(I)$ implies $h \in D^0(I)'$. □

On the group D^0 we introduce an order: $f > g$ if there is a number $a \in \mathbb{R}$ such that $(x)f = (x)g$, where $x \leqslant a$ and $(a+\varepsilon)f > (a+\varepsilon)g$ for all sufficiently small positive numbers $\varepsilon \in \mathbb{R}$. It is easy to check that the group D^0 with this order is totally ordered. Thus, $f > e$ in the totally ordered group D^0 if and only if $\eta_1 > 1$.

Theorem 4.4.2 (Chehata [63]). *The group D^0 and the groups $D^0(I)$ are order-simple.*

PROOF. By Propositions 4.4.1 and 4.4.2, each of the groups D^0 and $D^0(I)$ is an o-2-transitive group of order automorphisms of a totally ordered set. By definition, each of its elements has a bounded support. By Lemma 4.4.4, each of the groups D^0 and $D^0(I)$ coincides with its derived subgroup and, by Theorem 4.4.1, is simple. To complete the proof it suffices to introduce a two-sided total order on D^0. We set $f > g$ on D^0 if there exists a number $\alpha \in \mathbb{R}$ such that $f(\xi) = g(\xi)$, where $\xi \leqslant \alpha$ and $f(\alpha+\varepsilon) > g(\alpha+\varepsilon)$ for all sufficiently small positive $\varepsilon \in \mathbb{R}$. It is easy to check that D^0 is a totally ordered group. Thus, for $f \in D^0$ written in the form (4.4.1) we have $f > e$ if and only if $\theta_1 > 1$. □

It is not hard to show that the group D^0 is an l-group and the groups $D^0(I)$ are its l-subgroups with respect to the coordinate-wise order.

Theorem 4.4.2 with $\mathbb{R}$ replaced by any ordered field remains valid. The proof is similar to that of Theorem 4.4.2.

4.4.3. We denote by $\mathcal{S}$ the set of all monotonically increasing sequences

$$\Sigma = \{\xi_\iota \mid 1 \leqslant \iota \leqslant \tau,\ \iota, \tau \text{ are ordinals},\ \xi_\iota \in \mathbb{R}\}$$

where τ stands for an ordinal of a well-ordered sequence possessing the greatest element; moreover, $\xi_\sigma = \sup\{\xi_\iota \mid \iota < \sigma\}$ for every limit number $\sigma \leqslant \tau$. Let $D = D(\mathbb{R})$ be the set of all monotonically increasing continuous functions $f \in \mathcal{A}(\mathbb{R})$ such that there is a sequence $\Sigma_f = \{\xi_\iota \mid \iota \leqslant \tau\} \in \mathcal{S}$ such that $f(\xi) = \xi$ for all $\xi \notin [\![\xi_1, \xi_\tau]\!]$ and f is linear on every interval $[\![\xi_\iota, \xi_{\iota+1}]\!]$. It is clear that D is a subgroup of the group

$\mathcal{A}(\mathbb{R})$; moreover, the group D is o-2-transitive and the support of each of its elements is bounded.

Theorem 4.4.3 (Dlab [64]). *A group D is order-simple.*

PROOF. We first show that every element of the group D is a commutator in D. Let the interval $[\![\alpha, \beta]\!]$ be the convex closure of the support of an order automorphism $f \in D$ in $\mathbb{R}$, i.e., $\alpha = \inf\{\mathrm{supp} f\}$ and $\beta = \sup\{\mathrm{supp} f\}$. We choose points $\eta, \eta_i \in \mathbb{R}$, $i = 0, 1, 2, \ldots$, such that $\eta_1 = \alpha, \eta_2 = \beta$, $\eta_i < \eta_{i+1}$, $\eta = \sup\{\eta_i \mid i = 1, 2, \ldots\}$, and construct a function $g \in D$ as follows:

$$g(\xi) = \begin{cases} \xi, & \xi \notin [\![\eta_0, \eta]\!] \\ (\eta_2 - \eta_1)(\xi - \eta_0)/\eta_1 - \eta_0 + \eta_0, & \xi \in [\![\eta_0, \eta_1]\!] \\ \\ (\eta_{i+2} - \eta_{i+1})(\xi - \eta_i)/(\eta_{i+1} - \eta_i) + \eta_{i+1}, & \xi \in [\![\eta_i, \eta_{i+1}]\!], \quad i \geqslant 1 \end{cases}$$

We now consider the function f^* such that $f^*(\xi) = \xi$ for $\xi \notin [\![\eta_1, \eta]\!]$, and $f^*(\xi) = g^{-k} f^{\varepsilon} g^k(\xi)$ for $\xi \in [\![\eta_{k+1}, \eta_{k+2}]\!]$, $k = 0, 1, 2, \ldots$, where

$$\varepsilon = \begin{cases} 1 & \text{for even } k \\ -1 & \text{for odd } k \end{cases}$$

Then $f^* \in D$ and $f^*(\xi) = f(\xi)$ for $\xi \in [\![\eta_1, \eta_2]\!]$. At the same time, $g^{-1} {f^*}^{-1} g f^* = f$. Consequently, D coincides with its derived subgroup. By Theorem 4.4.1, the group D is simple.

The proof of the orderability of D is similar to that of the orderability of the Chehata group in Theorem 4.4.2. □

It is easy to check immediately that D is an l-group with respect to the coordinate-wise order.

4.5. Embeddings

In this section, in terms of right orders we describe partially ordered groups that are order-isomorphic to subgroups of l-groups or those of the Cartesian products of totally ordered groups.

4.5.1. We begin by proving the following theorem.

Theorem 4.5.1 (Hollister [65] and Conrad [66]). *Any lattice order on a group is an intersection of a family of right orders on the group.*

PROOF. Let G be an l-group and P be its semigroup of positive elements. By the Holland theorem (cf., for example, [36]), we can assume that G is an l-subgroup of an l-group $A(\Omega)$ of order automorphisms of a totally ordered set Ω and $g \in P$ if $(\alpha)g \geqslant \alpha$ for any $\alpha \in \Omega$.

For $g \notin P$ there is an element $\alpha_g \in \Omega$ such that $\alpha_g g < \alpha$. Starting from the element g, on the group G we construct a right order P_g such that $g \notin P_g$. To this end, we well order the set Ω so that its first element is α_g:

$$\alpha_1 = \alpha_g, \alpha_2, \dots, \alpha_i, \alpha_{i+1}, \dots \quad (i \in I)$$

and we introduce the right order P_g as in Example 1.4.5 by setting $g \in P_g$ if $\alpha_i g > \alpha_i$, where i is the least index among those $j \in I$ such that $\alpha_j \neq \alpha_j$. It is clear that the order P_g extends the order P and $g \notin P_g$. Consequently, $P = \bigcap_{g \notin P} P_g$. □

Let F be an l-group, G be a partially ordered group, and π be an order embedding of G into F, i.e., π is an isomorphism of the group G into the group F such that π and π^{-1} are order mappings of G and $\pi(G)$.

- The lattice-ordered group F with the order isomorphism $\pi : G \to F$ is called an l-group *free over the partially ordered group* G if

  (a) the l-subgroup of the l-group F generated by $\pi(G)$ coincides with the l-group F,

  (b) for any order homomorphism σ from the partially ordered group G into an l-group H there is an l-homomorphism $\tau : F \to H$ such that $\sigma = \pi\tau$.

A free l-group over a partially ordered group G is denoted by $F(G, \pi)$. It is evident that a free l-group over a partially ordered group exists only if the partial order on G is isolated.

Theorem 4.5.2 (Conrad [66]). *An l-group $F(G, \pi)$ free over a partially ordered group G exists if and only if the partial order P on the group G is an intersection of some collection of right orders on G.*

PROOF. Let $F = F(G,\pi)$ be an l-group free over the partially ordered group G. We can assume that $G \subseteq F$ and $P = G \cap Q$ provided that P is a partial order on the group G and Q is a lattice order on the l-group F. By Theorem 4.5.1, $Q = \bigcap_{\alpha\in\Lambda} Q_\alpha$, where $\{Q_\alpha | \alpha \in \Lambda\}$ is some set of right orders on the group F. Then

$$P = G \cap Q = G \cap \bigcap_{\alpha\in\Lambda} Q_\alpha = \bigcap_{\alpha\in\Lambda} (G \cap Q_\alpha)$$

where $\{G \cap Q_\alpha | \alpha \in \Lambda\}$ is a set of orders on the group G.

Conversely, let G be a partially ordered group and P be a partial order such that $P = \bigcap_{\alpha\in\Lambda} P_\alpha$, where $\{P_\alpha | \alpha \in \Lambda\}$ is a set of those right orders on the group G which extend the order P. For each $\alpha \in \Lambda$ we denote by G_α the group G with right order P_α. Then the representation R_α of the right-ordered group G_α by right translations is an isomorphism of the right-ordered group G_α into the group $A_\alpha = A(G_\alpha)$ of order automorphisms of the totally ordered set G_α. We denote by B_α an l-subgroup of the l-group A_α generated by $R_\alpha(G_\alpha)$ and by B the Cartesian product of the l-groups B_α : $\prod_{\alpha\in\Lambda} B_\alpha$. By π we denote the mapping from the group G into B defined by the following rule: $(\alpha)\pi(g) = R_\alpha(g)$ for the elements $g \in G$. It is easy to see that π is an isomorphism from the group G into the group B. We denote by F an l-subgroup of the l-group B generated by the set $\pi(G)$. We prove that F is an l-group that is free over the partially ordered group G. To this end, we first establish that π is an order isomorphism from the partially ordered group G with order P into the l-group F.

We note that if $g \in P$, then $R_\alpha(g) > e$ for any $\alpha \in \Lambda$. Otherwise there is an $\alpha \in \Lambda$ such that $(x)R_\alpha(g) < x$ for some element $x \in G_\alpha$ or $xg < x$ in G_α. Since G_α is a right-ordered group, we obtain $xgx^{-1} < e$ in G_α, i.e., $xgx^{-1} \notin P_\alpha$. However, $g \in P$ and, consequently, $xgx^{-1} \in P$, which contradicts the assumption $P \subseteq P_\alpha$. Therefore, $(\alpha)\pi(g) > e$ for any α, i.e., $\pi(g) > e$ in $B = \prod_{\alpha\in\Lambda} B_\alpha$.

Conversely, if $\pi(g) > e$ in B, then $R_\alpha(g) \geqslant e$ for any $\alpha \in \Lambda$. Since $g \neq e$, we have $R_\alpha(g) \neq e$ and $R_\alpha(g) > e$ in A_α. Consequently, $(e)R_\alpha(g) = e \cdot g = g$ in the totally ordered set G_α, i.e., $g \in P_\alpha$ for any α. Hence $g \in \bigcap_{\alpha\in\Lambda} P_\alpha$, $g \in P$. Thus, we have proved that π is an order isomorphism of the partially ordered group G in F, i.e., the condition (a) is valid.

We proceed to prove the condition (b). Let σ be an order homomorphism from the partially ordered group G into the l-group H. We find the required l-homomorphism of F into H by the formula $x \in F$, $x = \bigvee_{i\in I}\bigwedge_{j\in J} \pi(g_{ij})$, where $g_{ij} \in G$ and I, J are finite sets of indices. We set

$$\tau(x) = \bigvee_{i\in I}\bigwedge_{j\in J} \sigma(g_{ij})$$

To complete the proof it remains to show that $x \neq e$ in F if $\tau(x) \neq e$ in H. Indeed, this means that τ is a mapping; moreover, it is an l-homomorphism from the l-group F into the l-group H, and the equality $\sigma = \pi\tau$ follows from the construction of the mapping τ. Thus, we put $x = \bigvee_{i\in I}\bigwedge_{j\in J} \sigma(g_{ij}) \neq e$ in H for some elements $g_{ij} \in G$. We again apply the Holland theorem (cf. [36]) and find that H is an l-subgroup of an l-group $A(\Omega)$ of order automorphisms of a totally ordered set Ω. Since

$$\tau(x) = \bigvee_{i\in I}\bigwedge_{j\in J} \sigma(g_{ij}) \neq e$$

in Ω, there is an element ω_1 such that

$$\omega_1 \neq (\omega_1) \bigvee_{i\in IB}\bigwedge_{j\in J} \sigma(g_{ij})$$

As in Theorem 4.5.1, we well order the set Ω so that the element ω_1 is the least one in the set Ω:

$$\omega_1, \omega_2, \ldots, \omega_\eta, \omega_\eta+1, \ldots$$

We introduce the right order Q on the group H as follows: $h > e$ for $h \in H$ if $(\omega_\xi)h > \omega_\xi$, where ξ is the smallest index among ζ such that $(\omega_\zeta)h \neq \omega_\zeta$. With respect to this right order on the group H the inequality $(x)\tau > e$ holds and the right order Q is an extension of the lattice order on the l-group H and $\sigma(P) \subseteq \sigma(G)$. We note that each right order on the partially ordered group $\sigma(H)$ is induced by some right order on the partially ordered group G. Therefore, there is a right order P_α, $\alpha \in \Lambda$, on the group G with respect to which

$$\bigvee_{i\in I}^{\prime}\bigwedge_{j\in J}^{\prime} g_{ij} \neq e$$

where $\overset{\prime}{\bigvee}$ and $\overset{\prime}{\bigwedge}$ denote the operations of join and meet in the totally ordered set G_α. Then in the l-group $A(G_\alpha)$ of order automorphisms G_α the following equalities hold:

$$(e)x = \max_{i\in I}\{\min_{j\in J} e g_{ij}\} = \max_{i\in I}\{\min_{j\in J} g_{ij}\} = \overset{\prime}{\bigvee_{i\in B}} \overset{\prime}{\bigwedge_{j\in J}} g_{ij} \neq e$$

i.e., in $A(G_\alpha)$ the following relation is valid:

$$(e)\left(\bigvee_{i\in I} \bigwedge_{j\in J} R_\alpha(g_{ij})\right) \neq e$$

Consequently,

$$\bigvee_{i\in I} \bigwedge_{j\in J} R_\alpha(g_{ij}) \neq e, \quad x = \bigvee_{i\in I} \bigwedge_{j\in J} \pi(g_{ij}) \neq e$$

in F. □

Corollary 4.5.1 (Conrad [66]). *A partially ordered group G with order P is order-isomorphic to a subgroup of some suitable l-group if and only if P is an intersection of some family of right orders on the group G.*

Theorem 4.5.2 not only establishes the existence of the l-group $F(G,\pi)$ free over the partially ordered group G, but it also describes its structure. We note that if the partial order P on the group G is an intersection of right orders on the group G, then the l-group $F(G,\pi)$ free over G is uniquely defined up to an l-isomorphism. Indeed, if $F_1(G,\pi_1)$ and $F_2(G,\pi_2)$ are two l-groups that are free over G, then there are l-isomorphisms

$$\tau_1 : F_1(G,\pi_1) \to F_2(G,\pi_2)$$
$$\tau_2 : F_2(G,\pi_2) \to F_1(G,\pi_1)$$

such that $\pi_2 = \pi_1\tau_1$ and $\pi_1 = \pi_2\tau_2$. Consequently, $\pi_2 = \pi_2\tau_2\tau_1$ and $\pi_1 = \pi_1\tau_1\tau_2$. Since $\pi_1(G)$ and $\pi_2(G)$ generate the l-groups $F_1(G,\pi_1)$ and $F_2(G,\pi_2)$, we conclude that $\tau_1\tau_2$ is the identical mapping from the l-group $F_1(G,\pi_1)$ onto itself, $\tau_2\tau_1$ is the identical mapping from the l-group $F_2(G,\pi_2)$ onto itself, i.e., the l-group $F_1(G,\pi_1)$ is l-isomorphic to the l-group $F_2(G,\pi_2)$ and for fixed τ_1, τ_2 the corresponding l-isomorphism is unique.

4.5.2. The subgroups of the Cartesian products of totally ordered groups are of prime interest in the theory of partially ordered groups. A lot of partially ordered groups considered in different fields of mathematics, e.g., in functional analysis, are subgroups of the Cartesian products of additive groups of real numbers with the natural order.

- A partially ordered group is called a *vector group* if it is order-isomorphic to a subgroup of the Cartesian product of totally ordered groups.

The elements of a vector group

$$G \subseteq \overline{\prod_{\alpha\in\Lambda}} G_\alpha = \overline{G}$$

are functions f defined on the set Λ; moreover, $f(\alpha)$, $\alpha \in \Lambda$, belongs to the totally ordered group G_α. Every factor G_α of the Cartesian product $\overline{\prod_{\alpha\in\Lambda}} G_\alpha$ is called a *component* of the vector group G. The partial order on the vector group G is component-wise, i.e., $f \leqslant g$ if $f(\alpha) \leqslant g(\alpha)$ for any index $\alpha \in \Lambda$. It is easy to see that the class of vector groups is closed under taking Cartesian products and subgroups.

Theorem 4.5.3. *A partially ordered group is order-isomorphic to a vector group if and only if its partial order is an intersection of total orders.*

PROOF. Let G be a vector group

$$G \subseteq \overline{\prod_{\alpha\in\Lambda}} G_\alpha$$

and $P = \{f \in G | f(\alpha) \geqslant e,\ \alpha \in \Lambda\}$, be its semigroup of positive elements. With every component G_ζ we associate the total order P_ζ on the group $\overline{G} = \overline{\prod_{\alpha\in\Lambda}} G_\alpha$: we well order the set Λ so that the element ζ is the first one in Λ. According to this ordering of the set of indices, we order lexicographically the group $\overline{G}$. On $\overline{G}$ we define $f \in P_\zeta$ if $f(\beta) > e$, where β is the smallest index such that $f(\beta) \neq e$. It is clear that $P \subseteq P_\zeta$. For $g \in G$, $g \notin P \cup P^{-1} \cup \{e\}$, there are indices $\alpha, \beta \in \Lambda$ such that $g(\alpha) > e$ and $g(\beta) < e$. Hence $g \in P_\alpha$ and $g \notin P_\beta$, and consequently $g \notin P_\alpha \cap P_\beta$. This means that P is the intersection of total orders $G \cap P_\alpha$ and $\alpha \in \Lambda$ defined on the group G.

Conversely, let the partial order P of the group G be an intersection of total orders P_α and $\alpha \in \Lambda$. We consider the Cartesian product

$$H = \overline{\prod_{\alpha\in\Lambda}} G_\alpha$$

of totally ordered groups G_α that are isomorphic to the totally ordered group G with order P_α. It is easy to show that the mapping $\pi : G \to H$, $\pi(g)(\alpha) = g$, is an order mapping from the partially ordered group G into H. Consequently, the partially ordered group G is order-isomorphic to a vector group. □

Recall that the partial order P is called *strongly isolated* if the relation $x \cdot g_1^{-1}xg \ldots g_n^{-1}xg \in P$ implies $x \in P$.

Theorem 4.5.4 (Mal'cev [67]). *A partial order P of a fully orderable group G is an intersection of total orders if and only if the order P is strongly isolated.*

PROOF. Since any total order on the group is strongly isolated, the intersection of total orders is always strongly isolated. Conversely, let P be a strongly isolated order on a fully orderable group G and $a \notin P$. By Lemma 3.1.2, the semigroup $P \cdot S(a^{-1}, e)$ is a partial order on the group G. We extend this partial order to the total order Q_a defined on the group G. It is clear that $a \notin Q_a$. Hence $\bigcap_{a\in G\setminus P} Q_a = P$, which proves the theorem. □

Corollary 4.5.2 (Mal'cev [67]). *A partially ordered locally nilpotent group is order-isomorphic to a vector group if and only if its partial order is strongly isolated.*

Chapter 5

Relatively Convex Subgroups

In this chapter, we consider the properties of subgroups of right-ordered groups which are convex with respect to some right order. We will also describe groups with some restrictions on their subgroups. For this purpose the technique of groups of automorphisms of totally ordered sets is useful. Therefore, we first touch upon the development of the theory of the representations of right-ordered groups and its application to the problems of the relative convexity of subgroups.

5.1. Orderable Representations

5.1.1. In this section, we consider the representations of groups by automorphisms (permutations) of arbitrary not necessarily totally ordered sets. A group of automorphisms is denoted by a pair (G, Ω). We employ terms and notation that are usual for the theory of representations without assuming previously any structure of a set Ω or any properties of permutations of Ω. By tradition, for a representation of a group G we write briefly (G, Ω) instead of (G, Ω, ψ) because it is clear from the context what homomorphism ψ is meant. If H is a subgroup of a group G, then the right regular representation of G acting on the set of right cosets $R(G : H)$ will be denoted by $(G, R(G : H))$.

- A representation (G, Ω) of a group G by the automorphisms of an infinite set Ω is called
  - *partially ordered* if on the set Ω there is a partial order $\leqslant$ stable under the action of elements of G, i.e., $\alpha g \leqslant \beta g$ for any $g \in G$ and $\alpha, \beta \in \Omega$ such that $\alpha \leqslant \beta$,
  - *totally ordered* if the partial order on Ω is total,
  - *orderable* if on Ω we can introduce a total order which is stable under the action of elements of G.

In the case of a totally ordered representation, (G, Ω) is a group of order automorphisms of a totally ordered set Ω. Let (G, Ω) be a partially ordered representation.

- An element $g \in G$ such that $\alpha g > \alpha$ for some $\alpha \in \Omega$ is called *α-positive*; the set of all α-positive elements is called an *α-positive cone*; and the subset $\Upsilon(\alpha) = \{\gamma \in \Omega \mid \gamma \geqslant \alpha\}$ is called a *local α-cone*.

Proposition 5.1.1 (Tararin [68]). *A subset $S \neq \varnothing$ of a group G is an α-positive cone under some partially ordered representation (G, Ω), if and only if the following conditions hold:*

(a) *S is a pure subsemigroup of G and $e \notin S$,*

(b) $G_\alpha S G_\alpha \subseteq S$.

PROOF. Let (G, Ω) be a partially ordered representation and S be the set of all α-positive elements of the group G, $g_1, g_2 \in S$. From the inequality $\alpha g_1 > \alpha$ it follows that $\alpha g_1 g_2 = (\alpha g_1) g_2 > \alpha g_2 > \alpha$. Consequently, $g_1 g_2 \in S$. From the inequality $\alpha g > \alpha$, where $g \in S$, we obtain $\alpha g^{-1} < \alpha$, i.e., $g^{-1} \in S$. Let $x, y \in G_\alpha, g \in S$. Then the inequality $\alpha x g = (\alpha x) g = \alpha g > \alpha$ implies $\alpha x g y = (\alpha x g) y = (\alpha g) y > \alpha y = \alpha$, i.e., $x g y \in S$.

Conversely, let a subset $S \neq \varnothing$ of the group G satisfy the hypotheses of the proposition. On Ω we define an order by the following rule: for $\beta, \gamma \in \alpha G$ we set $\beta > \gamma$ if $\beta = \alpha g_1$, $\gamma = \alpha g_2$, and $g_1 g_2^{-1} \in S$, $g_1, g_2 \in G$; for $\beta, \gamma \notin \alpha G$ we set $\beta \geqslant \gamma$ if $\beta = \gamma$. We show that the definition of the order just introduced on the orbit αG is correct. Let $\alpha a = \alpha g_1$ and $\alpha b = \alpha g_2$, where $a, b \in G$ and $g_1 g_2^{-1} \in S$. Then

$$ab^{-1} = (ag_1^{-1})(g_1 g_2^{-1})g_2 b^{-1} \in S$$

since $ag_1^{-1}, g_2 b^{-1} \in G_\alpha$ and $g_1 g_2^{-1} \in S$. Let $\alpha g_1 > \alpha g_2$ and $\alpha g_2 > \alpha g_3$. From the equality $g_1 g_3^{-1} = g_1 g_2^{-1} g_2 g_3^{-1}$ we have $g_1 g_2^{-1}, g_2 g_3^{-1} \in S$ and $g_1 g_2 \in S$. Consequently, $\alpha g_1 > \alpha g_3$.

If $\alpha g_1 > \alpha g_2$ and $\alpha g_2 > \alpha g_1$, then $g_1 g_2^{-1}, g_2 g_1^{-1} \in S$, which contradicts the condition (a).

Finally, let $\alpha g_1 > \alpha g_2$ and $g \in G$. Then $g_1 g (g_2 g)^{-1} = g_1 g_2^{-1} \in S$. Consequently, $(\alpha g_1)g > (\alpha g_2)g$. □

- A subsemigroup S of the group G is called *α-closed* if $G_\alpha S G_\alpha \subseteq S$.

Proposition 5.1.2 (Tararin [68]). *A subset Υ of a set Ω in the transitive representation (G, Ω) is a local α-cone for $\alpha \in \Omega$ with respect to some nontrivial partial order if the set $S = \{g \in G \mid \alpha g \in \Upsilon \setminus \{\alpha\}\}$ is a pure α-closed subsemigroup of the group G.*

PROOF. If (G, Ω) is a partially ordered representation and Υ is a local α-cone, then, by Proposition 5.1.1, the set $S = \{g \in G \mid \alpha g \in \Upsilon \setminus \{\alpha\}\}$ is an α-positive cone and, consequently, a pure α-closed subsemigroup of the group G.

Let S be a pure α-closed subsemigroup of the group G. From the proof of Proposition 5.1.1, it follows that S defines the order with respect to which Υ is a local α-cone. □

Corollary 5.1.1. *A local α-cone defines a unique order on the transitive representation (G, Ω).*

Corollary 5.1.2. *A transitive representation (G, Ω) is orderable if and only if there is a pure α-closed subsemigroup S of G and an element $\alpha \in \Omega$ such that*

$$S \cup S^{-1} \cup G_\alpha = G_\alpha$$

Let a subsemigroup S of the group G have the following property:

$$G \setminus S \text{ is a semigroup} \tag{5.1.1}$$

It is clear that if a subsemigroup S of a group G has the property (5.1.1), then for any $x, y \in G$, the inclusion $xy \in S$ implies $x \in S$ or $y \in S$.

Lemma 5.1.1. *Let S be a pure subsemigroup of a group G and satisfy the property (5.1.1). Let $e \notin S$, and $H = G \setminus (S \cup S^{-1})$. Then H is a subgroup of G and $HSH \subseteq H$.*

PROOF. We first prove the fact that H is a subgroup of G. Let $x \in H$. Then $x \notin S \cup S^{-1}$. Consequently, $x^{-1} \in H$.

Let $x, y \in H$. Then $x, y \notin S \cup S^{-1}$. Since S, as well as S^{-1}, satisfy the condition (5.1.1), we have $xy \notin S$ and $xy \notin S^{-1}$. Therefore, $xy \in H$.

We now show that $HSH \subseteq S$. Let $x, y \in H$ and $p \in S$. Then $x, y \notin S \cup S^{-1}$. We assume that $u = xpy \notin S$. Then $p = x^{-1}uy^{-1}$. Since $u \notin S$ and $y^{-1}, x^{-1} \notin S$, we obtain the equality $p = x^{-1}uy^{-1}$, which contradicts the choice of the element p. Thus, $xpy \notin S$. □

Proposition 5.1.3 (Tararin [68]). *A subset S of the group G is an α-positive cone of some totally ordered representation (G, Ω) if and only if S is a pure subsemigroup of the group G satisfying the condition (5.1.1) and $e \notin S$.*

PROOF. Let S be an α-positive cone of a totally ordered representation (G, Ω). Then

$$S \cup S^{-1} \cup G_\alpha = G, \quad S \cap G_\alpha = \varnothing, \quad G_\alpha S G_\alpha \subseteq S$$

It is evident that $G \setminus S$ is a subsemigroup.

Conversely, let S be a pure subsemigroup of a group G satisfying the condition (5.1.1). By Lemma 5.1.1, the set $N = G \setminus (S \cup S^{-1})$ is a subgroup of the group G and $N \cap S = \varnothing$, $NSN \subseteq S$. The kernel G_N of the representation $(G, R(G : H))$ is equal to N. By Corollary 5.1.2, the representation $(G, R(G : H))$ is orderable with the α-positive cone S. The proposition is proved. □

Let (G, Ω) be a representation of a group G and $\alpha \in \Omega$. The intersections of α-closed subsemigroups of the group G with the group G are α-closed sets. Thus, we can discuss an α-closed subsemigroup $P_\alpha(A)$ generated by the set A. It is easy to see that

$$P_\alpha(A) = \{h_1 g_1 h_2 \ldots h_{2n-1} g_n h_{2n} \mid h_i \in G_\alpha, g_i \in A, n \geqslant 1\}$$
$$P_\alpha(P_\alpha(A \cup B) \cup C) = P_\alpha(A \cup B \cup C)$$

Proposition 5.1.4 (Tararin [68]). *In order that the α-positive cone of a partially ordered transitive representation (G, Ω) be extended to an α-positive cone of the totally ordered representation (G, Ω), it* *is necessary and sufficient that for any elements* $g_1, \dots, g_n$ *of the set* $G \setminus G_\alpha$ *there exist numbers* $\varepsilon_1, \dots, \varepsilon_n$ *equal to* ± 1 *such that*

$$e \notin P_\alpha(P, g_1^{\varepsilon_1}, \dots, g_n^{\varepsilon_n})$$

PROOF. Let an α-positive cone P of a transitive representation (G, Ω) be extended to an α-positive cone $\widehat{P}$ of some totally ordered representation (G, Ω). Then

$$G_\alpha \cap \widehat{P} = \varnothing$$
$$\widehat{P} \cup \widehat{P}^{-1} \cup G_\alpha = G$$

Since $g_1, \dots, g_n \notin G_\alpha$, there are numbers ε_i, $1 \leqslant i \leqslant n$, such that $g_i^{\varepsilon_i} \in \widehat{P}$. It is obvious that $P_\alpha(P, g_1^{\varepsilon_1}, \dots, g_n^{\varepsilon_n}) \subseteq \widehat{P}$ and $e \notin \widehat{P}$.

We assume that an α-closed subsemigroup S satisfies the hypotheses of the proposition. To show that $P_\alpha(S, g)$ or $P_\alpha(S, g^{-1})$ is a pure α-closed subsemigroup satisfying the hypotheses of the proposition we assume the contrary. Then in $G \setminus G_\alpha$ there are elements $g_1, \dots, g_n$ and $s_1, \dots, s_m$ such that

$$e \in P_\alpha(P_\alpha(S, g), g_1^{\varepsilon_1}, \dots, g_n^{\varepsilon_n}) = P_\alpha(S, g, g_1^{\varepsilon_1}, \dots, g_n^{\varepsilon_n})$$

$$e \in P_\alpha(P_\alpha(S, g^{-1}), s_1^{\mu_1}, \dots, s_m^{\mu_m}) = P_\alpha(S, g^{-1}, s_1^{\mu_1}, \dots, s_m^{\mu_m})$$

for any set of numbers $\varepsilon_1, \dots, \varepsilon_n$ and $\mu_1, \dots, \mu_m$ equal to ± 1. Consequently, $e \in P_\alpha(S, g^\varepsilon, g_1^{\varepsilon_1}, \dots, g_n^{\varepsilon_n}, s_1^{\mu_1}, \dots, s_m^{\mu_m})$ for any set of numbers $\varepsilon, \varepsilon_1, \dots, \varepsilon_n, \mu_1, \dots, \mu_m$ equal to ± 1, which contradicts our assumption. Thus, for some number ε equal to ± 1 the subsemigroup $P_\alpha(S, g^\varepsilon)$ satisfies the hypotheses of the proposition. In particular, $e \notin P_\alpha(S, g^\varepsilon)$, i.e., $P_\alpha(S, g^\varepsilon)$ defines a partial order on the representation (G, Ω).

Let $\widehat{P}$ stand for a maximal α-closed subsemigroup of G which contains P and satisfies the hypotheses of the proposition. For $g \in G \setminus G_\alpha$ either $P_\alpha(\widehat{P}, g)$ or $P_\alpha(\widehat{P}, g^{-1})$ defines a partial order on the representation (G, Ω). In view of the maximality of $\widehat{P}$, either g or g^{-1} is contained in $\widehat{P}$. This means that the order $\widehat{P}$ is total and, consequently, the representation (G, Ω) is orderable. □

Corollary 5.1.3. *A transitive representation* (G, Ω) *is orderable if and only if for any* $\alpha \in \Omega$ *and* $g_1, \dots, g_n \in G \setminus G_\alpha$ *there are numbers* $\varepsilon_1, \dots, \varepsilon_n$ *equal to* ± 1 *such that* $e \notin P_\alpha(g_1^{\varepsilon_1}, \dots, g_n^{\varepsilon_n})$.

Proposition 5.1.5 (Tararin [68]). *A subgroup H of the group G is a kernel of some orderable representation (G, Ω) if and only if for any $g_1, \dots, g_n \in G \setminus H$ there are numbers $\varepsilon_1, \dots, \varepsilon_n$ equal to ± 1 such that*

$$e \notin \mathrm{sgr}\,(H, g_1^{\varepsilon_1}, \dots, g_n^{\varepsilon_n})$$

PROOF. The necessity follows from Proposition 5.1.4 and the relation

$$\mathrm{sgr}\,(H, g_1^{\varepsilon_1}, \dots, g_n^{\varepsilon_n}) = P_\alpha(g_1^{\varepsilon_1}, \dots, g_n^{\varepsilon_n})$$

which is valid provided that H is the stabilizer of the element $\alpha \in \Omega$ in G.

To prove the sufficiency we consider the representation $(G, R(G : H))$. The subgroup H is the stabilizer of the coset H. Applying Proposition 5.1.4, we obtain the desired conclusion. □

In the sequel, we use the notation $P_H(X) = \mathrm{sgr}\,(H, X)$.

Proposition 5.1.6 (Tararin [68]). *A faithful transitive representation (G, Ω) of the group G by automorphisms of the set Ω is orderable if and only if the group G admits a right order with respect to which the stabilizer G_α of some element $\alpha \in \Omega$ is a convex subgroup.*

PROOF. The necessity follows from Theorem 3.4.1. Let a group G be right-orderable, $P = \{g \in G \mid g \geqslant e\}$ be its positive cone with respect to some right order, and G_α be a convex subgroup. Moreover, let $S = \{g \in P \mid \alpha g > \alpha\}$. We proceed to show that S is a pure α-closed subsemigroup of the group G.

Let $g_1, g_2 \in S$. Then $g_1 > e$, $g_2 > e$, $\alpha g_1 \neq \alpha$, $\alpha g_2 \neq \alpha$. It is obvious that $g_1 g_2 > e$. We assume that $\alpha g_1 g_2 = \alpha$. Then $g_1 g_2 \in G_\alpha$. Since G_α is a convex subgroup, we have $g_2 > g_1 g_2$. Consequently, we arrive at the inequality $e > g_1$, which contradicts our assumption that $g_1 > e$. Hence $g_1 g_2 \in S$. Thus, the purity of S follows from the purity of P.

Let $g \in S$, i.e., $g > e$, $\alpha g \neq \alpha$, and $x, y \in G_\alpha$. Then $xgy \notin G_\alpha$. Otherwise g belongs to G_α. It is clear that $gy > e$. Indeed, if $y > e$, then the inequality $g > e$ implies $gy > ey = y > e$. If $y^{-1} > e$, then from $g > y^{-1}$ it follows that $gy > e$. We assume that $xgy < e$. Then $x < y^{-1}g^{-1}$ and $y^{-1}g^{-1} < e$. In view of the convexity of G_α, we obtain

$y^{-1}g^{-1} \in G_\alpha$. Consequently, $g \in G_\alpha$, which contradicts the choice of the element g. Therefore, $xgy \in S$.

Thus, S is a pure α-closed subsemigroup of the group G and $S \cup S^{-1} \cup G_\alpha = G$. The last relation means that the representation (G, Ω) is orderable. □

- A subgroup H of a group G is called *right-relatively convex* if it is convex with respect to some right order on the group G.

Proposition 5.1.6 can be applied for finding a criterion for relative convexity of subgroups of a right-orderable group.

Corollary 5.1.4. *A representation (G, Ω), where G is a right-orderable group, is orderable if and only if the stabilizer G_α of an arbitrary element $\alpha \in \Omega$ is a right-relatively convex subgroup of the group G.*

Moreover, for any right-relatively convex subgroup H of a right-orderable group G there is a transitive representation (G, Ω) such that H coincides with the stabilizer of some element $\alpha \in \Omega$.

Corollary 5.1.5 (Todorinov [69]). *A subgroup H of a right-orderable group G is right-relatively convex if and only if for any elements $g_1, \ldots, g_n$ of the set $G \setminus H$ there are numbers $\varepsilon_1, \ldots, \varepsilon_n$ equal to ± 1 such that*

$$e \notin P_H(g_1^{\varepsilon_1}, \ldots, g_n^{\varepsilon_n})$$

Proof. If H is a convex subgroup of a right-orderable group G, then the representation $(G, R(G : H))$ is totally ordered with respect to the natural order on the set of right cosets $R(G : H)$.

To prove the sufficiency we consider the right regular representation $(G, R(G : H))$ of the group G by permutations of the set $R(G : H)$. The subgroup H is the stabilizer of the right coset H. By Proposition 5.1.6, the representation $(G, R(G : H))$ is orderable. Applying Corollary 5.1.4, we complete the proof. □

5.1.2. We use the above results to describe the properties of the system of right-relatively convex subgroups of a right-orderable group.

Proposition 5.1.7. *The union of a totally-ordered-by-inclusion system of right-relatively convex subgroups of a right-orderable group G is a right-relatively convex subgroup of G.*

The assertion follows from Corollary 5.1.5.

The above-mentioned results lead to the following useful criterion for right-orderability of groups.

Theorem 5.1.1 (Tararin [68]). *A group G is right-orderable if and only if there is a complete system $\mathfrak{C}(G)$ of subgroups satisfying the following condition:*

> *if $A_\alpha \prec \overline{A}_\alpha$ is a jump in $\mathfrak{C}(G)$, then the representation $(\overline{A}_\alpha, R(\overline{A}_\alpha : A_\alpha))$ is orderable.*

In addition, in the group G there is a right order with respect to which all subgroups of $\mathfrak{C}(G)$ are convex.

PROOF. The necessity is obvious. We show the sufficiency. In $\mathfrak{C}(G)$, any element $x \in G$, $x \neq e$, defines a jump of subgroups $A_\alpha \prec \overline{A}_\alpha$, where A_α is the union of all subgroups of $\mathfrak{C}(G)$ which do not contain the element x, and $\overline{A}_\alpha$ is the intersection of all subgroups of $\mathfrak{C}(G)$ containing the element x. By assumption, the representation $(\overline{A}_\alpha, R(\overline{A}_\alpha : A_\alpha))$ is orderable. Therefore, in G there is a subsemigroup Q_α satisfying the following conditions:

$$Q_\alpha \subset \overline{A}_\alpha, \qquad Q_\alpha \cap A_\alpha = \varnothing$$

$$Q_\alpha \cup Q_\alpha^{-1} \cup A_\alpha = \overline{A}_\alpha, \qquad A_\alpha Q_\alpha A_\alpha \subseteq Q_\alpha$$

Let $Q_0 = \bigcup_{\alpha \in I} Q_\alpha$, where the union is taken over all jumps in the system $\mathfrak{C}(G)$. We show that Q_0 is a pure linear subsemigroup of the group G. Let $a \in Q_\alpha$, $b \in Q_\alpha$, and $\overline{A}_\alpha \subseteq \overline{A}_\beta$. If $\overline{A}_\alpha = \overline{A}_\beta$, then $Q_\alpha = Q_\beta$ and $ab \in Q_\alpha$. If $\overline{A}_\alpha \subset \overline{A}_\beta$, then $\overline{A}_\alpha \subseteq A_\beta$. Therefore, $\overline{A}_\alpha Q_\beta \subseteq A_\beta Q_\beta \subseteq Q_\beta$. Consequently, $ab \in Q_\beta$ and Q_0 is a subsemigroup. It is evident that Q_0 is pure and linear.

From the inclusion $A_\alpha Q_\alpha A_\alpha \subseteq Q_\alpha$ it follows that A_α is convex in $\overline{A}_\alpha$. An arbitrary subgroup A of the system $\mathfrak{C}(G)$ is an intersection of lower subgroups of the jumps in $\mathfrak{C}(G)$. Hence it is convex. □

Proposition 5.1.8 (Plotkin [70]). *A subgroup of a torsion-free locally nilpotent group G is right-relatively convex if and only if it is isolated in G.*

PROOF (Tararin [68]). We assume that G is a torsion-free nilpotent group and H is its isolated subgroup. We consider the series of consecutive normalizers

$$H_0 \subseteq H_1 \subseteq H_2 \subseteq \ldots \subseteq H_n = G$$

where $H_{i+1} = N_G(H_i)$ is a normalizer in G of the subgroup H. Since G is a torsion-free nilpotent group, all subgroups H_i are isolated in G (cf. [71, Lemma 4.10] or [45]). Thus, all factors H_{i+1}/H_i and H_0 are torsion-free nilpotent groups. By Theorem 3.3.2, the group G admits a total right order with respect to which all subgroups H_i are convex.

Let H be an isolated subgroup of a torsion-free locally nilpotent group G. By Corollary 5.1.5, it suffices to show that for any elements $g_1, \ldots, g_n$ of the set $G \setminus H$ there exist numbers $\varepsilon_1, \ldots, \varepsilon_n$ equal to ± 1 such that $e \notin P_H(g_1^{\varepsilon_1}, \ldots, g_n^{\varepsilon_n})$. We assume the contrary. Then there are elements $g_1, \ldots, g_n$ of the set $G \setminus H$ such that $e \in P_H(g_1^{\varepsilon_1}, \ldots, g_n^{\varepsilon_n})$ for any numbers $\varepsilon_1, \ldots, \varepsilon_n$ equal to ± 1. In this case there is a finite collection of elements $\{h_1, \ldots, h_m\}$ of the subgroup H such that $e \in P_{H_1}(g_1^{\varepsilon_1}, \ldots, g_n^{\varepsilon_n})$ for any numbers $\varepsilon_1, \ldots, \varepsilon_n$ equal to ± 1, where H_1 is a subgroup of H generated by the elements $h_1, \ldots, h_m$. Let G_1 be a subgroup of G generated by the elements $h_1, \ldots, h_m, g_1, \ldots, g_n$. Let $H_2 = G_1 \cap H$. It is obvious that H_2 is an isolated subgroup of G_1 and, consequently, is a right-relatively convex subgroup. On the other hand, $H_2 \supseteq H_1$ and $g_1, \ldots, g_n \in G_1 \setminus H_2$. Consequently, $e \in P_{H_2}(g_1^{\varepsilon_1}, \ldots, g_n^{\varepsilon_n})$ for any numbers $\varepsilon_1, \ldots, \varepsilon_n$ equal to ± 1, which contradicts the right-relative convexity of the subgroup H_2 of H_1. □

Proposition 5.1.9 (Tararin [68] and Todorinov [69]). *A right-orderable group without proper right-relatively convex subgroups is an Abelian group of rank* 1.

Proof. We show that if G is a right-ordered group, $a \in G$, $a > e$, and the set $\{a^n \mid n \in \mathbb{Z}\}$ is bounded from above or from below, there exists a proper right-relatively convex subgroup of the group G containing the element a. For convenience, we set $a^n < b$ for all integers n. Let $L = \{x \in G \mid x < a^n \text{ for some } n, n \in \mathbb{N}\}$. Then $La = L$. Indeed, if $x \leqslant a^n$, then $xa \leqslant a^{n+1}$. Thus, $xa \in L$. Let $y = xa$, where $x \in L$ and, consequently, $x \leqslant a^n$, $n \in \mathbb{N}$. We obtain $xa \leqslant a^{n+1}$ and $y = xa \in L$. Since L is a convex subset of G, the subsets Lg are also convex. On the set $M = \{Lg \mid g \in G\}$ we can introduce a total order as follows:

$$Lg_1 \leqslant Lg_2 \Leftrightarrow Lg_1 \subseteq Lg_2$$

It is clear that this order is stable under the action of the group G. By Corollary 5.1.4, the stabilizer G_L of the set L is a right-relatively convex subgroup of G. Since $a \in G_L$, we have $G_L \neq E$, and since $b \notin L$, we conclude that $G_L \neq G$.

We now show that if G is a right-ordered group, $a > e$, and in G there are elements b and c such that $ca < c$ and $b < ba$, then there is a proper right-relatively convex subgroup of G containing the element a. Let $L = \{x \in G \mid xa^n \geqslant b \text{ for some } n, n \in \mathbb{N}\}$. Since $b \in L$, we have $L \neq \varnothing$. Therefore, L is a convex subset of the totally ordered set G. Indeed, if $x < y$ and $y \in L$, then $ya^n > xa^n \geqslant b$ for some natural number n. We show that $La = L$. Let $x \in L$. Then $xa^n \geqslant b$. If $n \geqslant 1$, then $(xa)a^{n-1} \geqslant b$. Hence $xa \in L$. If $x \geqslant b$, then $xa \geqslant ba \geqslant b$, and, moreover, $xa \in L$. We assume that $x \in L$. Then $xa^n \geqslant b$ and $(xa^{-1})a^{n+1} \geqslant b$. Thus, $xa^{-1} \in L$ and $x \in La$.

We show that $L \neq G$. In view of the inequalities $ca < c$ and $c < b$, we have $ca^2 < ca < c < b$. This means that for any natural number n the inequality $ca^n < b$ is fulfilled. Therefore, $c \notin L$. As above, we conclude that the representation (G, Ω), where $\Omega = \{Lg \mid g \in G\}$, is orderable and the stabilizer G_L is a proper convex subgroup of G.

Thus, if G is a right-ordered group without proper right-relatively convex subgroups, then for any $a \in G$, $a > e$, and $x, y \in G$, $x < y$, there is a natural number n such that $xa^n y$.

Let $a > e$ and $xa < x$ for some $x \in G$. Then $xa^2 < xa < x$, i.e., for any natural number n we have the inequality $xa^n < x$, which contradicts the above arguments. Consequently, $x < xa$ for any $a, x \in G$, $a > e$. Thus, G is a (two-sided) totally ordered group. Since G does not have nontrivial convex subgroups, by the Hölder theorem (cf. Theorem 2.2.1), it is Archimedean and, consequently, Abelian. It is easy to check that an Abelian group without right-relatively convex subgroups is an Abelian group of rank 1. □

Corollary 5.1.6 (Todorinov and Petrova [72]). *A group G has a unique (up to inverse) right order if and only if it is an Abelian group of rank* 1.

Corollary 5.1.7 (Tararin [68]). *Let G be a right-ordered group $a \in G$ and $a > e$. If the set $\{a^n \mid n \in \mathbb{Z}\}$ is bounded from above (from below) in G, then the element a belongs to a proper right-relatively convex subgroup of the group G.*

Corollary 5.1.8 (Tararin [68]). *Let G be a right-ordered group, $a \in G$, $a > e$. If in G there are elements b and c such that $c < b$, $ca < c$*

and $b < ba$, then the element a belongs to a right-relatively convex subgroup of the group G.

Proposition 5.1.10 (Tararin [68]). *The intersection of right-relatively convex subgroups of a right-orderable group G is a right-relatively convex subgroup of G.*

PROOF. Let $\{H_\alpha, \alpha \in I\}$ be right-relatively convex subgroups of the group G. For any $\alpha \in I$ we consider the transitive representation $(G, R(G : H_\alpha))$ of the group G on the set of right cosets $R(G : H_\alpha)$ of the group G by the subgroup H_α. It is evident that the stabilizer G_{H_α} of the element $H_\alpha \in R(G : H_\alpha)$ in G is equal to H_α. By Proposition 5.1.6, on the set $R(G : H_\alpha)$ it is possible to introduce a total order which is stable under the action of the group G. We now introduce a lexicographic order on the Cartesian product $\overline{\prod\limits_{\alpha \in I}} R(G : H_\alpha)$ of the totally ordered sets. It is evident that this order is stable under the action of the group G and the stabilizer of the element

$$f \in \overline{\prod_{\alpha \in I}} R(G : H_\alpha)$$

where $f_\alpha = H_\alpha$ for every $\alpha \in I$, coincides with

$$\bigcap_{\alpha \in I} H_\alpha$$

In view of Proposition 5.1.6, the subgroup $\bigcap\limits_{\alpha \in I} H_\alpha$ is a right-relatively convex subgroup of the group G. □

Due to Proposition 5.1.10, it is possible to introduce on the set of right-relatively convex subgroups of a right-orderable group G the operations of meet and join by the following rules:

$$X \wedge Y = X \cap Y$$
$$X \vee Y = \bigcap_{\alpha \in I} Z_\alpha$$

where $Z_\alpha (\alpha \in I)$ is the set of all right-relatively convex subgroups of G containing X and Y.

Theorem 5.1.2 (Tararin [68]). *The system $\mathfrak{C}(G)$ of all right-relatively convex subgroups of a right-orderable group G is a lattice.*

Let M be a subset of a right-orderable group G. We denote by $\mathrm{rc}\,(M)$ the intersection of all right-relatively convex subgroups of G containing M, i.e., the least right-relatively convex subgroup of G containing M. We refer to $\mathrm{rc}(M)$ as a *right-relatively convex subgroup of G generated by the subset M*. A subgroup H of the right-orderable group G will be called *bounded* in G if a right order can be introduced on the group G so that there is an element a such that $a > x$ for any $x \in H$ or $a < x$ for any $x \in H$. Otherwise H is called an *unbounded* subgroup of G.

Proposition 5.1.11 (Tararin [68]). *If H is a bounded subgroup of a right-orderable group G, then* $\mathrm{rc}\,(H) \neq G$.

PROOF. Let H be a bounded subgroup of G. On the group G we introduce a right order such that $x < a$ for any $x \in H$. The set

$$L = \{y \in G \mid \text{there exists an element } x \in H \text{ such that } y < x\}$$

is a convex subset of the totally ordered set G and $L \neq G$. Let $\Omega = \{Lg \mid g \in G\}$. It is obvious that Ω is a totally-ordered-by-inclusion set. Moreover, the group G regarded as the group of order automorphisms acts on Ω by the rule

$$(Lx)g = Lxg$$

The stabilizer G_L of the element $L \in \Sigma$ is a right-relatively convex subgroup of the group G. Since $H \subseteq G_L$, we obtain $\mathrm{rc}(H) \subseteq G_L$ and $\mathrm{rc}(H) \neq G$. □

5.2. A Finite Number of Right Orders

5.2.1. In this section, we consider right-orderable groups with only a finite number of distinct right orders.

Lemma 5.2.1. *Let there exist a series of subgroups of a right-orderable group G:*

$$E = H_0 \subseteq H_1 \subseteq \ldots \subseteq H_n = G$$

such that every subgroup H_i is a right-relatively convex subgroup of H_{i+1}. Then on the group G we can introduce a right order with respect to which all subgroups H_i of G are convex. The number of such different right orders is at least 2^n.

PROOF. By assumption, each of H_i is a right-relatively convex subgroup of H_{i+1}. Therefore, there is a right order $\leqslant$ and the semigroup of positive elements $\overline{P}_{i+1}$ of the group H_{i+1} such that the subgroup H_i is convex. Let

$$P_{i+1} = \{x \in \overline{P}_{i+1}, 1 \mid x \notin H_i\}$$

Then P_{i+1} is a semigroup. Indeed, for $a, b \in P_{i+1}$ we have $ab \in \overline{P}_{i+1}$, because $b < ab$. Since H_i is a convex subgroup of H_{i+1}, we have $ab \notin H_i$ and $ab \in \overline{P}_{i+1}$. The purity of the semigroup P_{i+1} is evident.

Further, if $a \in P_{i+1}, x, y \in H_i$, then $xay \in P_{i+1}$. Indeed, $xay \notin H_i$. To show that $xay \in \overline{P}_{i+1}$, i.e., $xay \geqslant e$, we assume the contrary, i.e., $xay < e$. Then $xa < y^{-1}$. In view of the convexity of H_i and the inequality $a > e$, we obtain $xa > e$. Therefore, $e < xa < y^{-1}$ and $xa \in H_i$, which contradicts the assumption $xa \notin H_i$. We also note that the semigroup P_{i+1}^{-1} has similar properties.

Let

$$P(\varepsilon) = \bigcup_{i=1}^{n} P_i^{\varepsilon_i} \cup \{e\}$$

where $\varepsilon_1, \ldots, \varepsilon_n = \pm 1$. We show that $P(\varepsilon)$ is a pure linear subsemigroup of the group G. It is obvious that $P(\varepsilon)$ is linear and pure. We consider $a, b \in P(\varepsilon)$. The inclusions $a \in P_i^{\varepsilon_i}$ and $b \in P_j^{\varepsilon_j}$ are valid for some i, j, $1 < i, j < n$. If $i = j$, then $ab \in P_i^{\varepsilon_i}$. Since $P_i^{\varepsilon_i}$ is a semigroup, the inclusion $ab \in P(\varepsilon)$ holds. If $i < j$, then $a \in H_{j-1}$ and $ab \in P_j^{\varepsilon_j}$ in view of the property of the semigroup $P_i^{\varepsilon_j}$. Consequently, $ab \in P(\varepsilon)$.

We now show that the subgroups H_i, $i = 1, \ldots, n$, are convex with respect to the right order $P(\varepsilon)$ defined on the group G. It suffices to show that for $e < x < a$, $a \in H_i$, $x \in H_{i+1}$, the inclusion $x \in H_i$ holds. The last inequalities imply the inclusion $x, ax^{-1} \in P(\varepsilon)$. We assume that $x \notin H_i$, $x, ax^{-1} \in P_{i+1}^{\varepsilon_{i+1}}$. Owing to the properties of the semigroup $P_{i+1}^{\varepsilon_{i+1}}$, the relations $x \in P_{i+1}^{\varepsilon_{i+1}}$ and $x^{-1} = a^{-1}ax^{-1} \in P_{i+1}^{\varepsilon_{i+1}}$ hold, which contradicts the purity of the semigroup $P_{i+1}^{\varepsilon_{i+1}}$. Hence $x \in H_i$.

Repeating the previous argument, we see that the orders $P(\varepsilon)$ are distinct for different sets of numbers $\varepsilon_1, \ldots, \varepsilon_n$ equal to ± 1. Thus, the number of distinct right orders with respect to which all subgroups H_i are convex is at least 2^n. □

From Lemma 5.2.1 we obtain the following lemma.

Lemma 5.2.2. *For any right-orderable group G the following assertions hold:*

(a) *if G has a finite number of right-relatively convex subgroups, then each such subgroup has the same property,*

(b) *if, on the group G, only a finite number of distinct right orders can be introduced, then for every right order on G the system of convex subgroups with respect to this order is finite,*

(c) *any right order P_N on a right-relatively convex subgroup N of the group G can be extended to a right order on G so that N will be a convex subgroup of G with respect to the extended order and $P_N = P \cap N$;*

(d) *if only a finite number of right orders can be introduced on G, then the same property holds for every right-relatively convex subgroup of G.*

Lemma 5.2.3. *Only a finite number of right orders can be introduced on an Abelian group (respectively, an Abelian group has only a finite number of right-relatively convex subgroups) if and only if the group is isomorphic to some subgroup of the additive group $\mathbb{Q}$ of rational numbers.*

PROOF. It is clear that any subgroup of $\mathbb{Q}$ has only two distinct right (total) orders. An Abelian group G of rank at least 1 has an infinite set of mutually distinct isolated subgroups. By Lemma 5.2.2, for any isolated subgroup H one can construct a right order on G with respect to which the subgroup H is convex. □

- A subnormal (normal) series

$$E = G_0 \subseteq G_1 \ldots \subseteq G_n = G$$

whose factors G_{i+1}/G_i are torsion-free Abelian groups of rank 1 is called *rational.*

Theorem 5.2.1 (Tararin [73]). *For a non-Abelian group G, only a finite number of distinct right orders can be introduced on G if and only if G has a normal rational series*

$$E = G_0 \subseteq G_1 \ldots \subseteq G_n = G$$

such that each of the factors G_{i+2}/G_i, $n-2 \geqslant i \geqslant 0$, does not admit nontrivial total order.

PROOF. We first assume that on a non-Abelian group G only a finite number of right orders can be introduced. Let P be a semigroup of positive elements corresponding to one of the right orders defined on G. It is clear that $x^{-1}Px$ is a pure linear semigroup for any element $x \in G$. Since the group G admits only a finite number of right orders, the number of distinct semigroups $x^{-1}Px$ is finite. Consequently, the subset $G^* = \{x \in G \mid x^{-1}Px = P\}$ is a subgroup of finite index of G. It is evident that G^* is a totally ordered group with respect to the induced order.

Let M be a proper convex subgroup of the group G. Then $M^* = G^* \cap M$ is also a proper convex subgroup of G^*. If N^* is a proper convex subgroup of G^*, then the subset

$$N = \{x \in G \mid a_1 \leqslant x \leqslant a_2 \text{ for some } a_1, a_2 \in N^*\}$$

is a proper convex subgroup of G. Thus, we have established a one-to-one correspondence between convex subgroups of the group G and of the group G^*.

Let $\mathfrak{C}(G)$ be the system of all convex subgroups of a right-ordered group G. By Lemma 5.2.1, $\mathfrak{C}(G)$ is finite:

$$E = G_0 \subsetneq G_1 \dots \subsetneq G_n = G$$

We consider the corresponding system of convex subgroups of the totally ordered group G^*:

$$\mathfrak{C}(G^*)\colon E = G_0^* \subsetneq G_1^* \dots \subsetneq G_n^* = G^*$$

The system $\mathfrak{C}(G^*)$ is normal in the group G^* and each of its factors G_{i+1}^*/G_i^* is an Archimedean totally ordered group with respect to the induced order. It is easy to show that the subgroup G_i^* has a finite index in G_i.

We take elements $a > e$ and $b > e$ of the group G. The elements ab and ba define the same jump of the convex subgroups $G_i \prec G_{i+1}$ in the system $\mathfrak{C}(G)$. By the finiteness of the index of the subgroup G_{i+1}^* of the group G_{i+1}, there are natural numbers k and m such that $(ab)^k, (ba)^m \in G_{i+1}^*$. Since G_{i+1}^*/G_i^* is an Archimedean totally ordered group, there is a natural number r such that $(ab)^r > ba$. By Theorem 2.4.1, the system $\mathfrak{C}(G)$ is subnormal and its factors G_{i+1}/G_i are Archimedean totally ordered groups. By Lemmas 5.2.1 and 5.2.3, the groups G_{i+1}/G_i, $i = 0, 1, \dots, n-1$, are Abelian groups of rank 1. Hence the system $\mathfrak{C}(G)$ is a subnormal rational series.

The task is now to show that the group G has a unique rational series. Let $E = H_0 \subsetneq H_1 \subsetneq \dots \subsetneq H_n = G$ be an arbitrary rational series of subgroups of the group G. Since G/H_{n-1} is isomorphic to a subgroup of the additive group $\mathbb{Q}$, by Lemma 5.2.1, H_{n-1} is a normal relatively convex subgroup of the group G. By Proposition 5.1.10, $N = G_{n-1} \cap H_{n-1}$ is a normal right-relatively convex subgroup of G. Since the group G/N is Abelian and admits only a finite number of distinct right orders, G/N is also isomorphic to a subgroup of the additive group $\mathbb{Q}$. Since the subgroups G_{n-1} and H_{n-1} are isolated in G, we have $H_{n-1} = G_{n-1}$. Using induction on the length of series, we can convince ourselves that the series $\mathfrak{C}(G)$ is unique and, consequently, is normal. Since the right order has been chosen arbitrarily and the series $\mathfrak{C}(G)$ is unique, all right-relatively convex subgroups of G are precisely the subgroups of the system $\mathfrak{C}(G)$.

We now show that none of the factors G_{i+2}/G_i, $i = 0, 1, \ldots, n-2$, of the system $\mathfrak{C}(G)$ is an orderable group. In accordance with what has been said above, on the group $S = G_{i+2}/G_i$ only a finite number of distinct right orders can be defined and the group $H = G_{i+1}/G_i$ is a unique proper right-relatively convex subgroup of S. We assume that S is a totally ordered group and consider the following natural representation φ of the group S by the inner automorphisms of S: for every $g \in S$ the mapping $g\varphi$ on S is defined as follows:

$$(x)g\varphi = g^{-1}xg \text{ for each } x \in S$$

By assumption, the order on S is two-sided. Therefore, φ is a representation of S by order automorphisms of the totally ordered set S. By Proposition 5.1.6, the stabilizer of an arbitrary element $g \in S$ in S is a right-relatively convex subgroup of S. Consequently, the centralizer $c(g) = \{x \in S \mid x^{-1}gx = g\}$ for any element $g \in S$, $g \neq e$, is equal to S or H. Hence $c(g)$ for $g \notin H$ is equal to S. But then the group S is Abelian and, by Lemma 5.2.3, is isomorphic to a subgroup of the additive group of rationals $\mathbb{Q}$, which contradicts our assumptions on the group S.

We now assume that there is a rational series

$$E = G_0 \subseteq G_1 \subseteq \ldots \subseteq G_n = G$$

all of whose factors G_{i+2}/G_i $(n-2 \geqslant i \geqslant 0)$ are not orderable groups. Sinse the group G_{i+1}/G_i is isomorphic to a subgroup $\mathbb{Q}$, it has only two mutually inverse orders, e.g., $\overline{P}_i$ and $\overline{P}_i^{-1}$. For any automorphism φ of the group G_{i+1}/G_i the set $\varphi(\overline{P}_i)$ is a pure linear subsemigroup that defines an order on G_{i+1}/G_i; consequently, $\varphi(\overline{P}_i) = \overline{P}_i$ or $\varphi(\overline{P}_i) = \overline{P}_i^{-1}$. We conclude that the automorphism φ either preserves the order on

G_{i+2}/G_i or inverts it. In particular, the automorphisms of the group G_{i+1}/G_i which are induced by the inner automorphisms of the group G_{i+2}/G_i preserve or invert the orders on G_{i+1}/G_i. Moreover, there exists an element $g_{i+2} \in G_{i+2}$ such that $g^{-1}\overline{P}_i g = \overline{P}_i^{-1}$; otherwise the group G_{i+2}/G_i is orderable. We fix such elements g_{i+2}, $i = 0, 1, \ldots, n-2$, and and arbitrary element $g_1 \neq e$, of the subgroup G_1. For every set of numbers $\varepsilon_1, \ldots, \varepsilon_n$ equal to ± 1 there is a right order $P(\varepsilon) = P(\varepsilon_1, \ldots, \varepsilon_n)$ on the group G with respect to which $g_i^{\varepsilon_i} \in P(\varepsilon)$ for every i, $i = 1, \ldots, n$. Indeed, ordering the factors G_i/G_{i-1} so that the elements $g_i^{\varepsilon_i} G_{i-1}$ $(i = 1, \ldots, n)$ are positive and setting $P(\varepsilon) = \bigcup\limits_{i=1}^{n} P_i(\varepsilon) \cup \{e\}$, where

$$P^i(\varepsilon) = \{x \in G_i \mid xG_{i-1} \in \overline{P}_i^{\varepsilon_i} \text{ in } G_i/G_{i-1}\}$$

we obtain the desired order.

We show that an order with the set of positive elements P of the group G coincides with one of the orders $P(\varepsilon)$. Without loss of generality, we can assume that $g_i \in P$ and $g_i \in P(\varepsilon)$ for all i, $i = 1, \ldots, n$. In view of the condition $g_1 \in P$, the set $\{g_1^k \mid k \in \mathbb{N}\}$ is contained in P. Since the subgroup G_1 is isomorphic to a subgroup of the group $\mathbb{Q}$ and P is an isolated subset of the group G, the set $\widetilde{P}_1 = \{u_1 \in G_1 \mid u_1 \in P(\varepsilon)\}$ is contained in P. Taking into account that P is a semigroup, we see that the inclusion $\{u_1 g_2 \mid u_1 \in G_1, u_1 \in P(\varepsilon)\} \subseteq P$ holds. On the other hand, $u_1 g = g_2(g_2^{-1} u_1 g_2)$. Hence

$$\{g_2 u_1' \in G_1, u_1' \in P^{-1}(\varepsilon)\} \subseteq P$$

Taking into account that P is a semigroup, we obtain

$$\{g_2 u_1' u_1 \mid u_1' \in P^{-1}(\varepsilon), u_1', u_2 \in G_1\} = g_2 G_1 \subseteq P$$

and $g_2^k G_1 \subseteq P$ for all natural numbers k. Since the quotient group G_2/G_1 is isomorphic to a subgroup of the group $\mathbb{Q}$, we obtain

$$\widetilde{P}_2 = \{u_2 v_1 \mid u_2 \in G_2 \setminus G_1, u_2 \in P(\varepsilon), v_1 \in G_1\} \subseteq P$$

Repeating the previous arguments word for word, we obtain

$$\widetilde{P}_n = \{u_n v_n \mid u_n \in G_n \setminus G_{n-1}, u_n \in P(\varepsilon), v_{n-1} \in G_{n-1}\} \subseteq P$$

Consequently, $P(\varepsilon) = \widetilde{P}_1 \cup \ldots \cup \widetilde{P}_n \subseteq P$ and $P = P(\varepsilon)$. □

Corollary 5.2.1. *An orderable group with a finite number of right orders is an Abelian group of rank* 1.

Theorem 5.2.2 (Tararin [73]). *If a group G admits a finite number of right orders, their number is equal to 2^n. For any natural number n there is a group on which the number of right orders is equal to 2^n.*

PROOF. The first assertion of the theorem follows from the proof of Theorem 5.2.1 because the number of distinct orders $P(\varepsilon)$ is equal to 2^n. To prove the second assertion, for any natural number n we construct a group which admits precisely 2^n distinct right orders. On the Cartesian power

$$\underbrace{\mathbb{Z} \times \ldots \times \mathbb{Z}}_{n} = G$$

of the additive group $\mathbb{Z}$ of integers, we define the multiplication operation as follows:

$$(a_n, \ldots, a_1) \cdot (b_n, \ldots, b_1) = (a_n + b_n, (-1)^{b_n} a_{n-1} + b_{n-1}, \ldots, (-1)^{b_2} a_1 + b_1)$$

Let

$$G_1 = \{(0, \ldots, 0, a_1) \mid a_1 \in \mathbb{Z}\}$$

$$\ldots\ldots\ldots\ldots\ldots$$

$$G_{n-1} = \{(0, a_{n-1}, \ldots, a_1) \mid a_{n-1}, \ldots, a_1 \in \mathbb{Z}\}$$

Then the series $E = G_0 \leqslant G_1 \ldots \leqslant G_n = G$ satisfies the hypotheses of Theorem 5.2.1. □

Theorem 5.2.3 (Tararin [73]). *A right-orderable group has a finite number of right-relatively convex subgroups if and only if it admits a finite number of right orders.*

PROOF. We apply induction on the number $r(G)$ of right-relatively convex subgroups. If $r(G) = 1$, then the group G has no proper right-relatively convex subgroups and is isomorphic to a subgroup of the group $\mathbb{Q}$ in virtue of Proposition 5.1.9. We assume that the assertion of the theorem is true for all groups G such that $r(G) < n$. Let $r(G) = n$. We denote by $H_1, \ldots, H_m$ the proper relatively convex subgroups of the group G. Then $r(H_i) < n$ for any $i = 1, \ldots, m$. We fix on the group G a right order with the semigroup P of positive elements. We set

$$P_0 = \{x > e \mid x \notin \bigcup_{i=1}^{m} H_i\}$$

$$P_i = P \cap H_i, \quad i = 1, \ldots, m$$

and note that P_0 can be empty. It is evident that $P = \bigcup_{i=0}^{m} P_i$. We prove that the number of distinct orders $g^{-1}Pg = \bigcup_{i=1}^{m} (g^{-1}P_i g) \cup g^{-1}P_0 g$ that are conjugate to P is finite.

The number of distinct sets $\bigcup_{i=1}^{m} g^{-1}P_i g$ is finite because the number of subgroups H_i is finite and, by the inductive assumption, the groups H_i, $i = 1, \ldots, m$, have only a finite number of distinct right orders. For $a \in P_0$ we have $g^{-1}ag \notin \bigcup_{i=1}^{m} H_i$, since the group $\bigcup_{i=1}^{m} H_i$ contains each subgroup H_i together with its conjugate group. The infinite cyclic subgroup (a) is not bounded in the group G, hence, by Corollary 5.1.7 and Proposition 5.1.9, the inequality $g^{-1}ag > e$ holds. Hence $g^{-1}P_0 g = P_0$ for every $g \in G$. Thus, we have proved the finiteness of the number of distinct right orders $g^{-1}Pg$ for $g \in G$.

Further consideration is similar to the proof of the necessity of the assumption in Theorem 5.2.1 with the remark that an Abelian group with a finite number of right-relatively convex subgroups is isomorphic to a subgroup of the group $\mathbb{Q}$. Arguing in the same way, for any right order on the group G its convex subgroups form a subnormal rational series and the latter satisfies the hypotheses of Theorem 5.2.1 provided that G is a non-Abelian group.

As was mentioned in the proof of Theorem 5.2.1, for a group with only a finite number of right orders, all of its right-relatively convex subgroups form a normal rational series. □

5.2.2. We consider the problem of cardinality of the set of distinct right orders on a group.

Theorem 5.2.4 (Zenkov). *Either the number of distinct right orders on an orderable group G is equal to 2 or the cardinality of the set of distinct right orders on G is uncountable.*

PROOF. For a totally ordered group G the system $\mathfrak{C}(G)$ of its convex subgroups is infrainvariant, totally-ordered-by-inclusion, subnormal, and with Archimedean factors:

$$\mathfrak{C}(G)\colon E \subseteq \ldots \subseteq G_\alpha \prec \overline{G_\alpha} \subseteq \ldots \subseteq G \tag{5.2.1}$$

where $\alpha \in I$. Moreover, if $G_\alpha \prec \overline{G_\alpha}$ is a jump of convex subgroups, then the group of automorphisms of the group $\overline{G_\alpha}/G_\alpha$ induced by the inner automorphisms of the normalizer $\mathfrak{n}_G(\overline{G}_\alpha)$ of the subgroup $\overline{G}_\alpha$ is isomorphic to a subgroup of the multiplicative group of positive real numbers.

We take some total order $\overline{P}_\alpha$ on $\overline{G}_\alpha/G_\alpha$. Then the set $P_\alpha = \{x \in \overline{G}_\alpha \mid \overline{x} > e$ for the order $\overline{P}_\alpha\}$ is a subsemigroup of the group $\overline{G}_\alpha$. We set $Q(P, I) = \bigcup_{\alpha \in I} P_\alpha \cup \{e\}$ and show that the subset $Q(P, I)$ defines a right order on G. Let $p_1 \neq e$, $p_2 \neq e$, $p_1, p_2 \in Q(P, I)$. There is an index $\alpha \in I$ such that $p_1, p_2 \in P_\alpha$. Therefore, $p_1 p_2 \in P_\alpha \subseteq Q(P, I)$, hence $Q(P, I)$ is a semigroup.

We assume that there is an element $x \in G$, $x \neq e$, such that $x \in Q(P, I) \cap Q(P, I)^{-1}$. Then $x \in P_\alpha$ for some index $\alpha \in I$ and $x \in \overline{G}_\alpha$, $x \notin G_\alpha$. On the other hand, $\overline{x} \in \overline{P}_\alpha \cap \overline{P}_\alpha^{-1}$. Therefore, $\overline{x} = e$ or, equivalently, $x \in G_\alpha$, which contradicts the choice of x. Consequently, $Q(P, I)$ is a pure set.

We now consider an arbitrary nonidentity element $g \in G$. This element determines some jump $G_\alpha \prec \overline{G}_\alpha$ in the system $\mathfrak{C}(G)$, i.e., $\overline{g} > e$ or $\overline{g} < e$ with respect to the order $\overline{P}_\alpha$. This means that $g \in Q(P, I) \cup Q(P, I)^{-1}$. Thus, we have proved that the set $Q(P, I)$ is linear.

By the construction of the order $Q(P, I)$, the cardinality of the set of right orders defined on G depends on the cardinality of the set I and the cardinality of the set of total orders on each of the factors $\overline{G}_\alpha/G_\alpha$ of jumps in the system $\mathfrak{C}(G)$.

In the case $|I| \geqslant \omega$, the cardinality of the set of distinct right orders on the group G is uncountable. Therefore, we assume that the series (5.2.1) has a finite length:

$$E = G_0 \subset G_1 \subset G_2 \subset \ldots \subset G_s = G \tag{5.2.2}$$

If the rank of at least one of the factors $\overline{G}_\alpha/G_\alpha$ is greater than 2, then this factor admits no less than the continuum of total orders and, consequently, the group G itself admits no less than the continuum of right orders. It remains to consider the case where the rank of $\overline{G}_\alpha/G_\alpha$ is equal to 1 for any $\alpha \in I$. This is equivalent to the fact that the factor $\overline{G}_\alpha/G_\alpha$ is isomorphic to a subgroup of the additive group $\mathbb{Q}$ of rational numbers. Thus, the series (5.2.2) is rational and has a finite length. If the length is 1, then the group G is isomorphic to a subgroup of the additive group $\mathbb{Q}$ and, thereby, admits only two distinct right orders.

We assume that the length of the series (5.2.2) is equal to 2. Then

$$E = G_0 \subset G_1 \subset G_2 = G$$

and G_1 is a normal subgroup of G. If $g^{-1}qg = q$ for any $g \in G \setminus G_1$ and $q \in G_1$, then G is an Abelian group of rank 2 and, consequently, admits the continuum of total orders. Thus, for any $g \in G$ the equality $g^{-1}qg = q^r$ is fulfilled, where $r \neq 1$ is some positive rational number. Therefore, the subgroup G_1 coincides with its centralizer in G and the quotient group G/G_1 is isomorphic to a subgroup of rank 1 of the group of automorphisms of an Abelian group of rank 1, i.e., of the multiplicative group of positive rational numbers. Thus, the quotient group G/G_1 is an infinite cyclic group. We note that the group G is isomorphic to a subgroup of the Smirnov group (cf. Example 2.1.1). Consequently, each right order on the Smirnov group constructed by means of the irrational number ε determines a right order on G and all such orders are distinct. Thus, on G the continuum of distinct right orders can be introduced.

If the length of the series (5.2.2) is at least 2, then the subgroup G_2 in the series is relatively convex and has an uncountable number of distinct right orders, as was found out above. Thus, if an orderable group G is not an Abelian group of rank 1, then it admits an uncountable set of distinct right orders. □

Theorem 5.2.5 (Zenkov). *Let G be a locally indicable group. Then either the number of distinct right orders is equal to 2^n for a suitable natural number n or the cardinality of the set of right orders on the group G is uncountable.*

PROOF. Since G is locally indicable, it admits a Conradian right order in view of Corollary 3.2.2. Then the system $\mathfrak{C}(G)$ of its convex subgroups is a soluble subnormal system of subgroups with Abelian torsion-free factors. Arguing as in the proof of Theorem 5.2.4, we see that we can restrict ourselves to consideration of a finite rational series:

$$E = G_0 \subset \ldots G_i \subset \ldots \subset G_n = G$$

The proof is by induction on the length n of the rational series. In the case $n = 2$, we have the series

$$E = G_0 \subset G_1 \subset G_2 = G$$

which is obviously normal. The group G is not orderable; otherwise, by Theorem 5.2.4, it would admit an uncountable set of distinct right orders. By Theorem 5.2.1, it has precisely four distinct right orders.

We assume that for $n = k-1$ the theorem is true, i.e., the group G admits no less than the continuum of right orders or precisely 2^{k-1} right orders, and the corresponding rational series is a unique normal series of subgroups which are convex with respect to any right order introduced on the group G.

Let $n = k$ and for some element $g \in G$ the relation $g^{-1}G_{n-2}g \neq G_{n-2}$ hold. Then $g^{-1}G_{n-2}g = G_i$ for some i, $0 \leqslant i \neq n-2 \leqslant n-1$, which is impossible. Therefore, G_{n-2} is a normal subgroup of G. If the quotient group G/G_{n-2} is an orderable group, then, as in the proof of Theorem 5.2.4, we deduce that G/G_{n-2} is a split extension of a subgroup, which is isomorphic to a subgroup of the additive group $\mathbb{Q}$ and is an infinite cyclic group. Therefore, G/G_{n-2} admits an uncountable set of right orders. Consequently, the group G itself amits an uncountable set of right orders. If the quotient group G/G_{n-2} is not orderable, then G admits precisely 2^n of right orders. □

5.3. A Finite Number of Total Orders

5.3.1. As already mentioned, every torsion-free Abelian group of rank 1 admits only one total order up to the inverse one. At present, it is not known whether for each finite number n there is a group admitting precisely $2n$ total orders. For soluble groups the problem has been solved. Moreover, all soluble groups with a finite number of orders have been described. This number is a multiple of 4. Before proving this assertion, we give some results on soluble totally ordered groups and their representations by automorphisms of totally ordered sets.

Lemma 5.3.1 (Kopytov [74]). *Let (G, Ω) be a group of order automorphisms of a totally ordered set Ω such that:*

(a) *for every element $g \in G$ there is an index $\alpha \in \Omega$ such that $\beta g = \beta$ for every $\beta \in \Omega$, $\beta \geqslant \alpha$;*

(b) *for any elements $\alpha, \beta \in \Omega$ there is an element $g \in G$ such that $\alpha g > \beta$.*

Then the group G is non-Abelian.

Proof. Let $F(g)$ be the set of elements $\alpha \in \Omega$ such that $\beta g = \beta$ for any $\beta \in \Omega$, $\beta \geqslant \alpha$, and $g \in G$, $g \neq e$, $\alpha \in \Omega$, $\alpha g > \alpha$. Then there are elements $\beta \in F(g)$ and $h \in G$ such that $\alpha h^{-1} = \gamma > \beta$. For g and

h the relations $\gamma h = \alpha < \gamma$ and $\gamma g h = \gamma h = \alpha$ are fulfilled. However, $\gamma h g = \alpha g > \alpha$, i.e., $hg \neq gh$. □

Theorem 5.3.1 (Kopytov [74]). *The derived subgroup of an order-simple totally ordered group is an order-simple group.*

PROOF. Let us assume the contrary, i.e., there is an order-simple totally ordered group G, the derived subgroup G' of which has a proper convex normal subgroup H of the group G'. It is clear that the group G is not Abelian. We denote by Ω the set of all normal subgroups of the group G' that are convex with respect to the induced order. The set Ω is totally-ordered-by-inclusion. We introduce a mapping φ from the group G into the group of order automorphisms of the totally ordered set Ω as follows:

$$H \cdot (g)\varphi = g^{-1}Hg \text{ for all } H \in \Omega,\ g \in G$$

If $H \in \Omega$, then $H = \overline{H} \cap G'$ for some convex subgroup $\overline{H}$ of the group G. Hence

$$\begin{aligned} H(g)\varphi &= g^{-1}(\overline{H} \cap G')g \\ &= g^{-1}(\overline{H} \cap G')g = H_1 \end{aligned}$$

and H_1 is a convex subgroup of the group G'. It is evident that H_1 is a normal subgroup of G', i.e., φ is a mapping of Ω into Ω. It is easy to check that φ is a mapping of Ω onto Ω.

Further, if $g, h \in G$ and $H \in \Omega$, then $H \cdot (gh)\varphi = h^{-1}g^{-1}Hgh = h^{-1}(H \cdot (g)\varphi)h = (H \cdot (g)\varphi)(h)\varphi = H \cdot (g\varphi)(h\varphi)$ and φ is a homomorphism of the group G into the group of permutations of the set Ω. It is easy to see that $A \cdot (g)\varphi \subseteq B \cdot (g)\varphi$ for any subgroups A, B of Ω such that $A \subseteq B$ and $g \in G$. Consequently, φ is a homomorphism of the group G into the group of order automorphisms of the totally ordered set Ω. We note that the mapping φ does not necessarily preserve the total order introduced on G.

We now show that $(G\varphi, \Omega)$ satisfies the assumptions of Lemma 5.3.1. Let $A = \overline{A} \cap G'$ and $B = \overline{B} \cap G'$, where $\overline{A}, \overline{B}$ are convex subgroups of G and $\overline{A} \subseteq \overline{B}$. Let $\overline{A}$ be a maximal convex subgroup of the group G such that $\overline{A} \cap G' = A$. If $\overline{A}^g \subseteq \overline{B}$ for any $g \in G$, then the subgroup $\bigcup_{g \in G} \overline{A}^g$ is a proper convex normal subgroup of G, which contradicts the order simplicity of G. Hence $\overline{A}^h \supset \overline{B}$ for some $h \in G$ and $A(h)\varphi \supseteq B$, $A(h)\varphi \neq B$. Thus, condition (b) is valid for the group $(G\varphi, \Omega)$.

Further, let $g \in G$, $(g)\varphi \neq e$, and let $\overline{A} \prec \overline{B}$ be a jump of convex subgroups of G defined by the element g. If $C \in \Omega$, $C = \overline{C} \cap G' \supseteq B$ for some convex subgroup $\overline{C}$ of the group G, then $C \cdot g\varphi = C$. In view of the order simplicity of the totally ordered group G, there is a subgroup C, Ω such that $C \supset \overline{B} \cap G'$. If $C_1 \in \Omega$, $C_1 = \overline{C}_1 \cap G'$ and $\overline{C}_1 \subset \overline{B}$, then there is an element $h \in G$ such that $h^{-1}\overline{C}_1 h \supseteq \overline{B}$. Therefore, $C_1 \cdot (h)\varphi \supseteq B$. In view of the order simplicity of the group G, we obtain the inequalities $\overline{B} \cap G' \neq E$ and $\overline{B} \cap G' \neq G'$, from which we conclude that for the group $(G\varphi, \Omega)$ condition (a) is valid.

Applying Lemma 5.3.1, we deduce that $G\varphi$ is not Abelian. On the other hand, the kernel of the representation φ contains G'. The contradiction obtained proves the theorem. □

Corollary 5.3.1 (Kopytov [74]). *Any soluble non-Abelian totally ordered group G has a proper normal convex subgroup.*

PROOF. On the contrary, let G be a soluble non-Abelian totally ordered group with minimal soluble length and G have no proper normal convex subgroups. Then $G' \neq E$. By Theorem 5.3.1, G' is an order-simple totally ordered group. Since the soluble length of the group G' is strictly less than that of the group G, by Theorem 5.3.1, the group G' is Archimedean. Therefore, for any convex subgroup C of the group G we have $G' \subset C$ or $C \cap G' = E$. The first inclusion contradicts the order simplicity of the group G because any subgroup of G containing G' is normal. In the second case, by the infrainvariance of the subgroup C, the subgroup C is contained in the center of the group G, which contradicts the order simplicity of the group G. □

We note that in Corollary 5.3.1 the condition of the solubility of the group G cannot be changed by the requirement of local solubility. A pertinent example was indicated by Kopytov [74]. The requirement of total orderability of the group G cannot be replaced by that of right orderability too. Thus, the group in Example 2.1.1 is a soluble non-Abelian right-ordered group without proper convex subgroups, including proper normal subgroups.

5.3.2. In this subsection, we prove the following theorem.

Theorem 5.3.2 (Kopytov [74]). *If a soluble group G admits a finite number of total orders and is not an Abelian group of rank 1, then the number of total orders is a multiple of 4. For any positive* *integer n there is a soluble group on which the number of distinct total orders is equal to* $4n$.

PROOF. We note that if the number of total orders on the group is finite, then it is even. We consider a soluble orderable group G with a finite number of total orders. With respect to any total order the group G has only a finite number of convex normal subgroups. By Corollary 5.3.1, this number is different from zero. Let A be a minimal convex (with respect to some order) subgroup of the group G. Then the group G/A, as well as the group G, has only $2k$ orders. Moreover, the group A has a finite number of orders invariant under the action of inner automorphisms of the group G. We assume that this number is $2s$. We note that the number of distinct orders on the group G, with respect to which A is a minimal convex subgroup, is $4s$. Indeed, let $P_1, \dots, P_{2s}$ be all total orders on the subgroup A which is invariant under the action of inner automorphisms of G and $\overline{Q}_1, \dots, \overline{Q}_{2k}$ be all total orders on the group G/A. Let $Q_i = \{x \in G \setminus A \mid xA \in \overline{Q}_i\}$. It is obvious that each of the sets $Q_i \cup P_j$, $1 \leqslant i \leqslant 2k$, $1 \leqslant j \leqslant 2s$, is a total order on the group G and all of these orders are distinct. Thus, the number of total orders on the group G, with respect to which the subgroup A is convex, is a multiple of 4. Since the group G has only a finite number of minimal convex subgroups, the number of all total orders on G is a multiple of 4.

We proceed to prove the second assertion of the theorem. For an arbitrary natural number n we construct a soluble group of a finite special rank such that exactly $4n$ distinct total orders can be introduced on it. Let $f(x)$ be a polynomial with integral coefficients which is irreducible over the field of rational numbers and has precisely n distinct positive real roots. As an example, we indicate the polynomial

$$f(x) = \sum_{i=0}^{n} \gamma_i x^{n-i} = \sum_{i=0}^{n} (x - x_0)^{n-i} p(p-1)\dots(p-i+1)C_n^{n-i}$$

where p is a prime number and x_0 is an integer exceeding all the roots of the polynomial

$$x^n + pC_n^{n-1} + \dots + p(p-1)\dots(p-n+1)$$

We consider an n-dimensional vector space A over the field $\mathbb{Q}$ and a linear transformation φ of this space such that the characteristic polynomial φ coincides with $f(x)$. We fix an element $a_0 \in A$ and construct an extension G of the group A by an infinite cyclic group

(b) such that the inner automorphism of the group G defined by the element b coincides with the linear transformation φ on A. In other words, G is a set of pairs (a, b^k), where $a \in A$, $k \in \mathbb{Z}$, on which the multiplication is defined by the following rule:

$$(a, b^k) \cdot (a', b^s) = (a + a'\varphi^k, b^{k+s})$$

We show that the group G is required, i.e., precisely $4n$ distinct total orders can be introduced on G.

Let $\{\alpha_i \mid 1 \leqslant i \leqslant n\}$ be the set of all roots of the polynomial $f(x)$. By the construction of $f(x)$, every number α_i is positive and real. We denote by $\mathbb{Q}(\alpha_i)$ an algebraic extension of the field $\mathbb{Q}$ by the integer α_i. The degree of this extension over $\mathbb{Q}$ is n. Furthermore, there is an isomorphism τ_i of the vector space $\mathbb{Q}(\alpha_i)$ over $\mathbb{Q}$ onto A such that $(1)\tau_i = a_0$, $(\alpha_i\xi)\tau_i = (\xi)\tau_i\varphi$, where a_0 is a fixed element of the space A, $\xi \in \mathbb{Q}(\alpha_i)$. Since α_i is a positive real number, $\mathbb{Q}(\alpha_i)$ is a subfield of the field $\mathbb{R}$ and the total order on $\mathbb{Q}(\alpha_i)$ induced by the natural order on the field $\mathbb{R}$ is stable under multiplication by α_i. Let R_i be a positive cone corresponding to the total order on A induced by the order on the field $\mathbb{Q}(\alpha_i)$ with the help of the isomorphism τ_i. Then the set R_i is stable under the action of φ.

Using the order R_i on A, we can construct two orders P_{+i} and P_{-i} on the group G by the following rule:

$$P_{+i} = \{(a, b^k) \mid k > 0 \text{ or } k = 0, a \in R_i\}$$

$$P_{-i} = \{(a, b^k) \mid k < 0 \text{ or } k = 0, a \in R_i\}$$

All the orders $P_{\pm 1}$, $i = 1, 2, \ldots, n$, are distinct. Indeed, $P_{+i} \neq P_{-j}$ for any indices i, j, $i \neq j$, since $b \in P_{+i}$ and $b \notin P_{-j}$. Moreover, every order R_{+i}, $1 \leqslant i \leqslant n$, is Archimedean, and the inner automorphism of the group G induced by the element b acts on A as the multiplication by the positive real number α:

$$\alpha_i = \frac{((a_0)\varphi)\tau_i^{-1}}{(a_0)\tau_i^{-1}}$$

Since $\alpha_i \neq \alpha_j$, where $i \neq j$, the total orders R_i, $i = 1, 2, \ldots, n$, are also distinct. Thus, G has at least $4n$ distinct orders $P_{\pm i}^{\pm 1}$.

We now prove that any total order on the group G coincides with one of the orders $P_{\pm i}^{\pm 1}$. Let P be a total order on the group G and $a_0 > e$ with respect to the order P. The group G is soluble and has a finite special rank. Therefore, every convex subgroup of G is normal. Let H be a proper convex (with respect to P) subgroup of G.

Since there are no isolated normal subgroups of G and H is a normal subgroup of G, we have $H = A$. Thus, A is convex with respect to any total order defined on the group G. Let τ be an order isomorphism of the totally ordered group A with order $P \cap A$ into the totally ordered additive group of the field of real numbers. Then

$$(b^{-1}a_0b)\tau \cdot ((a_0)\tau)^{-1} = (b^{-1}ab)\tau \cdot ((a)\tau)^{-1} = \alpha > 0$$

for any $a \in A$. Without loss of generality, we can assume that $(a_0)\tau = 1$. It is easy to see that the set $\{(1)\tau^{-1}, (\alpha)\tau^{-1}, \dots, (\alpha^{n-1})\tau^{-1}\}$ is a basis of the vector space A over $\mathbb{Q}$. Since the characteristic polynomial of the transformation φ is

$$f(x) = \sum_{j=0}^{n} \gamma_j x^{n-j}$$

where $\gamma_j \in \mathbb{Z}$, the following equality holds:

$$b^{-n}a_0b^n + \gamma_1 b^{-n+1}a_0b^{n-1} + \dots + \gamma_n a_0 = 0$$

acting on which by the isomorphism τ, we obtain

$$\begin{aligned}(b^{-n}a_0b^n &+ \gamma_1 b^{-n+1}a_0b^{n-1} + \dots \gamma_n a_0)\tau \\ &= \alpha^n + \gamma_1\alpha^{n-1} + \dots + \gamma_n = 0\end{aligned}$$

in the field $\mathbb{R}$. Hence α is a real root of the polynomial $f(x)$ and coincides with one of the numbers α_i, $1 \leqslant i \geqslant n$. Therefore, the order $P \cap A$ coincides with one of the orders R_i and P coincides with one of the orders $P_{\pm i}$. □

5.3.3. In addition to torsion-free Abelian groups of rank 1, there are also non-Abelian groups admitting precisely two total orders. The first example of a group of such a type was given by Dlab [64]. The proof of the fact that there are subgroups of the Dlab groups (cf. 4.4) which admit precisely two total orders can be found in [37]. Another example of such a type was presented by Bludov [75].

There are also totally ordered groups admitting a countable number of distinct total orders (cf. [76]).

5.4. Center of Right-Ordered Groups

In this section, for right-orderable groups we establish the relative convexity of its subgroups, which is often considered.

Lemma 5.4.1. *Let (G, Ω) be a transitive group of order automorphisms of a totally ordered set Ω. If the stabilizer G_α of some element $\alpha \in \Omega$ is a maximal right-relatively convex subgroup of the group G, then the group (G, Ω) is o-primitive.*

PROOF. Let (G, Ω) be a transitive group of order automorphisms, and let the stabilizer G_α of an element $\alpha \in \Omega$ be a maximal right-relatively convex subgroup of G. We assume that G is not o-primitive and θ is not a nontrivial convex G-congruence on Ω. Then the set $G_{\theta_\alpha} = \{g \in G \mid \alpha g \equiv \alpha \pmod{\theta}\}$ is a subgroup of the group G; moreover, $G_{\theta_\alpha} \neq G$ and $G_\alpha \subseteq G_{\theta_\alpha}$. We show that G_{θ_α} is a right-relatively convex subgroup of the group G. To this end, we well order the set Ω:

$$\alpha = \alpha_1 \prec \alpha_2 \prec \ldots \prec \alpha_i \prec \ldots$$

On the group G we introduce a relation $<$ by the following rule: $g_1 < g_2$ if $\alpha_{i(0)} g_1 < \alpha_{i(0)} g_2$, where, with respect to the order $\prec$, the element $\alpha_{i(0)} \in \Omega$ is the first one among the elements α_i such that $\alpha_i g_1 \neq \alpha_i g_2$. It is evident that the relation $<$ is a total right order on the group G. It is easy to show that G_{θ_α} is a convex subgroup of G. Hence G_{θ_α} strictly contains G_α, which contradicts the maximality of G_α. Therefore, (G, Ω) is an o-primitive group. □

Theorem 5.4.1 (Tararin [77]). *If a normal subgroup of a right-orderable group G is isomorphic to the additive group $\mathbb{Q}$, then it is a right-relatively convex subgroup of G.*

PROOF. Let G be a right-orderable group and Q be its normal subgroup isomorphic to the additive group $\mathbb{Q}$. We denote by $Q^* = \mathrm{rc}(Q)$ a right-relatively convex subgroup of G generated by Q. We prove that $Q^* = Q$. We first show that Q is a central subgroup of the group Q^*. To this end, we introduce some right order on Q^*. Let $a \in Q$, $a > e$, $x \in Q^*$. By Corollary 5.1.7, there are natural numbers k and r such that $a^k > x^{-1} > a^{-r}$. Therefore, $x^{-1} a^r a^k x > a^k x > e$ and $x^{-1} a x > e$, $x^{-1} a x \in Q$. Consequently, the total order on Q induced by the right order on the group Q^* is invariant under conjugation by elements of the group Q^*. Thus, the natural homomorphism of the group Q^* into the group of order automorphisms of the totally ordered set Q appears. The kernel of this representation is the centralizer C of the subgroup Q of the group Q^*. By Proposition 5.1.6, C is a right-relatively convex subgroup of the group G. Moreover, $Q \subseteq C$ because Q is an Abelian group and $C = Q^*$ because $Q^* = \mathrm{rc}(Q)$ and $Q \subseteq C$. Consequently, Q is a central subgroup of the group Q^*.

If B is an arbitrary proper right-relatively convex subgroup of the group Q^*, then $B \cap Q = E$. Indeed, B is an isolated subgroup of Q^*. Hence $B \cap Q$ is an isolated subgroup of the group Q. Therefore, $B \cap Q = E$ or $Q \subseteq B$. Since $\mathrm{rc}(Q) = Q^*$ and B is a proper right-relatively convex subgroup, we obtain $Q \not\subseteq B$. Hence $B \cap Q = E$. By Proposition 5.1.7, the group Q^* has maximal right-relatively convex subgroups, and any proper right-relatively convex subgroup of the group Q^* is contained in a maximal right-relatively convex subgroup of the group Q^*.

Let H be an arbitrary maximal right-relatively convex subgroup of the group Q^*. On Q^* we fix such a right order with respect to which the subgroup H is convex. Let

$$M = \{x \in Q^* \mid p^{-1} < x < q\}$$

for any $p, q \in Q$, $p, q > e$. We prove the equality $H = M$. The inclusion $H \subseteq M$ is evident. Conversely, if $x, y \in M$, $x, y \in Q^*$, $p^{-1} < x < q$, $p^{-1} < y < q_1$ for $p, p_1, q, q_1 \in Q$, $p, p_1, q, q_1 > e$, then $p^{-1}p_1^{-1} < xy < qq_1$, because Q is contained in the center of the group Q^*. Since the group Q is divisible, the elements of the form pp_1 and qq_1 "run through" the whole set of positive elements of the group Q provided that p, p_1, q, q_1 independently "run through" the whole set of positive elements of the group Q. Hence M is closed under multiplication. The fact that M is closed under taking the inverse element is obvious. Thus, the subgroup M of the group Q^* is bounded, contains H, and is contained in a proper right-relatively convex subgroup of the group Q^* by virtue of Proposition 5.1.11. In view of the maximality of the subgroup H, the following equality is fulfilled:

$$H = \{x \in Q^* \mid p^{-1} < x < q,\ p, q \in Q,\ p, q > e\} \tag{5.4.1}$$

We show that H is a normal subgroup of the group Q^*. To this end, we denote by Ω the set $R(Q^* : H)$ of the right cosets Hg, $g \in Q^*$, of the group Q^* by H with the total order induced by the above-fixed right order on the group Q^*. Since $Q \cap H = E$, the subgroup Q is embedded into Ω. We denote by $\widetilde{Q}$ the image Q under this embedding: $\widetilde{Q} = \{\widetilde{q} \mid \widetilde{q} = Hq,\ q \in Q\}$. We first prove that the set Ω is an order completion of its subset $\widetilde{Q}$, i.e.,

$$\widetilde{x} = \sup\{\widetilde{q} \mid \widetilde{x} > \widetilde{q}, q \in Q, x < q\} = \inf\{\widetilde{q} \mid \widetilde{x} < \widetilde{q}, q \in Q, x < q\} \quad (5.4.2)$$

for any $\widetilde{x} \in \Omega$. We assume the contrary, i.e., there are elements $\widetilde{a}, \widetilde{b} \in \Omega \setminus \widetilde{Q}$ such that $\widetilde{a} < \widetilde{b}$ and

$$\widetilde{a} < \widetilde{x} < \widetilde{b} \Rightarrow \widetilde{x} \notin \widetilde{Q} \quad (5.4.3)$$

for any $\widetilde{x} \in \Omega$. This means that there is a real number α such that $\widetilde{p} < \widetilde{a}$ for any rational number $p \in \mathbb{Q}$, $p \leqslant \alpha$, and $\widetilde{b} < \widetilde{q}$ for any rational number $q \in \mathbb{Q}$, $q > \alpha$. Let a and b be preimages in Q^* of the cosets $\widetilde{a}$ and $\widetilde{b}$ respectively: $\widetilde{a} = Ha$ and $\widetilde{b} = Hb$. Then the condition $\widetilde{a} < \widetilde{b}$ guarantees the validity of the relation

$$ba^{-1} \notin H \quad (5.4.4)$$

and the condition (5.4.3) assumes the form

$$p < a < b < q \text{ if } p, q \in Q, \quad (5.4.5)$$

Multiplying the latter inequality from the right by a^{-1}, we obtain $pa^{-1} < e < ba^{-1} < qa_{-1}$. Elements of the group Q commute with those of the group Q^*, and the inequalities $p < a < q$ hold. Therefore, $q^{-1} < a^{-1} < p^{-1}$. Hence $e < ba^{-1} < qa^{-1} = a^{-1}q < p^{-1}q$ for any elements $p, q \in Q$ such that $p \leqslant \alpha \leqslant q$. Since, for any $r \in Q$, there are elements $p, q \in Q$ such that $p \leqslant \alpha \leqslant q$ and $p^{-1}q < r$, we have $e < ba^{-1} < r$ for any $r \in Q$. In view of (5.4.1), we have $ba^{-1} \in H$, which contradicts the relation (5.4.4). The contradiction obtained proves that Eqs. (5.4.2) hold in Ω.

As usual, we denote by R the right regular representation of the group Q^* by automorphisms of a totally ordered set Ω. Since Q is contained in the center of the group Q^*, we have $\widetilde{p}R(h) = Hph = Hp$ for any $\widetilde{p} \in \widetilde{Q}$ and $h \in H$, i.e., the subgroup $R(H)$ is contained in the stabilizer of an element $\widetilde{p} \in \widetilde{Q}$. By Eqs. (5.4.2), the subgroup $R(H)$ coincides with the kernel of the representation R of the group Q^* by automorphisms of the totally ordered set Ω. Hence $R(H)$ is a normal subgroup of the group Q^*.

Since H is a maximal right-relatively convex subgroup of Q^*, the quotient group Q^*/H is a right-ordered group without nontrivial right-relatively convex subgroups and, by Proposition 5.1.9, Q^*/H is isomorphic to a torsion-free Abelian group of rank 1. Thus, $Q^* = Q \cdot H$. From what has been proved, it follows that $Q \cap H = E$ and that Q is contained in the center of the group Q^*. Therefore, $Q^* = Q \times H$, and the group Q is a right-relatively convex subgroup of the group Q^* and of the group G, i.e., $Q^* = Q$. □

Theorem 5.4.2 (Tararin [77]). *A divisible subgroup of the center of a right-orderable group is right-relatively convex.*

PROOF. Let A be a divisible subgroup of the center of a right-orderable group G. We assume that A is the direct product of the groups Q_α, $\alpha \in I$, that are isomorphic to the additive group $\mathbb{Q}$. We can also assume that the set of indices I is well-ordered. Taking into account the fact that the subgroup A of G is central and using Theorem 5.4.1, we see that every Q_α is a right-relatively convex subgroup of the group G for any index $\alpha \in I$. We assume that the assertion of the theorem has been proved for the groups $A_\beta = \prod_{\gamma \leqslant \beta} Q_\gamma$ and all indices $\beta < \alpha$. If α is a limit ordinal, then A_α is the union of the increasing sequence of the right-relatively convex subgroups A_β, $\beta < \alpha$. By Proposition 5.1.7, the subgroup A_α is also a right-relatively convex subgroup of G. If α is not a limit ordinal, then A_α is the direct product of the groups $A_{\alpha-1}$ and Q_α. By the induction assumption, $A_{\alpha-1}$ is a right-relatively convex subgroup of the group G. The quotient group $G/A_{\alpha-1}$ is right-orderable and contains a divisible central subgroup $\overline{Q}_\alpha = A_{\alpha-1}Q_\alpha/A_\alpha$. By Theorem 5.4.1, $\overline{Q}_\alpha$ is a right-relatively convex subgroup of the group G/A_α. Therefore, A_α is a right-relatively convex subgroup of the group $G/A_{\alpha-1}$.

Thus, all subgroups A_α are right-relatively convex. The subgroup A, being the union of the totally-ordered-by-inclusion right-relatively convex subgroups A_α, is right-relatively convex. □

Theorem 5.4.3 (Tararin [77]). *For any right-orderable group G, its normal divisible nilpotent subgroup of finite rank is right-relatively convex.*

PROOF. Let G be a right-orderable group and F be its divisible normal nilpotent subgroup of finite special rank n such that $F^* = \mathrm{rc}(F)$. We show that $F = F^*$. On G we fix a right order with respect to which F^* is convex. We proceed by induction on rational rank n. If $n = 1$, then F is a normal divisible subgroup of G which is isomorphic to the additive group $\mathbb{Q}$. By Theorem 5.4.1, F is a right-relatively convex subgroup of the group G.

We assume that the assertion of the theorem is true for every right-orderable group and each of its normal divisible nilpotent subgroups of rank at most n. By Theorem 2.4.2, the fixed right order on the group G induces a Conradian right order on F^*. Since the rank of F is finite, F has the greatest convex subgroup F_1. It is clear that F_1 is a divisible normal subgroup of F and its rank is at most n; moreover, the quotient group F/F_1 is an Archimedean totally ordered group.

Taking into account the unboundedness of F in F^* and the fact that the quotient group F/F_1 is Archimedean, for any $a \in F \setminus F_1$, $a > e$, and $g \in F^*$ we conclude that there are natural numbers k and r such that $a^k > g^{-1} > a^{-r}$. Hence $g^{-1}a^r a^k g > e$, which implies $g^{-1}ag > e$. Further, the inclusion $g^{-1}ag \notin F_1$ holds. Indeed, if $g^{-1}ag \in F_1$, then $(g^{-1}ag)^t = g^{-1}a^t g \in F_1$ for any natural number t. Consequently, $a > g^{-1}a^t g$, $ag^{-1}a^{-t}g > e$, and $ag^{-1}a^{-t}g \notin F_1$. By the above, $g(ag^{-1}a^{-t}g)g^{-1} = gag^{-1}a^t$ and $gag^{-1} > a^t$ for any natural number t, which contradicts the fact that the quotient group F/F_1 is Archimedean. Thus, F_1 is a normal subgroup of the group F^*. If $F_1 \neq E$, then, by the induction assumption, the subgroups F_1 and F/F_1 are right-relatively convex subgroups of the groups F^* and F^*/F_1 respectively. Therefore, F is a right-relatively convex subgroup of the group F^*. Hence $F = F^*$. We now set $F_1 = E$. Then F is a normal divisible Archimedean subgroup of the group F^*. In this case, F is a subgroup of the center of the group F^*. Indeed, since F is Archimedean for any $a \in F$ and $g \in F^*$, there are natural numbers k and r such that $a^k > g^{-1} > a^{-r}$. Consequently, $g^{-1}a^r a^k g > e$ and $g^{-1}ag > e$, $g^{-1}ag \in F$. Thus, the total order on F is invariant under conjugations by the elements of F^*. But in this case the centralizer C of the subgroup F is a right-relatively convex subgroup of the group F^*. Taking into account the commutativity of the group F and the equality $\mathrm{rc}(F) = F^*$, we obtain $C = F^*$. Therefore, F is a divisible subgroup of the center of the group F^*. By Theorem 5.4.2, F is a right-relatively convex subgroup of F^* and, consequently, a right-relatively convex subgroup of G. □

We now give an example that demonstrates the significance of the condition of the divisibility of the center in Theorem 5.4.2.

Example 5.4.1 (Mura and Rhemtulla [42]). It is easy to check that the group

$$G = \mathrm{gp}(x, y, z \mid x^2y^{-1}x^2y = y^2x^{-1}y^2x = z, \quad [x, z] = [y, z] = e)$$

is right-orderable, its center coincides with the infinite cyclic subgroup (z), and the quotient group $G/(z)$ is a torsion-free group and is not right-orderable by Theorem 3.2.2.

5.5. Criteria for Orderability

We begin with the definition of a relatively convex subgroup.

- A subgroup of an orderable group is called *relatively convex* if it is convex with respect to some total order on the group.

The notion of relatively convex subgroups of a group is close to the criterion for the orderability of the group. In particular, if a normal subgroup of a group is relatively convex, then the quotient group by this subgroup is orderable. Thus, in order to find a universal criterion for the relative convexity of a subgroup, one can start from necessary and sufficient conditions of the orderability of a group.

Theorem 5.5.1 (Kokorin and Kopytov [48]). *A subgroup of an orderable group G is relatively convex if and only if it is infrainvariant and for any set of elements $g_1, \ldots, g_n \in G \setminus H$ it is possible to select numbers $\varepsilon_1, \ldots, \varepsilon_n$ equal to ± 1 such that the set $S(g_1^{\varepsilon_1}, \ldots, g_n^{\varepsilon_n}) \setminus H$ is a semigroup.*

PROOF. Let H be a convex subgroup of a totally ordered group G. Then for any set of elements g_i, $1 \leqslant i \leqslant n$, we choose numbers ε_i subject to the condition $g_i^{\varepsilon_i} > e$. Let the subgroup H satisfy the hypotheses of the theorem. Considering the system of all subgroups conjugate to H, and joining to it the unions, intersections, the identity subgroup E, and the group G, we arrive at the system $\mathfrak{C}(G)$:

$$E \subset \overline{H} \subset \ldots \subset H_\alpha \subset H_{\alpha+1} \subset \ldots \widetilde{H} \subset G$$

where $\overline{H} = \bigcap_{g \in G} H^g$ and $\widetilde{H} = \bigcup_{g \in G} H^g$. In view of the infrainvariance of the subgroup H and the fact that the union and intersection of totally-ordered-by-inclusion subgroups are subgroups, we conclude that the subgroups of $\mathfrak{C}(G)$ satisfy the hypotheses of the theorem; moreover, the subgroups $\overline{H}$ and $\widetilde{H}$ are normal subgroups of the group G. To prove this assertion, for any g_i (or its conjugate) we can choose numbers $\varepsilon_i \pm 1$ the same as those that were chosen for H. If $H_\alpha \subset H_{\alpha+1}$ is a jump in $\mathfrak{C}(G)$, then the group $H_{\alpha+1}/H_\alpha$ is $\boldsymbol{n}(H_\alpha)/H_\alpha$-orderable, since for the group $H_{\alpha+1}/H_\alpha$ the hypotheses of the theorem become the orderability condition indicated in Theorem 3.3.2. Thus, the system $\mathfrak{C}(G)$ satisfies the hypotheses of Theorem 3.3.2. Applying this theorem, we complete the proof. □

Lemma 5.5.1 (Kokorin and Kopytov [48]). *Let M be a set of relatively convex totally-ordered-by-inclusion subgroups and their conjugate subgroups. Then the union and intersection of elements of M are relatively convex subgroups.*

PROOF. For normal subgroups subject to the assumptions of the lemma, the assertion is evident. Let $\mathfrak{C}(G)$ be the following system of subgroups of G:

$$E \subset \overline{H} \subset \ldots \subset H_\alpha \subset H_{\alpha+1} \subset \ldots \widetilde{H} \subset G$$

which consists of:

— a given set M of relatively convex subgroups,

— the set M' of thc unions and intersections of any family of subgroups of M,

— the set M'' of the unions and intersections of any families of subgroups of M' (in particular, $\overline{H}$ and $\widetilde{H}$ are the intersection and union of all subgroups of M'),

— the identity subgroup E,

— the whole group G.

It suffices to show that the system $\mathfrak{C}(G)$ satisfies the hypotheses of Theorem 3.3.2 because it is infrainvariant by definition. Indeed, let $H_\alpha \subset H_{\alpha+1}$ be a jump in $\mathfrak{C}(G)$. The subgroup H_α is the union of the subgroups H_γ, $H_\gamma \subseteq H_\alpha$, $H_\alpha \in M'$. If $H_\gamma \subseteq H_\alpha$, $H_\gamma \in M'$, and $x \in N(H_\alpha)$, then $x^{-1}H_\gamma x \subseteq H_\gamma$ and $x^{-1}H_\gamma x \in M'$. For any subgroup $H_\gamma \subseteq H_\alpha$ the subgroup $\overline{H}_\gamma = \bigcup_{x \in N(H_\alpha)} x^{-1}H_\gamma x$ is normal and relatively convex in $\boldsymbol{n}(H_\alpha)$; moreover, it is contained in $H_\alpha = \bigcup_{H_\gamma \subseteq H_\alpha} \overline{H}_\gamma$. But, as already mentioned, H_α is a relatively convex subgroup of $\boldsymbol{n}(H_\alpha)$. Therefore, the group $\boldsymbol{n}(H_\alpha)/H_\alpha$ is orderable. The case of the jumps $E \subset \overline{H}$ and $\widetilde{H} \subset G$ can be handled in a similar way. The application of Theorem 3.3.2 completes the proof. □

Theorem 5.5.2 (Kokorin [31]). *The intersection of relatively convex subgroups of an orderable group is relatively convex if and only if it is infrainvariant.*

PROOF. The necessity is obvious. Let $\{A_\gamma \mid \gamma \in I\}$ be a set of relatively convex subgroups of an orderable group G, and let the intersection

$$H = \bigcap_{\gamma \in I} A_\gamma$$

be infrainvariant.

If all subgroups A_γ are normal, the assertion of the theorem follows from the fact that a group which is approximated by orderable groups is also orderable.

In the general case, we consider the system $\mathfrak{C}(G)$ containing E, G, the subgroups $\{g^{-1}Hg \mid g \in G\}$, and all possible intersections and unions of subgroups from $\{g^{-1}Hg \mid g \in G\}$. We show that this system satisfies all the hypotheses of Theorem 3.3.2.

Let $H_\alpha \subset H_{\alpha+1}$ be a jump in $\mathfrak{C}(G)$. The subgroup H_α coincides with the union of all subgroups of the form $g^{-1}Hg$, where $g^{-1}Hg = \bigcap_{\gamma \in I} g^{-1}A_\gamma g$. We denote by $\widetilde{A}_\alpha$ the union of those subgroups $g^{-1}A_\gamma g$ for which $g^{-1}Hg \subseteq H_\alpha$. Then the subgroups $\widetilde{A}_\alpha \cap N(H_\alpha)$ are normal in the normalizer $\boldsymbol{n}(H_\alpha)$ of the subgroup H_α and relatively convex in $\boldsymbol{n}(H_\alpha)$ by Lemma 5.5.1. As was mentioned above, their intersection is equal to H_α and is relatively convex in $N(H_\alpha)$. Therefore, the group $H_{\alpha+1}/H_\alpha$ is $\boldsymbol{n}(H_\alpha)/H_\alpha$-orderable. The application of Theorem 3.3.2 to the system $\mathfrak{C}(G)$ completes the proof. □

Corollary 5.5.1. *The intersection of a totally-ordered-by-inclusion system of relatively convex subgroups of an orderable group is a relatively convex subgroup.*

The intersection of relatively convex subgroups unlike that of right-relatively convex subgroups is not always a relatively convex subgroup. Let $G = (a)\mathrm{wr}(b)$ be the wreath product of the infinite cyclic groups and $b^{-1}a_n b = a_{n+1}$. Every element $g \in G$ can be written uniquely in the form

$$g = b^k \cdot a_{n_1}^{k_1} a_{n_2}^{k_2} \dots a_{n_m}^{k_m} \tag{5.5.1}$$

where $k, k_1, \dots, k_m, n_1, \dots, n_m \in \mathbb{Z}$, $n_1 < n_2 < \dots < n_m$.

On G we consider total orders P and P' defined by the following rule:

P consists of elements of the form (5.5.1) with $k_m > 0$

P' consists of elements of the form (5.5.1) with $k_1 > 0$.

We consider the subgroup

$$A = \prod_{n=-\infty}^{0} (a_n)$$

which is convex with respect to the order P and the subgroup

$$A' = \prod_{n=0}^{\infty} (a_n)$$

which is convex with respect to the order P'. It is clear that their intersection $A \cap A'$ equal to the cyclic subgroup (a_0) is not an infrainvariant subgroup and therefore cannot be a relatively convex subgroup.

Theorem 5.5.3 (Kokorin and Kopytov [48]). *The union of a totally-ordered-by-inclusion system of relatively convex subgroups of an orderable group is relatively convex.*

PROOF. Let H be the union of the system of subgroups A_γ, $\gamma \in I$, satisfying the hypotheses of the theorem and $\mathfrak{C}(G)$ be the system of subgroups constructed by means of H as in the proof of Theorem 5.5.2. Let $H_\alpha \subset H_{\alpha+1}$ be a jump in $\mathfrak{C}(G)$. The union $\widetilde{A}_\gamma$ of all subgroups that are conjugate to A_γ and contained in H_α is a normal subgroup in the normalizer $\boldsymbol{n}(H_\alpha)$ and is a relatively convex subgroup in $\boldsymbol{n}(H_\alpha)$ by Lemma 5.5.1. Therefore, by Lemma 5.5.1 and the fact that H_α is the union of normal-in-$\boldsymbol{n}(H_\alpha)$ relatively convex subgroups $\widetilde{A}_\gamma$, we conclude that H_α is relatively convex in $\boldsymbol{n}(H_\alpha)$. The application of Theorem 3.3.2 completes the proof. □

The derived subgroup of an orderable group cannot be a relatively convex subgroup, since there is an orderable (moreover, nilpotent) group whose derived subgroup is not isolated. It is evident that the isolator of the derived subgroup is a relatively convex subgroup. The normalizer and centralizer of a relatively convex subgroup are not always relatively convex subgroups. In the lattice of subgroups of an orderable group, the join of two normal relatively convex subgroups is not always relatively convex, even if the join is a strongly isolated subgroup (cf. [48]).

5.6. Centers of "Small" Subgroups

5.6.1. To prove the principal statements of this section, we need some auxiliary results concerning the problem of the completion of a totally ordered group. These assertions can be proved by direct verification.

By a *completion* of a torsion-free Abelian or nilpotent group A we mean the smallest divisible group A^* containing A, i.e., the group such that for every one of its elements a and any natural number n the equation $x^n = a$ has a solution.

Lemma 5.6.1. *Let A be an Abelian orderable group and A^* be its completion. Then every total order on the group A can be uniquely extended to a total order on A^*. Every order automorphism φ of the totally ordered group A can be uniquely extended to an order automorphism φ^* of the group A^*.*

Lemma 5.6.2. *For any elements v and w of a totally ordered group G and a natural number k, the following implications hold:*

$$vw \leqslant wv \Rightarrow v^k w^k \leqslant (vw)^k \leqslant w^k v^k$$

$$vw > e \Rightarrow v^k w^k > e$$

$$v^k w^k > e \Rightarrow vw > e$$

Let G be a totally ordered group and A be its normal Abelian isolated subgroup. We denote by A^* the completion of the group A. We regard the group G as an extension of the group A by the quotient group G/H, corresponding to the system of Schreier factors M, the system of representatives

$$T = \{g_\alpha \mid \alpha \in G/H, g_\alpha \in G\}$$

and the set of automorphisms

$$\Phi = \{\varphi_g \mid g \in G, a\varphi_g = g^{-1}ag, a \in A\}$$

We denote by G^* the Schreier extension of A^* by the quotient group G/H, corresponding to the same system of Schreier factors M and the system of representatives T. As Φ we take the set of those automorphisms $\{\varphi_g^*\}$ of the group A^* which are the extensions of the automorphisms $\{\varphi_g\}$ of the group A (cf. [3, Ch. 15] or [45]). It is evident that the group G is a subgroup of the group G^*; moreover,

$$G^* = G \cdot A^*, \quad A = G \cap A^*, \quad G^*/A^* \cong G/A$$

For $a \in A^*$ we denote by $n(a)$ the smallest natural number such that $a^{n(a)} \in A$. Any element $x \in G^*$ can be represented (in general, not uniquely) in the form $x = ga$, where $g \in G$, $a \in A^*$. Let $\geqslant$ be a total order on the group G. We introduce the relation $\geqslant'$ on the group G^* as follows: $x = ga >' e$ if $g^n a^n > e$ in the group G, where $n = n(a)$.

Lemma 5.6.3. *The definition of the relation $\geqslant'$ on G^* is correct.*

PROOF. We assume that in G^* the inequality $x = ga = g'a'$ holds for some $g, g' \in G$ and $a, a' \in A^*$. From the definition of the group G^* it follows that $g' = gb$ and $a' = b^{-1}a$ for some $b \in A$. Let $ga >' e$ in the group G^*. Then $g^n a^n > e$ in G, where $n = n(a)$. Since $b \in A$, we have $n(b^{-1}a) = n(a)$. We assume that $gb \geqslant bg$ in the totally ordered group G. By Lemma 5.6.2,

$$(gb)^n b^{-n} a^n \geqslant b^n g^n$$

$$b^{-n} a^n (gb)^n \geqslant b^{-n} a^n b^n g^n = a^n g^n > e$$

in G. Consequently, $(gb)^n b^{-n} a^n > e$. Hence $g'a' >' e$ in G^*. The case $gb < bg$ in G is considered in a similar way, which completes the proof of the lemma. □

By Lemma 5.6.2, it also follows that if $ga >' e$ in the group G^* and $a^n \in A$, $a^m \in A$ for some positive integers n, m, then $g^m a^m > e$ and $g^n a^n > e$ in the group G.

Lemma 5.6.4. *In the group G^* the inequality $(ga)^{-1} >' e$ is valid if and only if $g^n a^n < e$ in the totally ordered group G.*

PROOF. Note that for any automorphism φ of the group A the equality

$$(\varphi_g^*(a))^n = ((a)^n)\varphi_g$$

holds in A^*, i.e., $(\varphi_g^*(a))^n \in A$. Therefore, $n(a) = n(\varphi_g^*(a))$. Since

$$(ga)^{-1} = g^{-1}\varphi_{g^{-1}}^*(a^{-1})$$

the inequality $(ga)^{-1} >' e$ is fulfilled if and only if $g^{-n}(\varphi_{g^{-1}}^*(a^{-1})^n) > e$ in the group G. On the other hand,

$$g^{-n}(\varphi_{g^{-1}}^*(a^{-1})^n) = g^{-n} g a^n g^{-1} = g^{-(n-1)}(g^n a^n)^{-1} g^{n-1}$$

Hence $(ga)^{-1} >' e$ in G^* if and only if in the group G the inequality $g^n a^n > e$ is fulfilled. □

Lemma 5.6.5. *The group G^* is a totally ordered group with order $\geqslant'$.*

Proof. It suffices to verify the relations O.1, O.2, O.5, and O.6 (cf. Chapter 1) for the group G^*.

The purity and linearity of the order $\geqslant'$ follow from Lemma 5.6.4.

Let in the group G^* the following inequalities hold: $ga >' e$ and $hb >' e$. Since $a^{n(a)n(b)} \in A$ and $b^{n(a)n(b)} \in A$, we can assume that $n = n(a) = n(b)$. By definition, the inequality $gahb >' e$ is valid in the group G^* if and only if the inequality $(gh)^n(h^{-1}a^nh)b^n > e$ is fulfilled in the group G. We now assume that in the group G the inequality $gh \geqslant hg$ holds. Conjugating it by the element h, we obtain $(h^{-1}gh)h \geqslant h(h^{-1}gh)$. Hence

$$(gh)^n = (h(h^{-1}gh))^n \geqslant h^n(h^{-1}gh)^n = h^n(h^{-1}g^nh)$$

Consequently, in the group G we have

$$\begin{aligned}(gh)^n(h^{-1}g^mh)b^n &\geqslant h^n(h^{-1}g^nh)(h^{-1}a^nh)b^n \\ &= b^n((b^nh^n)h^{-1}(g^na^n)h)b^n > e\end{aligned}$$

since $b^nh^n > e$ and $g^na^n > e$.

For $gh < hg$ in the group G the following relations hold:

$$\begin{aligned}(gh)^n(h^{-1}g^mh)b^n &> (h^{-1}g^nh)h^nb^n(h^{-1}a^nh) \\ &= h^{-1}a^{-n}h(h^{-1}(a^ng^n)h(h^nb^n))h^{-1}a^nh > e\end{aligned}$$

Thus, we have proved that the set of positive elements of the group G^* is a semigroup. We now show that this semigroup is normal. Let $ga > e$ in G^* and $hb \in G^*$. It suffices to show the stability of the relation $\geqslant'$ under conjugation by the elements $h \in G$ and $a \in A^*$. It is clear that $h \in G$ satisfies the following equality:

$$h^{-1}gah = (h^{-1}gh)\varphi_h^*(a)$$

Since $n = n(a) = n(\varphi_h^*(a))$, the inequality

$$(h^{-1}gh)\varphi_h^*(a) >' e$$

holds in the group G^* if and only if the following inequality is fulfilled:

$$h^{-1}g^nh(\varphi_h(a^n)) > e$$

On the other hand,

$$h^{-1}g^nh(\varphi_h(a^n)) = h^{-1}(g^na^n)h > e$$

in G. Therefore, $h^{-1}gah > e$. For $b \in A^*$ we have

$$b^{-1}gab = g(\varphi_g^*(b))^{-1}ab$$

Setting $n = n(a) = n(b)$, for simplicity, we obtain the inequality

$$g(\varphi_g^*(b))^{-1}ab > e$$

which is valid if and only if $g^n(g^{-1}b^{-n}g)a^n b^n > e$ or, equivalently, if $a^n g^n[g, b^n] > e$ in G. If $[g, b^n] \geqslant e$ in G, then $a^n g^n[g, b^n] \geqslant a^n g^n > e$. Consequently, the inequality $[g, b^n] < e$ is valid in G. Then

$$[g, b^n] < e \Rightarrow b^{-n}gb^n < g \Rightarrow b^{-n}g^{n-1}b^n = (b^{-n}gb^n)^{n-1} \leqslant g^{n-1}$$

This means that the order $\geqslant'$ on the group G^* is stable under the inner automorphisms of the group G^*. □

Lemmas 5.6.1–5.6.5 lead to the following theorem.

Theorem 5.6.1 (Fox [78]). *An orderable group G with a normal Abelian subgroup A is isomorphically embedded into a group G^* with a normal Abelian subgroup A^* such that the following conditions hold:*

— *the subgroup A^* of the group G^* contains a subgroup A of the group G,*

— $G^* = G \cdot A^*, \quad A = G \cap A^*, \quad G^*/A^* \cong G/A,$

— *any total order on the group G is extended uniquely to a total order on the group G^*.*

It is worth pointing out that a more general result for normal locally nilpotent subgroups was proved by Gurchenkov [79].

Corollary 5.6.1 (Kopytov [80]). *Any orderable group G is isomorphically embedded into a group G^* with divisible center Z^* such that the following assertions hold:*

— *the center Z^* of the group G^* contains the center Z of the group G;*

— $G^* = GZ^*, \quad Z = G \cap Z^*, \quad G^*/Z^* \cong G/Z;$

— *any total order on the group G can be extended uniquely to a total order on the group G^*.*

Lemma 5.6.6 (Kokorin [47]). *The product of a divisible subgroup of the center of an orderable group and its relatively convex subgroup is relatively convex.*

PROOF. Let Z^* be a divisible subgroup of the center of an orderable group G and H be a relatively convex subgroup of the group G. We take a total order on G with respect to which H is convex. Let $\mathfrak{C}(G) = \{H_\gamma \mid \gamma \in \Gamma\}$ be the system of convex subgroups of G with this order. We consider the system $\mathfrak{C}(G)'$ consisting of the products $H_\gamma Z^*$ of convex subgroups from $\mathfrak{C}(G)$ and Z^*. It suffices to show that the system $\mathfrak{C}(G)'$ satisfies the hypotheses of Theorem 3.3.1. We restrict ourselves to the proof of the strong isolation of the subgroups $H_\gamma Z^*$. The rest of the assertions are true because they are valid for $\mathfrak{C}(G)$.

Let

$$x \cdot g_1^{-1} x g_1 \cdot g_2^{-1} x g_2 \cdots g_n^{-1} x g_n = x^{e+g_1+\dots+g_n} = hz$$

where $h \in H_\gamma$, $z \in Z^*$, $z^{n+1} = z$, $z_1 \in Z$. Then $(xz_1^{-1})^{e+g_1+\dots+g_n} = h$. In view of the strong isolation of the subgroup H_γ, we have $xz^{-1} \in H_\gamma$. Hence $x \in H_\gamma Z^*$, which proves the strong isolation of the subgroup $H_\gamma Z^*$. □

Lemma 5.6.6 and Theorem 5.6.1 imply the following theorem.

Theorem 5.6.2 (Kokorin [47] and Kopytov [80]). *The quotient group of an orderable (fully orderable) group by its center is orderable (fully orderable). Any term of the upper central series of an orderable group is a relatively convex subgroup of this group.*

5.6.2. For an arbitrary orderable group we consider the problem of the relative convexity of a normal isolated soluble subgroup of finite special rank.

Lemma 5.6.7. *Let G be an orderable group and A be its isolated Abelian normal subgroup of finite rank which is Archimedean ordered with respect to some induced total order. Then A is a relatively convex subgroup of the group G.*

PROOF. We denote by $C = c(A)$ the centralizer of a subgroup A of a group G subject to the hypotheses of the lemma. Since A is a normal subgroup of G, we conclude that C is a normal subgroup of G. The quotient group G/C is isomorphic to a subgroup of order automorphisms of the Archimedean ordered group A. Therefore, G/C is a torsion-free Abelian group. Hence C is a relatively convex subgroup of the group G. By Theorem 3.3.1, there exists a system $\mathfrak{C}(G)$ of convex subgroups of G which contains the subgroup C:

$$G \supset \ldots \supset C \supset \overline{C}_\alpha \supset C_\alpha \supset \ldots \supset E$$

In view of Theorem 5.6.1, we can assume that A is a divisible Abelian group. We note that A is a strongly isolated subgroup of G. Indeed, let

$$x^{g_1+\ldots+g_n} = x^{g_1} \cdot x^{g_2} \ldots x^{g_n} = a \in A$$

It is clear that the mapping φ: $y \to y^{g_1+\ldots+g_n}$ is an order monomorphism of A. Since A is a divisible Abelian group of finite rank, it is easy to see that $A = \varphi(A)$ and φ are order automorphisms of the group A. Therefore, there is an element $y \in A$ such that $a = y^{g_1+\ldots+g_n}$. The quotient group G/C is orderable and $A \subseteq C$. Hence $x \in C$. Consequently,

$$x^{g_1+\ldots+g_n} = y^{g_1+\ldots+g_n}$$

By the commutativity of the elements x^{g_i} and y^{g_j}, we obtain the equality

$$(xy^{-1})^{g_1+\ldots+g_n} = e$$

and deduce that $xy^{-1} = e$ and $x = y = e \in A$.

We now construct a new soluble infrainvariant system

$$\mathfrak{C}'(G)\colon G \supset \ldots \supset C \supset \overline{C}_\alpha A \supset C_\alpha A \supset \ldots \supset A \supset E$$

We show that the system $\mathfrak{C}'(G)$ satisfies all the hypotheses of Theorem 3.3.1. To this end, we consider an arbitrary jump $\overline{C}_\alpha A \supset C_\alpha A$ in the system $\mathfrak{C}'(G)$ such that $\overline{C}_\alpha A$ is contained in C. We note that the normalizers $\boldsymbol{n}_G(C_\alpha)$ and $\boldsymbol{n}_G(C_\alpha A)$ of the subgroups C_α and $C_\alpha A$ coincide. For an arbitrary element $g \in \boldsymbol{n}_G(C_\alpha A) \setminus \boldsymbol{n}_G(C_\alpha)$ only one of the following inclusions holds:

$$C_\alpha^g \supset C_\alpha, \quad C_\alpha^g \subset C_\alpha$$

For convenience, we set $C_\alpha^g \supset C_\alpha$. Conjugating the last inclusion by the element g a sufficient number of times, we obtain the following chain of subgroups:

$$\ldots C_\alpha^{g^n} \supset C_\alpha^{g^{n-1}} \supset \ldots \supset C_\alpha \qquad (5.6.1)$$

On the other hand, in view of the equalities $(C_\alpha A)^g = C_\alpha^g A = C_\alpha A$, we have

$$(C_\alpha)^{g^n} A = \ldots = C_\alpha^g A = C_\alpha A$$

Taking the intersection of the subgroups of the system (5.6.1) with the subgroup A, we arrive at a system of isolated subgroups of the group A:

$$(C_\alpha \cap A) \subseteq (C_\alpha^g \cap A) \subseteq \ldots (C_\alpha^{g^n} \cap A) \subseteq \ldots \tag{5.6.2}$$

Since every C_α is a relatively convex subgroup of G and the rank of the group A is finite, every $C_\alpha^{g^k} \cap A$ is an isolated and normal subgroup of A. Consequently, the system (5.6.2) is stabilized at some finite step, i.e., there is a natural number n such that

$$C_\alpha^{g^n} \cap A = C_\alpha^{g^{n+1}} \cap A = C_\alpha^{g^{n+t}} \cap A = \ldots$$

for any natural t.

We show that, in fact, for any natural number k the following equality holds:

$$C_\alpha^{g^k} \cap C_\alpha^{g^{k-1}} A = C_\alpha^{g^{k-1}} (C_\alpha^{g^k} \cap A)$$

Indeed, let $x = yz$, where $x \in C_\alpha^{g^k}$, $y \in C_\alpha^{g^{k-1}}$, $z \in A$. Then $z = y^{-1}x \in C_\alpha^{g^k} \cap A$. Hence $x \in C_\alpha^{g^{k-1}}(C_\alpha^{g^k} \cap A)$. Furthermore,

$$\begin{aligned} C_\alpha^{g^{n+1}} &= C_\alpha^{g^{n+1}} \cap C_\alpha^{g^{n+1}} A = C_\alpha^{g^{n+1}} \cap C_\alpha^{g^n} A \\ &= C_\alpha^{g^n} (C_\alpha^{g^{n+1}} \cap A) = C_\alpha^{g^n} (C_\alpha^{g^n} \cap A) = C_\alpha^{g^n} \end{aligned}$$

which is impossible.

We show that the subgroup $C_\alpha A$ is a strongly isolated subgroup of $\boldsymbol{n}_G(C_\alpha A)$. Indeed, if $x^{g_1+\cdots+g_n} = cb$ for some elements $g_1, \ldots g_n, x$ of $\boldsymbol{n}_G(C_\alpha A)$, we take an element $a \in A$ such that $b = a^{g_1+\cdots+g_n}$. Since a and x commute, we have $x \in C \cap N_G(C_\alpha)$ and $(xa^{-1})^{g_1+\cdots+g_n} = c \in C_\alpha$. We note that if $x \notin C$, the equality $x^{g_1+\cdots+g_n} = cb$ is impossible by the orderability of the quotient group G/C. However, $xa^{-1} \in \boldsymbol{n}_G(C_\alpha)$, which contradicts the fact that C_α is a strong isolated subgroup of $N_G(C_\alpha)$. It is evident that the inner automorphisms of $\boldsymbol{n}_G(C_\alpha A)/C_\alpha A$ commute on the quotient group $\overline{C}_\alpha A/C_\alpha A$. □

Theorem 5.6.3 (Medvedev [81]). *Let an isolated normal subgroup H of an orderable group G be a soluble group of a finite rank. Then H is a relatively convex subgroup of the group G.*

PROOF. We use induction on the rank of H. If the rank of H is 1, then H is Abelian and Archimedean ordered with respect to any total order. By Lemma 5.6.7, H is a relatively convex subgroup. We now assume that the rank of H is equal to r. Let $\leqslant$ be some total order on the group G. Then every subgroup H which is convex with respect to the induced order $\leqslant$ is a normal and isolated subgroup of the group G. Since the rank of H is finite, there is the smallest nonidentity convex subgroup A_1 of H. The convex subgroup A_1 is Archimedean ordered and satisfies the hypotheses of Lemma 5.6.7. Thus, the quotient group G/A_1 is orderable and the subgroup H/A_1 is normal, isolated, and soluble; moreover, the rank of H/A_1 is strictly less than r. By the induction assumption, H/A_1 is a relatively convex subgroup of G/A_1. Therefore, the quotient group $(G/A_1)/(H/A_1)$ is orderable. By the theorem on homomorphisms of groups, G/H is isomorphic to $(G/A_1)/(H/A_1)$. Hence the quotient group G/H is orderable and H is a relatively convex subgroup of the group G. □

Corollary 5.6.2 (Mura and Rhemtulla [82]). *A subgroup of an orderable soluble group of finite rank is relatively convex if and only if it is normal and isolated.*

The following example shows that the condition of finiteness of the rank in Theorem 5.6.3 is essential.

EXAMPLE 5.6.1. Let $G = \text{gp}\ (a, b \mid b^{-1}ab = a^{-1})$ be a group with generators a, b and the relation $b^{-1}ab = a^{-1}$. Let also F/A be isomorphic to G, where F is a free group of rank 2. We note that the group G has a normal series $G \supseteq (a) \supseteq E$ with infinite cyclic factors. By Theorem 3.5.1, the group F/A' is orderable and the group F/A is not even a group with unique roots. Thus, the normal Abelian subgroup A/A' is not a relatively convex subgroup of F/A'.

Chapter 6

Orders on Free Products

6.1. The Vinogradov Theorem

In this section, we prove that the free product of orderable (right-orderable) groups is an orderable (right-orderable) group. This result by Vinogradov [83] is critical in the theory of totally ordered groups and in that of right-ordered ones.

Let $\{G_\alpha \mid \alpha \in I\}$ be a set of groups, and let every group G_α be the quotient group of the free group $F_\alpha = \mathrm{gp}\,(\mathrm{x}_{\alpha\beta} \mid \beta \in \mathrm{B}_\alpha)$ with free generators $x_{\alpha\beta}$ by the normal subgroup N_α.

- The *free product* $G^* = \prod\limits_{\alpha \in I}{}^* G_\alpha$ of groups G_α is the quotient group of the free group F with the set of free generators $\{\overline{x}_{\alpha\beta} \mid \alpha \in I, \beta \in B_\alpha\}$ by the normal subgroup N which is the normal closure in F of the set of all words $w_\alpha = w_\alpha(\overline{x}_{\alpha\beta})$, $\beta \in B_\alpha$, where $w_\alpha(x_{\alpha\beta}) \in N_\alpha$.

The mapping φ_α such that $(x_{\alpha\beta}N_\alpha)\varphi_\alpha = \overline{x}_{\alpha\beta}$, as is known (cf. [45] and [3]), is an isomorphism of the group G_α onto the subgroup $\overline{G}_\alpha$ of the group $G^* = \mathrm{gp}\,(\overline{G}_\alpha \mid \alpha \in I)$. If a group H is generated by subgroups

H_α, $\alpha \in I$, such that for every $\alpha \in I$ there is a homomorphism ψ_α of G_α onto H_α, there exists a homomorphism ψ of G^* onto H such that

$$a\psi_\alpha = a\varphi_\alpha\psi$$

for any $\alpha \in I$ and $a \in G_\alpha$.

In the sequel, we identify the group G_α with the subgroup $G_\alpha\varphi_\alpha$ of the group G^* and the elements $x_{\alpha\beta}N_\alpha$ with the elements $\overline{x}_{\alpha\beta}N$. As is known, every element $g \in G^*$, $g \neq e$, can be uniquely represented in the form

$$g = g_{\alpha(1)}g_{\alpha(2)}\cdots g_{\alpha(n)} \qquad (6.1.1)$$

where $g_{\alpha(i)} \in G_{\alpha(i)}$ and $\alpha(i) \in I$, $g_{\alpha(i)} \neq e$, $\alpha_i \neq \alpha_{i+1}$, for any $i = 1, 2, \ldots n-1$. The number n is called the *length of* the *word* g.

To prove the Vinogradov theorem on the orderability of the free product of groups, we need some auxiliary results on ordered groups of matrices over an associative ring with unit and without zero divisors. We note that the proof presented below was suggested by Bergman [84]. A similar proof was obtained by Strojnowski and Wojas [85].

Let R be an associative ring with unit and without zero divisors. Throughout this chapter, we use the following notation:

$R[t]$ is the ring of polynomials over R with central indeterminate t,

$M_2(R[t])$ is the ring of all 2×2-matrices over $R[t]$,

$GL_2(R[t])$ is the group of invertible 2×2-matrices of the ring $M_2(R[t])$.

Lemma 6.1.1 (Bergman [84]). *Let G and H be subgroups of the group $GL_2(R[t])$ such that for every matrix $g \in G \setminus E$ the degree of its entry g_{12} regarded as a polynomial is strictly greater than the degrees of the remaining entries g_{ij}, $(i,j) \neq (1,2)$. If for every matrix $h \in H \setminus E$ the degree of its entry h_{21} is strictly greater than the degrees of h_{ij}, $(i,j) \neq (2,1)$, then the subgroup of the group $GL_2(R[t])$ generated by the subgroups G and H is isomorphic to the free product of G and H.*

PROOF. It suffices to show that the product of the form (6.1.1)

$$a_1a_2a_3\ldots a_n$$

is distinct from the identity matrix; here $a_{2i+1} \in G$, $a_{2i} \in H$, and $2i+1 \leqslant n$. For a matrix a_i we denote by $\deg(a^i_{rs})$ the degree of its entry located in the ith row and the jth column. It is easy to show that the coefficient of t^k, where

$$k = \deg(a_{12}^1) + \deg(a_{21}^2) + \deg(a_{12}^3) + \cdots + \deg(a_{rs}^n)$$

is not zero. In the last summand, $(r, s) = (1, 2)$ if n is odd and $(r, s) = (2, 1)$ if n is even. □

Corollary 6.1.1. *The subgroup of $GL_2(R[t])$ generated by the matrices*

$$\begin{bmatrix} 1 & t \\ 0 & 1 \end{bmatrix}, \quad \begin{bmatrix} 1 & 0 \\ t & 1 \end{bmatrix}$$

is a free group.

Let G and H be groups constituted by invertible elements of an associative ring R with unit and without zero divisors. It is obvious that the mapping

$$\varphi\colon g \to \begin{bmatrix} 1 & t \\ 0 & 1 \end{bmatrix}^{-1} \begin{bmatrix} g & 0 \\ 0 & 1 \end{bmatrix} \begin{bmatrix} 1 & t \\ 0 & 1 \end{bmatrix} = \begin{bmatrix} g & t(g-1) \\ 0 & 1 \end{bmatrix}$$

is an isomorphic embedding of the group G into the group $GL_2(R[t])$. It is also evident that the mapping

$$\psi\colon h \to \begin{bmatrix} 1 & 0 \\ t & h \end{bmatrix}^{-1} \begin{bmatrix} 1 & 0 \\ 0 & h \end{bmatrix} \begin{bmatrix} 1 & 0 \\ t & 1 \end{bmatrix} = \begin{bmatrix} 1 & 0 \\ t(h^{-1}-1) & h \end{bmatrix}$$

is an isomorphic embedding of the group H into the group $GL_2(R[t])$.

From Lemma 6.1.1 we obtain the following assertion.

Corollary 6.1.2. *The subgroup of the group $GL_2(R[t])$ generated by $G\varphi$ and $H\psi$ is isomorphic to the free product $G * H$ of G and H.*

- An associative ring R with unit is called *totally ordered* if on R there is a total order $\leqslant$ such that $a+c \leqslant b+c$ and $au \leqslant bu$, $ua \leqslant ub$ for any $a, b, c, u \in R$ such that $a \leqslant b$ and $u \geqslant 0$.

Let a ring R be totally ordered. Every matrix $g \in GL_2(R[t])$ can be uniquely represented as $g = c + t \cdot h$, where c and h are matrices over R. We denote by U the subgroup of the group $GL_2(R[t])$ formed by matrices $g = c + th$ such that c is a diagonal matrix with positive entries.

Lemma 6.1.2 (Bergman [84]). *The group U is orderable.*

Proof. Let $a, b \in U$, $a \neq b$, and

$$a - b = z^{(0)} + \cdots + z^{(i)}t^i + \cdots + z^{(n)}t^n$$

where

$$z^{(i)} = \begin{bmatrix} z_{11}(i) & z_{12}(i) \\ z_{21}(i) & z_{22}(i) \end{bmatrix}$$

and $z_{11}(i), z_{12}(i), z_{21}(i), z_{22}(i) \in R$. We set $a > b$ in U if the first nonzero term of the sequence

$$z_{11}(0), z_{12}(0), z_{21}(0), z_{22}(0), \ldots, z_{11}(i), z_{12}(i), z_{21}(i), z_{22}(i), \ldots$$

is a positive element of R. It is evident that the relation $\leqslant$ just introduced is a total order on U.

Further, for an arbitrary element $c \in U$, $c = c^{(0)} + \ldots c^{(i)}t^i \ldots$, the following equalities hold:

$$ca - cb = c(a - b) = (c^{(0)} + \ldots)(z^{(i)} + \ldots) = c^{(0)}z^{(i)}t^i + \ldots$$

here

$$c^{(0)} = \begin{bmatrix} c_{11}(0) & 0 \\ 0 & c_{22}(i) \end{bmatrix}$$

and $c_{11}(0), c_{22}(0) > 0$ in the totally ordered ring R. Since $c_{11}(0) > 0$ and $c_{22}(0) > 0$ in R, the signs of elements of the matrix $c^{(0)}z^{(i)}$ are the same as those of the matrix $z^{(i)}$. Hence $ca > cb$. We can show that $ac > bc$ in a similar way. □

Let G and H be orderable groups and $\mathbb{Z}[G \times H]$ denote the integral group ring of the direct product $G \times H$ of G and H. We introduce a total order on the group $G \times H$ and extend it to an order on $\mathbb{Z}[G \times H]$ assuming that the element $c = \sum k_i c_i$, where $k_i \in \mathbb{Z}$, $c_i \in G \times H$, is positive if the first nonzero coefficient of c is positive (i.e., the coefficient of the least c_i with respect to the order introduced on the group $G \times H$). The ring $\mathbb{Z}[G \times H]$ with this order is totally ordered. We apply the above results to $\mathbb{Z}[G \times H]$ to convince ourselves that the following theorem holds in the case of two free factors. By transfinite induction, we can conclude that Theorem 6.1.1 remains valid for an arbitrary number of free factors.

Theorem 6.1.1 (Vinogradov [83]). *The free product G^* of orderable groups $\{G_\alpha \mid \alpha \in I\}$ is an orderable group, and for every group G_α its total order can be extended to a total order on the group G^*.*

Corollary 6.1.3. *The free product G^* of right-orderable groups $\{G_\alpha \mid \alpha \in I\}$ is a right-orderable group, and for every group G_α its right total order can be extended to a right total order on the group G^*.*

PROOF. We consider the natural homomorphism

$$\varphi\colon G^* = \prod_{\alpha\in I}{}^{*} G_\alpha \to \prod_{\alpha\in I} G_\alpha$$

of the free product $G^* = \prod_{\alpha\in I}{}^{*} G_\alpha$ onto the Cartesian product $\prod_{\alpha\in I} G_\alpha$. Its kernel D is called the *Cartesian subgroup* of the group G^*. It is known that the Cartesian subgroup is a free group. By Theorem 6.1.1, the normal subgroup D is orderable and the quotient group G^*/D is a right-orderable group. In view of Theorem 1.6.2, the group G^* also admits a right order which is an extension of the right orders on free factors. □

6.2. Free Products with Amalgamation

In this section, we establish the orderability and right-orderability of some (far from all!) free products of orderable groups with amalgamation.

6.2.1. Let groups G_α, $\alpha \in I$, be given, and let a subgroup H_α isomorphic to the group H be selected. We denote by φ_α an isomorphism of H_α onto H. Then a mapping $\psi_{\alpha\beta} = \varphi_\alpha\varphi_\beta^{-1}$ is an isomorphism between H_α and H_β.

- By the *free product of groups G_α, $\alpha \in I$, with an amalgamated subgroup H* we mean the quotient group G of the free product of the groups G_α, $\alpha \in I$, by the normal subgroup generated by all elements of the form $h_\alpha h_\beta^{-1}$, where $h_\beta = h_\alpha\psi_{\alpha\beta}$, h_α "run through" the subgroup H_α, and α, β are all admissible pairs of indices from I.

As is well known (cf. [3]), the constructed mapping θ_α of G_α into G is an isomorphism. Moreover, for all pairs of indices $\alpha, \beta \in I$ the following equalities hold:

$$G_\alpha\theta_\alpha \cap G_\beta\theta_\beta = H_\alpha\theta_\alpha = H_\beta\theta_\beta$$

and for any element $x \in G_\alpha$ the following relation is fulfilled:

$$x\theta_\alpha = x\psi_{\alpha\beta}\theta_\beta$$

In the sequel, we identify the groups G_α, $\alpha \in I$, with subgroups $G_\alpha\theta_\alpha$ of the group G and use the following notation for the free product G of the groups G_α with an amalgamated subgroup H and the "identifying" isomorphisms $\psi_{\alpha\beta}$:

$$G = \prod_{\alpha \in I} G_\alpha(H_\alpha \overset{\psi_{\alpha\beta}}{=} H_\beta)$$

We recall how to construct G from the groups G_α, $\alpha \in I$, and the isomorphisms $\psi_{\alpha\beta}$. For each index $\alpha \in I$ we fix a system $X_{\alpha\gamma}$, $\gamma \in \Gamma_\alpha$, constituted by the representatives of right cosets of G_α by the subgroup H_α, which are distinct from H_α. Then every element $g \in G$ can be uniquely represented as follows:

$$g = x_{\alpha(1)}x_{\alpha(2)} \cdots x_{\alpha(n)} \cdot h$$

where $h \in H$, $x_{\alpha(i)} \in X_{\alpha(i)}$, and $\alpha(i+1) \neq \alpha(i)$ for $i = 1, 2, \ldots, n-1$ and $\alpha(i) \in I$ for $i = 1, 2, \ldots, n$. Furthermore, $g \in H$ if and only if $n > 0$ and $g = e$ if and only if $n = 0$ and $h = e$ in H.

The following example shows that, in general, the free product of orderable groups with an amalgamated subgroup is not orderable, even in the case of an amalgamated subgroup which is normal and relatively convex in each of its free factors.

Let A_1 be the group of all matrices of the form

$$\begin{bmatrix} 2^{k/2} & r_1 + r_2\sqrt{2} \\ 0 & 1 \end{bmatrix}$$

where k is an integer and r_1, r_2 are rational numbers. The subgroup H_1 of A constituted by matrices of the form

$$\begin{bmatrix} 1 & r_1 + r_2\sqrt{2} \\ 0 & 1 \end{bmatrix}$$

is the direct product of additive groups of rational numbers. Moreover, the subgroup H_1 is normal in A_1. Since the group A_1 has a finite special rank (equal to 3) and is soluble, we conclude that, by Theorem 3.3.1, each of its relatively convex subgroups is normal. But the group A_1 has the unique nontrivial isolated normal subgroup H_1. Consequently, H_1 is convxxH i»harxsËeyt to any total order on A_1

and has no proper convex subgroups. But in this case, any total order on A_1 induces an Archimedean order on H_1.

Let A_2 be the group of all matrices of the form

$$\begin{bmatrix} 2^n & r & 0 & 0 \\ 0 & 1 & 0 & 0 \\ 0 & 0 & 3^n & s \\ 0 & 0 & 0 & 1 \end{bmatrix}$$

where n is an integer and r, s are rational numbers. It is evident that the group A_2 is orderable and the subgroup H_2 of all matrices of the form

$$\begin{bmatrix} 1 & r & 0 & 0 \\ 0 & 1 & 0 & 0 \\ 0 & 0 & 1 & s \\ 0 & 0 & 0 & 1 \end{bmatrix}$$

is normal and relatively convex in A_2. It is easy to show that no total A_2-order on H_2 is Archimedean. Then the free product

$$G = A_1 * A_2(H_1 \overset{\varphi}{=} H_2)$$

of the groups A_1 and A_2 with the amalgamated subgroup $H = H_1 \overset{\varphi}{=} H_2$ is not orderable for any isomorphism φ identifying H_1 with H_2.

Proposition 6.2.1. *The free product*

$$G = A_1 * A_2(H_1 \overset{\varphi}{=} H_2)$$

of orderable groups A_1 and A_2 with the amalgamated subgroup $H = H_1 \overset{\varphi}{=} H_2$, which is relatively convex and normal in each of its free factors, is orderable if and only if the identifying isomorphism φ is an order isomorphism under some total A_1-order on H_1 and A_2-order on H_2. Moreover, if the group G is orderable, then the subgroup H is relatively convex in it.

PROOF. The necessity is obvious. We now prove the sufficiency. We first show that H is a G-orderable subgroup of the group G. Let P_1 be a total A_1-order on H_1 and $P_2 = \varphi(P_1)$ be a total A_2-order on H_2. Then P_1 is a total G-order on $H = H_1$. Since the group G/H, being isomorphic to the group $A_1/H_1 * A_2/H_2$, is orderable, we can conclude that the group G is orderable and H is its relatively convex subgroup. □

6.2.2. To present further results on the orderability of the free product of orderable groups with an amalgamated subgroup, we give some auxiliary assertions on an ordered group of matrices over an associative ring with unit and without zero divisors. The following lemma is a generalization of Lemma 6.1.1.

Lemma 6.2.1 (Bergman [84]). *Let G and F be multiplicative subgroups of the ring $M_2(R[t])$ such that the degree of the entry x_{12} of the matrix*

$$x = \begin{bmatrix} x_{11} & x_{12} \\ x_{21} & x_{22} \end{bmatrix}, \qquad x \in F \setminus G$$

is greater than the degrees of the entries x_{11}, x_{21}, x_{22}. Moreover, the degree of the entry y_{21} of the matrix

$$y = \begin{bmatrix} y_{11} & y_{12} \\ y_{21} & y_{22} \end{bmatrix}$$

where $y \in G \setminus F$, is greater than the degrees of the entries y_{11}, y_{12}, y_{22}. Then the group generated in $M_2(R[t])$ by the subgroups G and F is the free product $G * F(G \cap F = G \cap F)$ of G and F with the amalgamated subgroup $G \cap F$.

Proof. Let $H = G \cap F$. We denote by X_G and X_F the systems of representatives of cosets of groups G and F by the subgroup H. We consider an element $w \in M_2(R[t])$ of the form $w = z_t z_2 \ldots z_1 u$, where $u \in H$, and elements z_i, $i = 1, \ldots, t$, of X_G and X_F. Moreover, the elements z_i and z_{i+1}, $i = 1, \ldots, t-1$, belong to different groups G or F. It suffices to prove (cf., for example, [86]) that w is distinct from the identity matrix in $M_2(R[t])$. Indeed, if $w = z_t z_2 \ldots z_1 u = e$, then

$$u \begin{pmatrix} x \\ y \end{pmatrix} = \begin{pmatrix} 1 \\ 1 \end{pmatrix}$$

for some $x, y \in R$. Taking into account the hypotheses of the lemma and calculating the product, we obtain

$$w \begin{pmatrix} 1 \\ 1 \end{pmatrix} = z_t z_2 \ldots z_1 u \begin{pmatrix} 1 \\ 1 \end{pmatrix} \neq \begin{pmatrix} 1 \\ 1 \end{pmatrix}$$

which completes the proof. □

Let M be a group constituted by invertible elements of the ring R, $N = M \times M$, and $\Delta M = \{(m, m) \mid m \in M\} \subseteq N$. It is possible to identify ΔN with the group of diagonal 2×2-matrices over R.

Conjugating this diagonal subgroup by the matrices

$$\begin{bmatrix} 1 & t \\ 0 & 1 \end{bmatrix} \quad \begin{bmatrix} 1 & 0 \\ t & 1 \end{bmatrix}$$

we arrive at the pair of isomorphic embeddings of $M \times M$ into $M_2(R[t])$:

$$\varphi_1: (m, m') \to \begin{bmatrix} m & t(m - m') \\ 0 & m' \end{bmatrix}$$

$$\varphi_2: (m, m') \to \begin{bmatrix} m & 0 \\ t(m' - m) & m' \end{bmatrix}$$

It is clear that

$$(M \times M)\varphi_1 \cap (M \times M)\varphi_2 = \Delta M$$

By Lemma 6.2.1, the following assertion is true.

Proposition 6.2.2 (Bergman [84]). *Let M be a group of invertible elements of the ring R, $N = M \times M$, and $\Delta M = \{(m, m) \mid m \in M\} \subseteq M \times M$. Then the subgroup of the group $GL_2(R[t])$ generated by the subgroups $(M \times M)\varphi_1$ and $(M \times M)\varphi_2$ is isomorphic to the free product $(M \times M) * (M \times M)(\Delta M = \Delta M)$ with the amalgamated subgroup ΔM.*

Proposition 6.2.2 and Lemma 6.1.1 imply

Proposition 6.2.3 (Bergman [84]). *Let M be an orderable group. Then the free product $(M \times M) * (M \times M)(\Delta M = \Delta M)$ with the amalgamated subgroup ΔM is an orderable group.*

Let $\overline{\prod\limits_{x \in X}} M_x$ be the Cartesian product of the family $\{M_x \mid x \in X\}$ of groups, each of which is isomorphic to the group M, and let the natural isomorphisms of the group M onto each of the groups M_x be fixed. The subgroup

$$\Delta^{(X)} M = \{f \in \overline{\prod_{x \in X}} M_x \mid f(x_1) = f(x_2) \text{ for any } x_1, x_2 \in X\}$$

is called the *diagonal of the Cartesian product* $\overline{\prod\limits_{x \in X}} M_x$.

The following theorem is a generalization of Proposition 6.2.2.

Theorem 6.2.1 (Bergman [84]). *For any orderable group H and its subgroup T the following conditions are equivalent:*

(a) *there exists an orderable group $M \supseteq H$ and a subset $X \subseteq M$ such that T coincides with the centralizer $c_H(X)$ of the subset X in H,*

(b) *there exists an orderable group $M \supseteq H$ and an element $x \in M$ such that T coincides with the centralizer $c_H(x)$ of the element $x \in H$,*

(c) *there exists an orderable group M and a pair of homomorphisms α, β from H into M such that $T = \{h \in H \mid h\alpha = h\beta\}$,*

(d) *the free product with the amalgamated subgroup $H * H(T = T)$ is orderable.*

PROOF. We show that the condition (a) implies the condition (b). Consider a group $\bar{\prod}_{x \in X} M_x$ that is the Cartesian product of the copies of the group M indexed by $x \in X$. Let f denote an element of $\bar{\prod}_{x \in X} M_x$ such that $f(x) = x$ for any $x \in X$. It is obvious that the centralizer $c_{\Delta^{(X)}M}(f)$ of the element f in the diagonal subgroup $\Delta^{(X)}M$ is a subgroup isomorphic to $c_M(X)$. Taking into account that $\Delta^{(X)}M \supseteq \Delta^{(X)}H$ and identifying H with $\Delta^{(X)}H$, we obtain $c_H(f) = T$.

It is clear that the condition (b) implies the condition (a). Hence they are equivalent.

We assume that the condition (b) is valid and establish the validity of the condition (c). Let $h\alpha = h$ and $h\beta = h^x$ for any $h \in H$. Then α and β are isomorphic embeddings of the group H into the group M and

$$\{h \in H \mid h\alpha = h = h^x = h\beta\} = T$$

We now assume that the condition (c) is valid and prove the validity of the condition (b). We suppose that there are homomorphisms α and β satisfying the condition (c). We also may assume that the kernels of the homomorphisms α and β are trivial groups E; otherwise we can insert $M \times H$ instead of M and α', β', where $h\alpha' = (h, h\alpha)$, $h\beta' = (h, h\beta)$, instead of α, β. Consequently, the group H can be embedded into the multiplicative group $GL_2(R[t])$ of invertible 2×2-matrices over the ring $R[t]$, where $R = \mathbb{Z}[M]$, by the following rule:

$$\varphi\colon h \to \begin{bmatrix} h\alpha & 0 \\ 0 & h\beta \end{bmatrix}$$

It is evident that the intersection of $H\varphi$ and the centralizer of the matrix

$$\begin{bmatrix} 1 & t \\ 0 & 1 \end{bmatrix}$$

from $GL_2(R[t])$ coincides with the image of the subgroup T.

We now prove that the condition (c) follows from the condition (d). We assume that the homomorphisms α and β satisfy the condition (c). As above, we can also assume that the kernels of the homomorphisms α and β coincide and are equal to E. Then the mapping φ, $h\varphi = (h\alpha, h\beta)$, is an isomorphic embedding of H into $N = M \times M$.

By Lemma 6.2.1, the group $(M \times M) * (M \times M)(\Delta M = \Delta M)$ is orderable. It is evident that the subgroup generated by the isomorphic images $H\varphi$ of free factors of $M \times M$ is isomorphic to the free product

$$H * H(\Delta M \cap H\varphi) = (\Delta M \cap H\varphi)$$

and, by Proposition 6.2.2, is an orderable group.

We now prove that if the condition (d) is valid, then the condition (c) is valid too. It suffices to set α and β to be equal to the canonical embeddings of the group H into the free factors $H * H(T = T)$. □

- A subgroup T of an orderable group H subject to the condition (c) is called an *equalizer subgroup* of the group H.

6.2.3. Let G be an arbitrary orderable group, A be its subgroup, and H be an arbitrary orderable homomorphic image of G. We denote by $D(A)$ a subset of G formed by elements $g \in G$ such that $g\alpha = g\beta$ for any pair of homomorphisms $\alpha\colon G \to H$ and $\beta\colon G \to H$ which map G onto an orderable group H and satisfy the equality $a\alpha = a\beta$ for any $a \in A$.

It is easy to check that $D(A)$ is a subgroup of the orderable group G. In the case $D(A) = A$, the subgroup A is said to be *dominion.* Since the class of orderable groups is closed under taking the Cartesian products, any dominion subgroup of an orderable group G is an equalizer subgroup. From the conditions (c) and (d) of Theorem 6.2.1 it follows that for an orderable group H and its subgroup T the group $H * H(T = T)$ is orderable if and only if T is a dominion subgroup of H.

The following lemma shows that not every subgroup of an orderable group is dominion.

Lemma 6.2.2 (Bergman [84]). *A dominion subgroup T of any orderable group G has the following properties:*

(a) *T is an isolated subgroup,*

(b) *for any elements $g_1, \ldots, g_m$ of $G \setminus T$ there are numbers $\varepsilon_1, \ldots, \varepsilon_m$ equal to ± 1 and such that*

$$e \notin P_T(g_1^{\varepsilon_1}, \ldots, g_m^{\varepsilon_m}) = \operatorname{sgr}(T, g_1^{\varepsilon_1}, \ldots, g_m^{\varepsilon_m})$$

PROOF. Since the condition (a) follows from the condition (b) in an obvious way, we restrict ourselves to the proof of the validity of (b). Let $g_1, \ldots, g_m \in G \setminus T$. Since T is a dominion subgroup, there are homomorphisms α and β which map the group G into an orderable group M, coincide on T, and satisfy the relation $x\alpha \neq x\beta$ for every element $x \in G \setminus T$. We fix some total order on the group M. Since the homomorphisms α and β are different on elements $g_1, \ldots, g_m$, for $i = 1, \ldots, m$ we set $\varepsilon_i = +1$ if $g_i\alpha < g_i\beta$ and $\varepsilon_i = -1$ otherwise. Then for any $g \in P_T(g_1^{\varepsilon_1}, \ldots, g_m^{\varepsilon_m})$ we have $g\alpha < g\beta$. Hence $g \notin T$. □

In view of Proposition 5.1.5, we arrive at the following assertion.

Corollary 6.2.1. *A dominion subgroup of any orderable group is right-relatively convex.*

Theorem 6.2.2 (Bergman [84]). *Let H be an orderable group and T be its subgroup. The free product with the amalgamated subgroup $H * H(T = T)$ is orderable if and only if T is a right-relatively convex subgroup of the group H.*

PROOF. It suffices to show that the conditions (a)–(d) of Theorem 6.2.1 are equivalent to the requirement that T is right-relatively convex. Let H be a subgroup of some totally ordered group M and let the subgroup T be the centralizer in H of some element $x \in M$. We consider the following representation $(H\varphi, M)$ of the group H by order automorphisms of the totally ordered set M:

$$x \cdot h\varphi = x^h = h^{-1}xh$$

where $x \in M$ and $h \in H$. The subgroup T coincides with the stabilizer of the element $x \in M$ and, by Corollary 5.1.4, is a right-relatively convex subgroup of the group H.

To prove the converse assertion we consider a right-relatively convex subgroup T of an orderable group H. In view of Proposition 5.1.6, we can assume that H is a group of order automorphisms of a totally ordered set Ω and the subgroup T is the stabilizer of some element $\alpha \in \Omega$. Let K be an arbitrary totally ordered ring and $U = \sum_{\alpha\in\Omega} Ka_\alpha$ be a free k-module with the set of free generators $\{a_\alpha \mid \alpha \in \Omega\}$. We totally order U by extending the order given on Ω. It is clear that H with this order is naturally represented by order (K-linear) automorphisms of the module U.

Further, we will denote by M the split extension of the additive group U by the group H defined by this representation:

$$M = \{(h, u) \mid h \in H, u \in U\}$$

and the multiplication in M is defined by the following rule:

$$(h, u)(h', u') = (hh', u \cdot h' + u)$$

By construction, the group M is orderable and the centralizer of the element a_α in H coincides with the stabilizer of the element $\alpha \in \Omega$ in H. □

6.2.4. We now consider the right-orderability of free products with amalgamation.

Theorem 6.2.3 (Bergman [84]). *Let T be a group and G_1, G_2 be groups containing T. Then the following conditions are equivalent:*

(a) *there is a right-orderable group H and isomorphic embeddings $\varphi_1\colon G_1 \to H$ and $\varphi_2\colon G_2 \to H$ such that $T\varphi_1 = T\varphi_2$ and $t\varphi_1 = t\varphi_2$ for any $t \in T$,*

(b) *the free product $G_1 * G_2(T = T)$ is right-orderable.*

Proof. Let H be a right-orderable group satisfying the condition (a). Then there is a homomorphism φ of the group $G_1 * G_2(T = T)$ into H, which is an extension of φ_1 and φ_2.

The kernel N of the homomorphism φ has a trivial intersection with each of the free factors G_1, G_2 and, consequently, with all subgroups of $G_1 {*} G_2(T = T)$ conjugate to G_1 and G_2. In this case, from the description of subgroups of free products with amalgamation (cf., for example, [86]) it follows that N is a free group. Therefore, the group N is orderable and, by Theorem 1.6.2, the free product $G_1 * G_2(T = T)$ is a right-orderable group. The converse assertion is obvious. □

Corollary 6.2.2. *Let* (a) *be an infinite cyclic group. Then the groups* $(a) * (a)((a^m) = (a^m))$ *and* $(a^m) * (a^n)((a^{mn}) = (a^{mn}))$ *are right-orderable.*

Corollary 6.2.3. *Let* G *be an arbitrary right-orderable group and* T *be its arbitrary subgroup. Then the free product* $G * G(T = T)$ *is also a right-orderable group.*

6.3. Right-Orderable Groups with Amalgamation

In this section, we show that the classes of right-orderable groups and of locally indicable groups do not have the amalgamation property.

- A formal system $\{G_1, G_2, H, \sigma_1, \sigma_2\}$, where G_1, G_2, H are groups of the class $\mathfrak{K}$ and σ_i, $i = 1, 2$, are isomorphic embeddings of H into G_i, $i = 1, 2$, is called a *V-formation* in $\mathfrak{K}$.
- A V-formation is *amalgamated* by $\{\varphi_1, \varphi_2, G\}$ if $G \in \mathfrak{K}$ and φ_i, $i = 1, 2$, are isomorphic embeddings of G_i, $i = 1, 2$, into G; moreover, $\sigma_1\varphi_1 = \sigma_2\varphi_2$.
- A class of groups $\mathfrak{K}$ has the *amalgamation property* if any V-formation is amalgamated in some system $\{\varphi_1, \varphi_2, G\}$, where $G \in \mathfrak{K}$.

Theorem 6.3.1 (Passman [87], Medvedev [88], and Bergman [89]). *The classes of right-orderable and of locally indicable groups do not have the amalgamation property.*

PROOF. We consider the right-orderable groups

$$G_1 = \mathrm{gp}\,(a_1, b_1 \mid a_1^{-1} b_1 a_1 = b_1^{-1})$$
$$G_2 = \mathrm{gp}\,(a_2, b_2 \mid b_2^{-1} a_2 b_2 = a_2^{-1})$$

and their subgroups $H_1 = \mathrm{gp}\,(a_1^2, b_1)$ and $H_2 = \mathrm{gp}\,(b_2^2, a_2)$. It is evident that H_1 and H_2 are free Abelian groups of rank 2. We also consider the free Abelian group $H = (x) \times (y)$ generated by x and y. Let $\sigma_1 \colon H \to H_1$ and $\sigma_2 \colon H \to H_2$ be isomorphisms of H onto H_1 and H_2 respectively and $x\sigma_1 = b_1$, $y\sigma_1 = a_1^2$, $x\sigma_2 = b_2^2$, $y\sigma_2 = a_2$. We assume that the V-formation $\{G_1, G_2, H, \sigma_1, \sigma_2\}$ is amalgamated in the class of right-orderable groups. Then there exist a group G and isomorphisms φ_1 of G_1 into G and φ_2 of G_2 into G such that

$$x\sigma_1\varphi_1 = b_1\varphi_1 = x\sigma_2\varphi_2 = b_2^2\varphi_2 \tag{6.3.1}$$

$$y\sigma_1\varphi_1 = a_1^2\varphi_1 = y\sigma_2\varphi_2 = a_2\varphi_2 \tag{6.3.2}$$

We show that the group G is not right-orderable. To this end, we assume the contrary. Let the group G admit a right order. We assume that

$$a_1\varphi_1 > e, \quad b_2^\varepsilon\varphi_2 > e$$

where $\varepsilon = \pm 1$. Consequently,

$$\begin{gathered} a_1\varphi_1 \cdot b_2^\varepsilon\varphi_2 > e \\ b_2^\varepsilon\varphi_2 \cdot a_1\varphi_1 > e \end{gathered}$$

Taking into account the defining relations of the groups G_1 and G_2 and Eqs. (6.3.1) and (6.3.2), we obtain

$$\begin{aligned} e &< (a_1\varphi_1 \cdot b_2^\varepsilon\varphi_2)^2(b_2^\varepsilon\varphi_2 \cdot a_1\varphi_1)^2 \\ &= a_1\varphi_1 b_2^\varepsilon\varphi_2 a_1\varphi_1 \cdot (b_2^\varepsilon)^2\varphi_2 \cdot a_1\varphi_1 b_2^\varepsilon\varphi_2 a_1\varphi_1 \\ &= a_1\varphi_1 b_2^\varepsilon\varphi_2 \cdot a_1\varphi_1^2 \cdot a_1^{-1}\varphi_1 b_1^\varepsilon\varphi_1 a_1\varphi_1 \cdot b_2^\varepsilon\varphi_2 a_1\varphi_1 \\ &= a_1\varphi_1 b_2^\varepsilon\varphi_2(a_1\varphi_1)^2 b_1^{-\varepsilon}\varphi_1 b_2^\varepsilon\varphi_2 a_1\varphi_1 \\ &= a_1\varphi_1 b_2^\varepsilon\varphi_2(a_1\varphi_1)^2 b_2^{-2\varepsilon}\varphi_2 b_2^\varepsilon\varphi_2 a_1\varphi_1 \\ &= a_1\varphi_1 \cdot b_2^\varepsilon\varphi_2 \cdot a_2\varphi_2 \cdot b_2^{-\varepsilon}\varphi_2 \cdot a_1\varphi_1 \\ &= a_1\varphi_1 \cdot a_2^{-1}\varphi_2 \cdot a_1\varphi_1 \\ &= a_1\varphi_1 \cdot a_1^{-2}\varphi_1 \cdot a_1\varphi_1 = e \end{aligned}$$

The contradiction obtained completes the proof. □

The following theorem is another statement of Theorem 6.2.3.

Theorem 6.3.2. *The free product $G = G_1 * G_2(H_1 \overset{\varphi}{=} H_2)$ of right-orderable groups G_1 and G_2 with an amalgamated subgroup $H = H_1 \overset{\varphi}{=} H_2$ is right-orderable if and only if the V-formation defined by this free product is amalgamated in the class of right-orderable groups.*

Chapter 7

Quasivarieties

7.1. Properties of Quasivarieties

7.1.1. We introduce some definitions.

- A formula φ of the first-order predicate calculus is called a *quasi-identity* of signature σ if it assumes the form

$$(\forall x_1)\dots(\forall x_n)((A_1 = B_1)\&\dots\&(A_k = B_k) \Rightarrow (A = B))$$

  where $A_1, \dots, A_k$, $B_1, \dots, B_k$, A, B are terms in the variables $x_1, \dots, x_n$ in the signature σ.

- A class $\mathfrak{K}$ of algebraic systems is called a *quasivariety of signature* σ if there exists a set Φ of quasi-identities such that $\mathfrak{K}$ consists of all algebraic systems of signature σ on which all the quasi-identities of Φ are true.

Let I be a set. In the sequel, we omit quantifiers.

- A system $\mathcal{F}$ of subsets of a set I is called a *filter* if $\mathcal{F}$ satisfies the following three conditions:

  — $A, B \in \mathcal{F} \Rightarrow A \cap B \in \mathcal{F}$

— $A \in \mathcal{F}, \quad A \subseteq B \subseteq I \Rightarrow B \subseteq \mathcal{F}$

— $\varnothing \neq \mathcal{F}$

- A filter over I is called the *Fréchet filter* if it consists of all subsets $M \subseteq I$ whose complements $I \setminus M$ are finite.

It is clear that the set of all filters over I is partially ordered by inclusion.

- A maximal filter over I with respect to the partial order by inclusion is called an *ultrafilter*.

As is known (cf., for example, [90]), a filter $\mathcal{F}$ is an ultrafilter if and only if for any set $A \subseteq I$ either A or $I \setminus A$ is contained in $\mathcal{F}$.

Let $\{A_\alpha \mid \alpha \in I\}$ be a set of algebraic systems of signature σ, and $\mathcal{F}$ be a filter over I. On the Cartesian product $\bar{\prod\limits_{\alpha \in I}} A_\alpha$ we introduce the relation $\equiv_{\mathcal{F}}$ by setting $f \equiv_{\mathcal{F}} g$ if the set K of indices $\alpha \in I$ such that $f(\alpha) = g(\alpha)$ is contained in $\mathcal{F}$. It is easy to verify that $\equiv_{\mathcal{F}}$ is an equivalence relation. Moreover, it is stable under all operations in signature σ, i.e., if

$$f_1 \equiv_{\mathcal{F}} g_1, \ldots, f_k \equiv_{\mathcal{F}} g_k$$

then

$$\varphi_\gamma(f_1, \ldots, f_k) \equiv_{\mathcal{F}} \varphi_\gamma(g_1, \ldots, g_k)$$

for any operation φ_γ in σ. This means that on the set of equivalence classes of $\equiv_{\mathcal{F}}$ all operations φ_γ in σ can be defined in a natural way. The algebraic system $A/_{\equiv_{\mathcal{F}}}$ just constructed is called the *reduced product* of the algebraic systems A_α, $\alpha \in I$, by the filter $\mathcal{F}$ and is denoted by

$$\bar{\prod_{\alpha \in I}} A_\alpha/\mathcal{F}$$

- For an ultrafilter $\mathcal{F}$ over I the algebraic system $\bar{\prod\limits_{\alpha \in I}} A_\alpha/\mathcal{F}$ is called the *ultraproduct* of the algebraic systems A_α, $\alpha \in I$, by the ultrafilter $\mathcal{F}$.

- A class $\mathfrak{K}$ of algebraic systems is called *ultraclosed* if any ultraproduct of systems of $\mathfrak{K}$ is contained in $\mathfrak{K}$.

The following Mal'cev theorem expresses one of the basic results in the theory of quasivarieties of algebraic systems.

Theorem 7.1.1 (Mal'cev [90]). *A class $\mathfrak{K}$ of algebraic systems is a quasivariety if and only if it satisfies the following conditions:*

(a) *$\mathfrak{K}$ is ultraclosed,*

(b) *$\mathfrak{K}$ is closed under taking subsystems,*

(c) *$\mathfrak{K}$ is closed under taking the Cartesian products,*

(d) *$\mathfrak{K}$ contains a single-element subsystem.*

Proposition 7.1.1. *The class of right-ordered groups is ultraclosed.*

PROOF. Let G_α, $\alpha \in I$, be a set of right-ordered groups regarded as algebraic systems of signature $\{\cdot, ^{-1}, e, \leqslant\}$ (here $\leqslant$ stands for a total right order), and let $\mathcal{F}$ be an arbitrary ultrafilter over I. It is evident that the algebraic system $\bar{\prod\limits_{\alpha\in I}} G_\alpha/\mathcal{F} = G$ is a group. On G we introduce the relation $\leqslant$ as follows: $f \leqslant g$ for $f, g \in G$ if the set K of indices $\alpha \in I$ such that $f(\alpha) \leqslant g(\alpha)$ is contained in the ultrafilter $\mathcal{F}$. It is clear that $\leqslant$ is a total right order on the group G. □

Proposition 7.1.2. *The class $\mathfrak{O}_{\mathrm{r}}$ of right-orderable groups is a quasivariety.*

PROOF. The conditions (b)–(d) for right-orderable groups are obvious. The condition (a) follows from Proposition 7.1.1. □

The class of all locally indicable groups is denoted by $\mathfrak{J}$.

Proposition 7.1.3. *The class $\mathfrak{J}$ of locally indicable groups is a quasivariety.*

PROOF. Let G_α, $\alpha \in I$, be locally indicable groups, $G = \bar{\prod\limits_{\alpha\in I}} G_\alpha/\mathcal{F}$ be an ultraproduct of the groups G_α, $\alpha \in I$, by the ultrafilter $\mathcal{F}$, and $f_1\mathcal{F}, \dots, f_k\mathcal{F}$ be a finite set of elements of $\bar{\prod\limits_{\alpha\in I}} G_\alpha/\mathcal{F}$. We denote by $\widehat{G}_\alpha$ a subgroup of G_α generated by the elements $f_1(\alpha), \dots, f_k(\alpha)$. It is evident that the subgroup $\mathrm{gp}\,(f_1\mathcal{F}, \dots, f_k\mathcal{F})$ is contained in the ultraproduct of the groups $\bar{\prod}_{\alpha\in I} \widehat{G}_\alpha/\mathcal{F}$. By assumption, for each index $\alpha \in I$ there is a homomorphism $\varphi_\alpha : G_\alpha \to Z$ of the group G_α onto the infinite cyclic group Z and $f_i(\alpha)\varphi_\alpha \neq e$ at least for one index $i \in \{1, \dots, k\}$.

It is easy to show that the family $\{\varphi_\alpha \mid \alpha \in I\}$ of homomorphisms defines a homomorphism τ of the group $\bar{\prod}_{\alpha\in I} \widehat{G}_\alpha/\mathcal{F}$ into the ultraproduct $\bar{\prod}_{\alpha\in I} Z_\alpha/\mathcal{F}$ of the infinite cyclic groups Z_α, $\alpha \in I$, by the ultrafilter $\mathcal{F}$, and for some index $i \in \{1, \dots, k\}$ the relation $f_i\mathcal{F}\tau \neq e$ holds. Therefore, the subgroup $H = \operatorname{gp}(f_1\mathcal{F}\tau, \dots, f_k\mathcal{F}\tau)$ of the group $\bar{\prod}_{\alpha\in I} Z_\alpha/\mathcal{F}$ generated by the elements $f_1\mathcal{F}\tau, \dots, f_k\mathcal{F}\tau$ is a torsion-free nontrivial finitely generated Abelian group. Consequently, there is a homomorphism ψ of H onto Z such that

$$\psi f_i\mathcal{F}\tau\psi \neq e$$

for some i, $1 \leqslant i \leqslant k$. In this case the composition $\tau\psi$ maps the group H onto an infinite cyclic group. Consequently, the class of locally indicable groups is ultraclosed.

The conditions (a)–(d) for the class of locally indicable groups are obviously satisfied. □

7.1.2. We now introduce some definitions.

- An abstract group class $\mathfrak{R}$ is called *superaccessible* if every group which has a normal system of subgroups with the factors of $\mathfrak{R}$ belongs to $\mathfrak{R}$.

For example, the class of right-orderable groups admitting the Conradian right orders (cf. Chapter 2) is superaccessible. We denote this class by $\mathfrak{O}_c$.

Let $\mathfrak{R}$ be an arbitrary abstract group class.

- A group G is called
  - $\mathfrak{R}$-*projectable* if there exists a homomorphism of G onto a nontrivial group of $\mathfrak{R}$,
  - *locally* $\mathfrak{R}$-*projectable* if for any nontrivial finite-generated subgroup of G there exists a nontrivial homomorphism of G onto a nontrivial group of $\mathfrak{R}$.

Let $\Gamma(\mathfrak{R})$ stand for the set of all formulas

$$(\forall x_1)\dots(\forall x_n)(\mathop{\&}_{j=1}^{m} u_j(x_1,\dots,x_n)=e) \Rightarrow \mathop{\&}_{i=1}^{n} x_i = e \tag{7.1.1}$$

such that the group

$$G = \mathrm{gp}\,(x_1,\dots,x_n \mid u_1(x_1,\dots,x_n)=e,\dots,u_m(x_1,\dots,x_n)=e) \tag{7.1.2}$$

is not $\mathfrak{R}$-projectable. We note that formulas of $\Gamma(\mathfrak{R})$ are not quasi-identities, but any formula of $\Gamma(\mathfrak{R})$ is equivalent to some system of quasi-identities in a group class.

We denote by $\mathrm{q}\,(\Gamma(\mathfrak{R}))$ the group quasivariety that is defined by the system $\Gamma(\mathfrak{R})$ of formulas or the equivalent system of quasi-identities. We also denote by $\mathrm{q}(G)$ the group quasivariety that is generated by the group G or by the least group quasivariety containing the group G.

Lemma 7.1.1. *For an arbitrary group quasivariety $\mathfrak{R}$ the quasivariety $\mathrm{q}\,(\Gamma(\mathfrak{R}))$ is contained in the class of locally $\mathfrak{R}$-projectable groups.*

Proof. We assume that the quasi-identities of $\Gamma(\mathfrak{R})$ hold in the group G. Let H be a nontrivial finitely generated subgroup of the group G. If H is a finitely presented group, then H is also $\mathfrak{R}$-projectable; otherwise the validity of formulas of $\Gamma(\mathfrak{R})$ in the group H implies $H = E$.

We now consider the following infinitely presented group

$$H = \mathrm{gp}\,(x_1,\dots,x_n \mid u_1(x_1,\dots,x_n)=e,\dots,u_m(x_1,\dots,x_n)=e,\dots)$$

For an arbitrary natural number m we set

$$H_m = \mathrm{gp}\,(x_1,\dots,x_n \mid u_1(x_1,\dots,x_n)=e,\dots,u_m(x_1,\dots,x_n)=e)$$

Since for every natural number m the group H is a homomorphic image of H_m, arguing as above, we can show that H_m is $\mathfrak{R}$-projectable.

Let $H^* = \overline{\prod_{m\in\mathbb{N}}} H_m/\mathcal{F}$ be the reduced product of the groups H_m, $m \in \mathbb{N}$, by the Fréchet filter $\mathcal{F}$. It is obvious that the subgroup of $H^* = \overline{\prod_{m\in\mathbb{N}}} H_m/\mathcal{F}$ generated by elements $x_1,\dots,x_n$ (from now on, it is convenient to regard the symbols $x_1,\dots,x_n$ as signature constants) is isomorphic to H. We identify H with the above-mentioned subgroup and consider the group

$$\widehat{H}^* = \overline{\prod_{m\in\mathbb{N}}} \widehat{H}_m/\mathcal{F}$$

which is the reduced product of the groups $\widehat{H}_m$, $m \in \mathbb{N}$, by the Fréchet filter $\mathcal{F}$, where for any natural number m the group $\widehat{H}_m$ is a nontrivial homomorphic image of H_m belonging to the quasivariety $\mathfrak{R}$.

Since for all $m \in \mathbb{N}$ the group $\widehat{H}_m$ satisfies the formula $\underset{i=1}{\overset{n}{\&}} x_i \neq e$, the group $\widehat{H}^*$ also satisfies this formula. This means the nontriviality of the subgroup $\mathrm{gp}(x_1, \ldots, x_n) \subseteq \widehat{H}^*$. On the other hand, this subgroup is a homomorphic image of the group H. Hence H is $\mathfrak{R}$-projectable. □

Lemma 7.1.2. *For any group quasivariety $\mathfrak{R}$ the class of locally $\mathfrak{R}$-projectable groups is contained in the class of groups possessing subnormal systems of subgroups and factors from $\mathfrak{R}$.*

PROOF. We consider a locally $\mathfrak{R}$-projectable group G. As is known, there exists a subnormal system of subgroups of G with factors belonging to the quasivariety $\mathfrak{R}$ provided that on the group G it is possible to define a binary predicate R such that the algebraic system $G[R]$, obtained from G by enrichment of the group signature with the predicate R, satisfies the universal formulas of the following system Σ_0 (quantifiers are omitted):

$$\neg xRx$$
$$xRy \& yRz \Rightarrow xRz$$
$$xRz \& \neg yRz \Rightarrow xRy$$
$$xRz \& yRz \Rightarrow xy^{-1}Rz$$
$$\neg x = e \Rightarrow eRx$$
$$xRy \Rightarrow y^{-1}xyRy$$

moreover, the factors of the subnormal system which are defined by the predicate R belong to $\mathfrak{R}$. The last assertion can be expressed by means of the system of formulas of the first-order language. In fact, for any formula

$$\sigma = \underset{j=1}{\overset{m}{\&}} u_j(x_1, \ldots, x_n) = e \Rightarrow u_0(x_1, \ldots, x_n) = e$$

of the system Σ of quasi-identities defining the quasivariety $\mathfrak{R}$, the formula

$$\mathop{\&}_{i=1}^{n} \neg yRx_i \& (\mathop{\&}_{i=1}^{m} u_j(x_1, \ldots, x_n)Ry) \Rightarrow u_0(x_1, \ldots, x_n)Ry$$

expresses the fact that the factors of our system satisfy the quasi-identity σ.

We denote by Σ_1 the following system of formulas:

$$\{\sigma \mid \sigma \in \Sigma\} \cup \{\sigma \in \Sigma_0\}$$

The above reasoning shows that the existence on G of a predicate R such that on the algebraic system $G[R]$ the formulas of Σ_1 are valid is equivalent to the existence of a subnormal system of subgroups of G with factors from $\mathfrak{R}$.

Let T be an arbitrary finite subset of the group G. We construct a strictly decreasing chain of sets

$$T = T_0 \supset T_1 \supset \ldots$$

such that $T_{i+1} = T_i \cap N_i$ for any $i \in \mathbb{N}$, where N_i is the kernel of some homomorphism of the subgroup $\mathrm{gp}\, T_i$ onto a nontrivial group of $\mathfrak{R}$ if such a homomorphism exists. Since G is a locally $\mathfrak{R}$-projectable group, the chain is finite and ends with the empty set or with the trivial subgroup.

We now enrich the group G with predicates R and D, assuming that for elements $a, b \in G$ the formula aRb is true if and only if there exists an index m such that $f \in T_m$ and $b \in T_{m-1} \setminus T_m$, and assuming that Da is true if and only if $a \in T$. The algebraic system just obtained will be denoted by G^T.

It is easy to check that the system G^T satisfies the systems Σ_2 each of whose formulas is obtained from the corresponding formula of the σ_1 by imposing the restriction Dt on each term t which occurs in this formula.

Let Δ denote the family of all finite subsets of the group G. For every $T \in \Delta$ we set

$$K_T = \{S \in \Delta \mid T \subseteq S\}$$

Let $\mathcal{F}$ be an arbitrary ultrafilter on the set Δ containing the family of subsets $\{K_T \mid T \in \Delta\}$. Since the family $\{K_T \mid T \in \Delta\}$ is closed under intersections, such an ultrafilter exists.

We consider the ultraproduct

$$G^* = \overline{\prod_{T\in\Delta}} G^T/\mathcal{F}$$

of the algebraic systems G^T, $T \in \Delta$, by the ultrafilter $\mathcal{F}$. It is clear that the system Σ_2 of formulas is valid on G^*.

Finally, let G^0 be a subsystem of the system G^* whose underlying set consists of the equivalence classes modulo $\mathcal{F}$, defined by the constant functions. The reduct of the algebraic system G^0 to a group is isomorphic to the group G. Since all the formulas of the system Σ_2 are universal, they are also valid on G^0. For every element $a \in G$ the set

$$\{T \in \Delta \mid Da \text{ is true on } G^T\} = K$$

contains the set $K_{\{a\}}$. We see that K belongs to the ultrafilter $\mathcal{F}$. Therefore, the system G^0 satisfies the formula $\forall x Dx$. Consequently, the validity of the system Σ_2 on G^0 implies the validity of the system Σ_1 on G^0. But this means that the group G has a subnormal system of subgroups with factors in the quasivariety $\mathfrak{R}$, i.e., G has a subnormal system of subgroups with factors in $\mathfrak{R}$. □

In the statement of the following theorem we use the notation below:

— $\mathfrak{R}_s$ is the least superaccessible quasivariety containing $\mathfrak{R}$,

— $\mathfrak{R}_p$ is the class of all locally $\mathfrak{R}$-projectable groups,

— $\mathfrak{R}_q$ is the class of all groups having subnormal systems of subgroups with factors in $\mathfrak{R}$.

Theorem 7.1.2 (Brodskiĭ [91]). *For an arbitrary group quasivariety $\mathfrak{R}$ the following equalities hold:*

$$\mathfrak{R}_s = \mathfrak{R}_p = \mathfrak{R}_q = \mathrm{q}\,(\Gamma(\mathfrak{R}))$$

PROOF. By Lemmas 7.1.1 and 7.1.2, the following inclusions hold:

$$\mathfrak{R}_p \subseteq \mathfrak{R}_q \subseteq \mathrm{q}\,(\Gamma(\mathfrak{R}))$$

Let a group G be locally $\mathfrak{R}$-projectable and some formula from the system (7.1.1) fail to be true in G. Then there exist $a_1, \ldots, a_n$ such that

$$\mathop{\&}_{j=1}^{m} u_j(a_1, \ldots, a_n) = e$$

and the subgroup $\mathrm{gp}\,(a_1, \ldots, a_n)$ is nontrivial. Since $\mathrm{gp}\,(a_1, \ldots, a_n)$ is $\mathfrak{R}$-projectable, we conclude that the group defined by (7.1.2) is also $\mathfrak{R}$-projectable, which contradicts the definition of the system $\Gamma(\mathfrak{R})$. The contradiction obtained proves the inclusion $\mathfrak{R}_{\mathrm{q}} \subseteq \mathfrak{R}_{\mathrm{p}}$.

Let a group G have a subnormal system $\mathfrak{C}(G)$ with factors in $\mathfrak{R}$, and let H denote a nontrivial finitely presented subgroup of the group G. We consider the jump $N_1 \prec N_2$ in $\mathfrak{C}(G)$ such that N_2 is an intersection of all subgroups of $\mathfrak{C}(G)$ containing H, and let N_1 be the union of all subgroups of $\mathfrak{C}(G)$ which do not contain H. Then the restriction on H of the natural homomorphism of N_2 onto N_2/N_1 is a homomorphism of H onto a nontrivial group of $\mathfrak{R}$. Consequently, $\mathfrak{R}_{\mathrm{q}} \subseteq \mathfrak{R}_{\mathrm{p}}$. Thus, we have proved the equalities

$$\mathfrak{R}_{\mathrm{p}} = \mathfrak{R}_{\mathrm{q}} = \mathrm{q}\,(\Gamma(\mathfrak{R}))$$

The corresponding class will be denoted by $\mathfrak{R}_{\mathrm{p}}$. The equality $\mathfrak{R}^0 = \mathrm{q}\,(\Gamma(\mathfrak{R}))$ shows that $\mathfrak{R}^0$ is a quasivariety, and the equality $\mathfrak{R}^0 = \mathfrak{R}_{\mathrm{q}}$ implies that the class $\mathfrak{R}^0$ is contained in every superaccessible class containing $\mathfrak{R}$. It is not hard to see that for the group G there is a subnormal system of subgroups with factors in $\mathfrak{R}^0$ if and only if G is locally $\mathfrak{R}^0$-projectable. It is evident that a locally $\mathfrak{R}_0$-projectable group G is $\mathfrak{R}$-projectable. The above assertion and the equality $\mathfrak{R}^0 = \mathfrak{R}_{\mathrm{p}}$ imply that $\mathfrak{R}^0$ is a superaccessible class. Thus, $\mathfrak{R}^0 = \mathfrak{R}_{\mathrm{p}}$. □

Corollary 7.1.1. *The class $\mathfrak{O}_{\mathrm{c}}$ of groups admitting Conradian right orders coincides with the class $\mathfrak{J}$ of locally indicable groups.*

PROOF. Let $\mathfrak{A}$ denote the class of all torsion-free Abelian groups. Since for any finitely presented group of $\mathfrak{A}$ there is a homomorphism of the group onto an infinite cyclic group, the class of locally $\mathfrak{A}$-projectable groups coincides with the class of all locally indicable groups. The application of Theorem 7.1.1 to the class $\mathfrak{A}$ completes the proof. □

Thus, from Theorem 7.1.1 we obtain $\mathfrak{O}_{\mathrm{c}} = \mathfrak{J} \subseteq \mathfrak{O}_{\mathrm{r}}$. The question on the coincidence of varieties $\mathfrak{O}_{\mathrm{r}}$ and $\mathfrak{J}$ (cf. Theorems 7.4.1 and 7.4.2) will be considered below.

7.2. Model Theory

In this section, following Ehrenfeucth and Mostowski [92] and Rabin [93], we establish the relationship between right-orderable groups and the model theory.

Let $\mathfrak{X}$ be an arbitrary axiomatizable class of models, M be a model of $\mathfrak{X}$, and $\operatorname{Aut} M$ be the group of all automorphisms of the model M.

- We say that a group G
  - — has a *faithful representation by automorphisms* of the model M if G is isomorphic to some subgroup G_1 of the group $\operatorname{Aut} M$,
  - — is called *universal* if for every axiomatizable class $\mathfrak{X}$ containing an infinite model there is a model $M \in \mathfrak{X}$ such that G has a faithful representation by automorphisms of M.

We now describe how to construct a model $M^{(J,\langle I,\mathcal{D}\rangle)}$ starting from a given model M and a pair $(J, \langle I, D\rangle)$, where J is a totally ordered set and $\langle I, \mathcal{D}\rangle$ is a filter over some set I. With every element $j \in J$ we associate one copy $\langle I_j, \mathcal{D}_j\rangle$ of the filter $\langle I, \mathcal{D}\rangle$. Let $M^{\langle I,\mathcal{D}\rangle}$ designate the reduced power of the model M by the filter $\langle I, \mathcal{D}\rangle$. Let us agree to identify the constant elements of the reduced power $M^{\langle I,\mathcal{D}\rangle}$ by a filter $\langle I, \mathcal{D}\rangle$ with the corresponding elements of the model M. In other words, the model M is regarded as a submodel of the model $M^{\langle I,\mathcal{D}\rangle}$. Moreover, if N is a submodel of the model M and $\langle I, \mathcal{D}\rangle$ is a filter, we assume that $N^{\langle I,\mathcal{D}\rangle}$ is a submodel of the model $M^{\langle I,\mathcal{D}\rangle}$. We also assume that the nonconstant elements of the reduced power $M^{\langle I(j),\mathcal{D}(j)\rangle}$ of the model M by the filter $\langle I(j), \mathcal{D}(j)\rangle$ are obtained from the corresponding elements of the model $M^{\langle I,\mathcal{D}\rangle}$ by attaching the index j, and elements with distinct indices are not equal.

In view of the above assumptions, for distinct indices $j_1, j_2 \in J$ the following relations hold:

$$M^{\langle I(j_1),\mathcal{D}(j_1)\rangle} \cap M^{\langle I(j_2),\mathcal{D}(j_2)\rangle} = M$$

$$M^{\langle I(j_1),\mathcal{D}(j_1)\rangle} \cup M^{\langle I(j_2),\mathcal{D}(j_2)\rangle} \subseteq M^{\langle I(j_1),\mathcal{D}(j_1)\rangle\cdot\langle I(j_2),\mathcal{D}(j_2)\rangle}$$

where the expression

$$M^{\langle I(j_1),\mathcal{D}(j_1)\rangle\langle I(j_2),\mathcal{D}(j_2)\rangle}$$

stands for the reduced power of the model $M^{\langle I(j_1),\mathcal{D}(j_1)\rangle}$ by the filter $\langle I(j_2),\mathcal{D}(j_2)\rangle$.

We fix an element b of $M^{\langle I,\mathcal{D}\rangle}$, which does not belong to a submodel of M, and identify every element b^j with its index $j \in J$.

Let $T = \{j_1 < \ldots < j_n\}$ be a finite subset of a totally ordered set J. We denote by M_T the n-reduced power

$$M^{\langle I(j_1),\mathcal{D}(j_1)\rangle \ldots \langle I(j_n),\mathcal{D}(j_n)\rangle}$$

of the model M by the filters

$$\langle I(j_1),\mathcal{D}(j_1)\rangle \ldots \langle I(j_n),\mathcal{D}(j_n)\rangle$$

and let $M^{(J,\langle I,\mathcal{D}\rangle)}$ denote the set-theoretical sum of the models M_T, $T \in S_\omega(J)$, where $S_\omega(J)$ is the set of all finite subsets of J:

$$M^{(J,\langle I,\mathcal{D}\rangle)} = \bigcup_{M_T, T\in S_\omega(J)} M_T$$

The model $M^{(J,\langle I,\mathcal{D}\rangle)}$ is called the *reduced limit* of the model M by $(J,\langle I,\mathcal{D}\rangle)$.

Lemma 7.2.1 *Let M be an infinite model and φ: $J(1) \to J(2)$ be an isomorphism of the totally ordered set $J(1)$ onto $J(2)$. Then for any nonprincipal filter $\langle I,\mathcal{D}\rangle$ there is an isomorphism*

$$\widehat{\varphi} : M^{(J(1),\langle I,\mathcal{D}\rangle)} \to M^{(J(2),\langle I,\mathcal{D}\rangle)}$$

such that the restriction on $J(1)$ of $\widehat{\varphi}$ coincides with φ.

PROOF. We construct an isomorphism between $M^{(J(1),\langle I,\mathcal{D}\rangle)}$ and $M^{(J(2),\langle I,\mathcal{D}\rangle)}$ as follows. Let a be an arbitrary element of the model $M^{(J(1),\langle I,\mathcal{D}\rangle)}$. There is a finite subset $T = \{j_1 < \ldots < j_n\}$ of the set J_1 such that $a \in M_T$. We assume that the image of the set T under the isomorphism φ is the set $T' = \{j_1' < \ldots < j_n'\}$. Let $\widehat{\varphi}(a) = a'$, where a' is obtained by replacing the indices $j_1, \ldots, j_n$ associated with a by the indices $j_1', \ldots, j_n'$. It is easy to check that the mapping $\widehat{\varphi}$ is the required isomorphism. □

Corollary 7.2.1. *Let M be an infinite model, J be a totally ordered set, and $\langle I,\mathcal{D}\rangle$ be not a principal filter. Then any automorphism of the totally ordered set J is extended to an automorphism of the model $M^{(J,\langle I,\mathcal{D}\rangle)}$.*

We introduce the following definition (cf. [92]). A group of automorphisms $\operatorname{Aut} M$ of the model M *strongly contains* a group (G,Ω)

of order automorphisms of a totally ordered set Ω if the latter is the universe of M and there is a subgroup F of the group Aut M such that the restriction $F|_\Omega$ coincides with G.

Theorem 7.2.1 (Ehrenfeucth and Mostowski [92]). *For an arbitrary totally ordered set Ω, in an axiomatizable model class with an infinite model there is a model of which the group of automorphisms strongly contains the group of order automorphisms of Ω.*

Theorem 7.2.2 (Rabin [93]). *The class of universal groups coincides with the class $\mathfrak{O}_r$ of right-orderable groups.*

PROOF. Since the class of totally ordered sets is axiomatizable and contains infinite totally ordered sets, any universal group G is a subgroup of the group of automorphisms of a totally ordered set. By Theorem 3.4.1, the group G is right-orderable.

Let an axiomatizable class $\mathfrak{X}$ of models of signature Σ contain an infinite model, T be the elementary theory of $\mathfrak{X}$, and G be a right-orderable group such that it is a group of automorphisms of a totally ordered set Ω.

To construct a model $B \in \mathfrak{X}$ of which the group of automorphisms contains a subgroup isomorphic to G, we apply the Skolemization method (cf., for example, [94]). We can assume that any sentence σ of the theory T is written in the prenex normal form

$$\sigma = (\forall x_1)\dots(\forall x_n)(\exists y_1)\dots$$
$$(\exists y_m)(\forall x_{n+1})\dots(\exists y_{m+1})\dots\Phi(x_1,\dots,x_t,y_1,\dots,y_{\rho(\sigma)})$$

We extend the signature Σ by adding, for each variable y_i, a new symbol of k-ary operation f_i^σ, $1 \leqslant i \leqslant \rho(\sigma))$, where k denotes the number of universal quantifiers preceding $\exists y_i$ in the formula σ. We denote by $\overline{\Sigma}$ the signature obtained (it is called the *Skolemization of signature* Σ). Let

$$\overline{\sigma} = (\forall x_1)\dots(\forall x_t)\Phi(x_1,\dots,x_t,f_1^\sigma(x_1,\dots,x_n),\dots,f_{\rho(\sigma)}^\sigma(x_1,\dots,x_t))$$

$$\overline{\mathrm{T}} = \{\overline{\sigma} \mid \sigma \in \mathrm{T}\}$$

As is well known (cf., for example, [94]),

— every model $M = \langle A, \Sigma\rangle$ of signature Σ is a reduct of some model $\overline{U}$ of the theory $\overline{T}$ of signature $\overline{\Sigma}$,

— if $\overline{M}$ is a model of the theory $\overline{\mathrm{T}}$ of signature $\overline{\Sigma}$, then the reduct $\overline{M}|_{\Sigma} = M_1$ of the system $\overline{M}$ to the signature Σ is a model of the theory T,

— if an algebraic system $\overline{M} = \langle A, \overline{\Sigma} \rangle$ of signature $\overline{\Sigma}$ is a model of the theory $\overline{\mathrm{T}}$, B is a subset of the set A, and $\overline{B}$ is a subsystem of the algebraic system $\overline{M} = \langle A, \overline{\Sigma} \rangle$; generated by the set B; then the reduct of the system $\overline{B}$ to the signature Σ is a model of the theory T.

Thus, the set of formulas

$$\overline{\mathrm{T}} = \{\overline{\sigma} \mid \sigma \in \mathrm{T}\}$$

has an infinite model. Let $\overline{M} = \langle M, \overline{\Sigma} \rangle$ be a model of the theory $\overline{\mathrm{T}}$ such that $\Omega \subseteq M$, and let $\mathrm{Aut}\, M$ strongly contain the group G. Let $\overline{B} = \langle \overline{\Omega}, \overline{\Sigma} \rangle$ be a subsystem of the system $\overline{M}$ generated by the set Ω and $g \in G$, $\varphi \in \mathrm{Aut}\, \overline{M}$ be such that $g = \varphi|_{\Omega}$. We assert that $\overline{g} = \varphi|_{\overline{\Omega}}$ is a mapping of $\overline{\Omega}$ onto itself. Indeed, if $\omega \in \overline{\Omega}$, then $\omega = w(\omega_1, \dots, \omega_n)$, where w is a term constructed from a finite set of operations $f \in \Sigma$, $f_i^{\sigma} \in \overline{\Sigma} \setminus \Sigma$, and $\omega_1, \dots, \omega_n \in \Omega$. Then

$$\omega\overline{g} = \omega\varphi = w(\omega_1\varphi, \dots, \omega_n\varphi) = w(\omega_1 g, \dots, \omega_n g)$$

Since $\omega_1 g, \dots, \omega_n g \in \Omega$ and $\overline{\Omega}$ is closed with respect to f_i^{σ}, we obtain $\omega\overline{g} \in \overline{\Omega}$. Furthermore, the element

$$c = w(\omega_1\varphi^{-1}, \dots, \omega_n\varphi^{-1}) = w(\omega_1 g^{-1}, \dots, \omega_n g^{-1})$$

belongs to $\overline{\Omega}$; moreover, $c\overline{g} = c\varphi = b$. Thus, g maps $\overline{\Omega}$ onto $\overline{\Omega}$.

We now assume that $\overline{g}_1$ and $\overline{g}$ are two automorphisms of the model $\overline{B}$ such that

$$\overline{g}|_{\Omega} = \overline{g}_1|_{\Omega} = g \in G$$

For any element $\omega \in \Omega$ we have

$$\omega\overline{g} = w(\omega_1\overline{g}, \dots, \omega_n\overline{g}) = w(\omega_1 g, \dots, \omega_n g) = w(\omega_1\overline{g}_1, \dots, \omega_n\overline{g}_1) = \omega\overline{g}_1$$

Therefore, $\overline{g} = \overline{g}_1$. Thus, every element $g \in G$ is the restriction on Ω of only one automorphism $\overline{g} \in \mathrm{Aut}\, \overline{B}$. Hence the group $\mathrm{Aut}\, \overline{B}$ contains a subgroup isomorphic to the group G.

Let $B_1 = \overline{B}|_{\Sigma}$ be the reduct of the system $\overline{B}$ to the signature Σ. As already mentioned, B_1 is a model of the theory T and the group of automorphisms $\mathrm{Aut}\, B_1$ contains a subgroup that is isomorphic to the group G. □

7.3. Axiomatic Rank

We begin with the definition of axiomatic rank.

- A quasivariety $\mathfrak{K}$ has *axiomatic rank n* if it is prescribed by a system of quasi-identities in n variables but cannot be described by a system of quasi-identities in fewer variables. We say that a quasivariety $\mathfrak{K}$ has an *infinite axiomatic rank* if such a natural number n does not exist.

The following groups are of great importance in finding axiomatic ranks of the basic quasivarieties of right-ordered groups:

$$G_{m,k} = \mathrm{gp}\,(x_1, \dots, x_{2m}, c \mid [x_i, x_j, x_l] = e,\ i, j, l = 1, \dots, 2m$$
$$c^{-1}x_i c = x_i^{-1}, i = 1, \dots, 2m;\ c^{4k} = \prod_{i=1}^{m}[x_i, x_{m+i}]) \tag{7.3.1}$$

where $m, k \geqslant 1$.

Lemma 7.3.1. *For any natural numbers m and k the groups $G_{m,k}$ are torsion-free.*

PROOF. We note that any element $g \in G_{m,k}$ is uniquely represented in the form

$$g = c^{k_0}x_1^{k_1}\dots x_{2m}^{k_{2m}} \prod_{1\leqslant i<j\leqslant 2m} [x_i, x_j]^{l_{ij}}$$

where $k_0 \in \{0, 1, \dots, 4k-1\}$, $k_1, \dots, k_{2m}, l_{ij} \in \mathbb{Z}$, $i, j \in 1, \dots, 2m$. Therefore, the subgroup $K = \mathrm{gp}\,(x_1, \dots, x_{2m})$ is a free nilpotent group of class 2 with free generators $x_1, \dots, x_{2m}$. We assume that g has a finite order in $G_{m,k}$. Taking instead of g its power, we can assume that $k_0 = 2k$. Since $g^2 \in K$ and K is a torsion-free group, we have $g^2 = e$. Thus,

$$\prod_{i=1}^{m}[x_i, x_{m+i}](x_1^{k_1}\dots x_{2m}^{k_{2m}})^2 \prod_{1\leqslant i<j\leqslant 2m} [x_i, x_j] = e$$

However, the last relation is impossible for a free nilpotent group of class 2. □

Lemma 7.3.2. *For any natural numbers m and k the group $G_{m,k}$ is not right-orderable.*

PROOF. The assertion follows immediately from Theorem 3.2.2. As a subgroup K we must take the subgroup $K = \mathrm{gp}(x_1, \dots, x_{2m})$. The index of the subgroup K in $G_{m,k}$ is finite, and any right order on K is Conradian by Theorem 2.4.2. Since the quotient group $G_{m,k}/G'_{m,k}$ by its derived subgroup $G'_{m,k}$ is a finite group, the group $G_{m,k}$ is not right-orderable. □

Lemma 7.3.3 (Fedorov [95]). *Let $K = \mathrm{gp}(x_1, \dots, x_{2m})$ be a free nilpotent group of class 2 with free generators $x_1, \dots, x_{2m}$. If the relation*

$$h^s = \prod_{i=1}^{t} [g_i, g_{t+i}]$$

holds for some elements $h, g_1, \dots, g_{2t}$ of the group K and the integer s, then the element h can be represented as a product of C^2_{2t} commutators.

PROOF. Let $A = \mathrm{gp}(g_1, \dots, g_{2t})$, and let I be the isolator of the subgroup AK'/K' of the quotient group K/K'. We choose the system of elements $b_1K', \dots, b_rK'$ of K/K', where $b_1, \dots, b_r \in K$, $r \leqslant 2t$, are free generators of a free Abelian group I, and complete this system so as to arrive at the system of free generators $b_1K', \dots, b_{2m}K'$ of the group K/K'. Applying Theorem 42.51 of [39], we conclude that $\{b_1, \dots, b_{2m}\}$ are free generators of the group K. Thus, for some $l(ij) \in \mathbb{Z}$, where $i, j \in \{1, \dots, 2m)\}$, the following equality holds:

$$h = \prod_{1 \leqslant i < j \leqslant 2m} [b_i, b_j]^{l(ij)}$$

Since $h^s \in \mathrm{gp}(b_1, \dots, b_r)$, we obtain $l(ij) = 0$, where $j > r$, $1 \leqslant i \leqslant 2m$. Thus,

$$h^s = \prod_{1 \leqslant i < j \leqslant r} [b_i, b_j^{sl(ij)}]$$

which completes the proof. □

Let a group variety $\mathfrak{M}_p$ be given by the following identities:

$$x^p = e, \quad [x, y, z] = e$$

where $p \geqslant 3$ is a prime number. Throughout this section, F_m stands for a free group of rank m from $\mathfrak{M}_p$ and F'_m is the derived subgroup of F_m.

Lemma 7.3.4 (Fedorov [95]). *An element*

$$a = [a_1, a_2] \dots [a_{2n-1}, a_{2n}] \in F_m$$

can be represented as a product of less than n commutators if and only if the elements $a_1, \dots, a_{2n}$ *are linearly dependent, modulo* F'_m.

PROOF (Budkin). We first assume that $a_1, \dots, a_{2n}$ are linearly dependent, modulo F'_m, i.e., $a_{1^{k_1}} \dots a_{2n}^{k_{2n}} \in F'_m$ and $k_i \not\equiv 0 \pmod p$ for some integer i, $1 \leqslant i \leqslant 2n$. For definiteness, we set $i = 1$. Then $a_1 = a_2^{m_2} \dots a_{2n}^{m_{2n}} c$ for suitable integers $m_2, \dots, m_{2n}$ and $c \in F'_m$. Therefore,

$$\begin{aligned} a &= [a_{2^{m_2}} \dots a_{2n}^{m_{2n}}, a_2][a_3, a_4] \dots [a_{2n-1}, a_{2n}] \\ &= \prod_{s=2}^{n} [a_{2s-1}, a_2]^{m_{2s-1}} [a_{2s-1}, a_{2s}][a_{2s}, a_2]^{m_{2s}} \end{aligned}$$

In any nilpotent group of class 2 the identity

$$[x, z]^k [x, y][y, z]^l = [xyz^{k-l}, xy^2 z^{2k-l}]$$

holds. Hence the element a can be represented as a product of $n - 1$ commutators.

Conversely, let $a_1, \dots, a_{2n}$ be linearly independent, modulo F'_m. It is clear that $2n \leqslant m$. We complete this system so as to obtain the system $a_1, \dots, a_m$ of free generators of the group F_m. We assume that a can be represented as a product of less than n commutators, i.e.,

$$[a_1, a_2] \dots [a_{2n-1}, a_{2n}] = \prod_{i=1}^{k} [t_i(a_1, \dots, a_m), t_{k+i}(a_1, \dots, a_m)]$$

for some $t_1, \dots, t_{2k}$ of the group signature and $k < n$. Any relation connecting free generators of the group F_m is an identity in $\mathfrak{M}_p$, i.e., the equality just obtained can be regarded as an identity in $\mathfrak{M}_p$. Its validity in an arbitrary group $G \in \mathfrak{M}_p$ implies that any element $g \in G'$ can be written as a product of at most k commutators. Using this fact, we estimate the order of the derived subgroup of an arbitrary free group $F_s \in \mathfrak{M}_p$.

In the group F_s the number of elements of the form $[u, v]$ is at most p^{2s}. Hence in the group F_s, the number of elements of the form $[u_1, v_1] \dots [u_k, v_k]$ is at most p^{2ks}. Since any element of F'_s can be written in such a form, we have $| F'_s | \leqslant p^{2sk}$. But since $| F'_s | = p^{s(s-1)/2}$, for $s > 4k + 1$ a contradiction arises, which proves the lemma. □

Corollary 7.3.1. *Let $K = \text{gp}(x_1, \ldots, x_m)$ be a free nilpotent group of class 2 with free generators $x_1, \ldots, x_m$. If $g_1, \ldots, g_{2n}$ are independent, modulo K', then the element $g = [g_1, g_2] \cdots [g_{2n-1}, g_{2n}]$ cannot be represented as a product of less than n commutators.*

Theorem 7.3.1 (Budkin [96]). *The quasivarieties $\mathfrak{O}_r$ of right-orderable groups and $\mathfrak{J}$ of locally indicable groups have infinite axiomatic rank.*

PROOF. It suffices to construct, for each natural number n, a group B_n such that each of its n-generated subgroups belongs to $\mathfrak{O}_r(\mathfrak{J})$, but the group B_n itself does not belong to $\mathfrak{O}_r(\mathfrak{J})$ or $\mathfrak{J}$. As B_n we can take the group (7.3.1) with a natural number $k \geqslant 1$ and $m = n^4$. Indeed, by Lemma 7.3.2, the group $G_{m,k}$ is not right-orderable. In view of Corollary 3.2.2, $G_{m,k}$ is not locally indicable. Hence $G_{m,k} \notin \mathfrak{O}_r$ and $G_{m,k} \notin \mathfrak{J}$.

Let A be an arbitrary n-generated subgroup of the group $G_{m,k}$. We show that $A \in \mathfrak{J}$ and, consequently, $A \in \mathfrak{O}_r$. For definiteness, we assume that $A = \text{gp}(c^{e_1}a_1, \ldots, c^{e_n}a_n)$, where

$$a_i \in K = \text{gp}(x_1, \ldots, x_{2m})$$
$$e(i) \in \{0, 1, \ldots, 4k - 1\}$$

We also introduce the following notation:

$$C = \text{gp}(a_1, \ldots, a_n, a_1^c, \ldots, a_n^c, c^{4k})$$

I is the isolator of the subgroup C' of the group K, and $B = \text{gp}(C, I, c)$. Our next purpose is to construct a homomorphism ψ of the group B onto an infinite cyclic group such that its kernel is contained in the subgroup $\text{gp}(c^2, K)$. If $C \subseteq K'$, the group B is a torsion-free Abelian group by Lemma 7.3.1. Therefore, it is locally indicable. Since $A \subseteq B$, the group A is also locally indicable. Hence we can assume that $C \nsubseteq K'$.

We choose elements $b_1, \ldots, b_t \in C$ such that $b_1K', \ldots, b_tK'$ are free generators of the nontrivial free Abelian group CK'/K'. From the inclusions $a_ia_i^c \in K'$, $i = 1, 2, \ldots, n$, it follows that $1 \leqslant t \leqslant n$. For an element $g \in B$ we denote by $\overline{g}$ its image under the action of the natural homomorphism $B \to B/I$ and show that $\overline{M} = \text{gp}(\overline{b}_1, \ldots, \overline{b}_t)$ is an isolated group of CI/I.

Let $\overline{x}^s = \overline{b}_1^{r_1} \ldots \overline{b}_t^{r_t}$ for $x \in CI$ and $s \neq 0$. Then

$$x^sK' = b_1^{r_1} \ldots b_t^{r_t}K'$$

Since $b_1K', \ldots, b_tK'$ are generators of the free Abelian group CK'/K' and $xK' \in CK'/K'$, all the numbers r_i are divisible by s for any $i = 1, \ldots, t$.

Let $r_i' = r_i \cdot s^{-1}$. Since CI/I is a torsion-free Abelian group, we obtain

$$\overline{x} = \overline{b}_1^{r_1'} \ldots \overline{b}_t^{r_t'}$$

i.e., $\overline{x} \in \overline{M}$. Therefore, $\overline{M}$ is an isolated subgroup of the group CI/I. As is known, any isolated subgroup of a finitely generated Abelian group can be singled out by means of a direct factor. Therefore, $CI/I = \overline{M} \times \overline{N}$ for some subgroup $\overline{N}$ of CI/I.

We show that $\overline{c}^{4k} \neq e$. To this end, we assume the contrary: $\overline{c}^{4k} = e$, i.e., $c^{4k} \in I$. For $t = 1$, the group CK' is contained in the group $\mathrm{gp}(b_1, K')$. Thus, $C' = I = (e)$. Consequently, $c^{4k} = e$, which is not true.

Let now $t \geqslant 2$. Since CK'/K' is a free Abelian group of rank t, any element of C' can be represented as a product of C_t^2 commutators. In view of Lemma 7.3.3, c^{4k} can be represented as a product of C_{2l}^2 commutators, where $l = C_t^2$. But, by Lemma 7.3.4, the element $\prod_{i=1}^{m} [x_i, x_{i+m}]$ cannot be written as a product of less than m commutators. The inequality $C_{2l}^2 < m$ implies $\overline{c}^{4k} \neq e$.

Since the elements $b_1K', \ldots, b_tK'$ are linearly independent, modulo K', the following inclusion holds: $\overline{c}^{4k} \in \overline{N}$. We now show that the cyclic subgroup $(\overline{c}^{4k})$ is singled out in the group $\overline{N}$ by a direct factor.

By the theorem on subgroups of free finitely generated Abelian groups (cf., for example, [53, Theorem 3.2.1]), we can choose free generators $\overline{c}_1, \ldots, \overline{c}_q$ of the group $\overline{N}$, $c_i \in C$, $i = 1, \ldots, q$, such that $\overline{c}^{4k} = \overline{c}_1^d$ for some natural number d. We now show that $d = 1$. To this end, we assume the contrary: $d \neq 1$. The equality $\overline{c}^{4k} = \overline{c}_1^d$ implies the equality

$$\prod_{i=1}^{m} [x_i, x_{m+i}] = c_1^d h$$

for a suitable element $h \in I$. As already mentioned, for $t = 1$ we have $C' = I = E$. Consequently, $h = e$. It remains to note that the relation

$$\prod_{i=1}^{m} [x_i, x_{m+i}] = c_1^d$$

fails in K for $d \geqslant 2$. Thus, we can assume that $t \geqslant 2$.

We now take an element $g \in C'$ such that $h^s = g$ for some integer $s > 0$. Since C' is a free Abelian group of rank C_t^2, any of its elements (in particular, g) can be represented as a product of C_t^2 commutators. By Lemma 7.3.3, the element h can be represented as a product of $r = C_{2l}^2$ commutators, where $l = C_t^2$.

Thus, we can assume that

$$h = \prod_{i=1}^{r} [g_i, g_{r+i}]$$

where $g_1, \ldots, g_{2r}$ are some elements of K.

We first study the case of even d. Let F_2 denote a free group of the variety defined by the identities $x^4 = e$, $[x, y]^2 = e$, $[x, y, z] = e$:

$$F_2 = \mathrm{gp}\,(y_1, \ldots, y_{2m})$$

It is evident that the mapping $x_i \to y_i$, $i = 1, \ldots, 2m$, can be extended to a homomorphism φ of the group K into F_2. Let $g_i\varphi = f_i(y_1, \ldots, y_{2m})$, $i = 1, \ldots, 2r$. Since $2 \leqslant t \leqslant n$ and $r = C_{2C_t^2}^2 < n^4 = m$, in F_2 the following relation holds:

$$\prod_{i=1}^{m} [y_i, y_{m+i}] = \prod_{i=1}^{r} [f_i(y_1, \ldots, y_{2m}), f_{r+i}(y_1, \ldots, y_{2m})]$$

The latter is an identity in F_2 and implies that every element of F_2' can be represented as a product of r commutators. However, the group F_2 contains at most 2^{4m} distinct elements of the form $[u, v]$. Therefore, from the above identity it follows that $|F_2'| \leqslant 2^{4mr}$. Indeed, $|F_2'| = 2^{2(2m-1)}$. However, by the choice of m we have $4mr < m(2m-1)$. Hence d is odd.

Let d be a multiple of some prime number $p > 2$. We can argue as above, but instead of F_2 we must consider the free group F_p of the variety defined by the identities $x^p = e$, $[x, y, z] = e$. Then $d = 1$ and, consequently, the cyclic subgroup $(\overline{c}^{\,4k})$ is singled out in the group $\overline{N}$ by a direct factor.

Thus, for some subgroup $\overline{X}$ of the group $\overline{N}$ the following decomposition takes place:

$$CI/I = (\overline{b}_1) \times \ldots \times (\overline{b}_t) \times (\overline{c}^{\,4k}) \times \overline{X}$$

Since $b_i^c = b_i^{-1} b_i b_i^c$ and $b_i b_i^c \in K'$, there exist elements $\overline{f}_1, \ldots, \overline{f}_t$ of $\overline{X}$ and integers $l_1, \ldots, l_t$ such that

$$\overline{b}_i^{\,\overline{c}} = \overline{b}_i^{\,-1} \overline{c}^{\,4kl_i} \overline{f}_i, \quad i = 1, \ldots, t$$

Let

$$\overline{L} = \mathrm{gp}\,(\overline{b}_1^{-1}\overline{c}^{2kl_1}, \ldots, \overline{b}_t^{-1}\overline{b}_t^{-1}\overline{c}^{2kl_t}, \overline{X})$$

It is easy to see that $\overline{L}$ is a normal subgroup of the quotient group $\overline{B} = B/I$. Moreover, $(\overline{c}) \cap \overline{L} = E$. Hence $\overline{B}/\overline{L}$ is an infinite cyclic group.

Let ψ be a natural homomorphism of B onto $\overline{B}/\overline{L}$. It is evident that its kernel is contained in the group $\mathrm{gp}\,(c^2, K)$. Since $\mathrm{gp}\,(c^2, K)$ is a torsion-free nilpotent group, the kernel of the homomorphism ψ is a locally indicable group. Hence the group B is also locally indicable and, consequently, right-orderable by Corollary 3.3.1. □

7.4. Locally Indicable Groups

As has already been indicated, any locally indicable group is right-orderable. For a long time it was not known whether every right-orderable group is locally indicable. The first example that had a negative answer was constructed by Bergman [89]. Another example was given by Tararin [97]. In this section, we present both examples.

7.4.1. In this subsection, we discuss the example constructed by Tararin [97]. Let $\mathcal{A}(\mathbb{Q})$ be the group of all order automorphisms of a totally ordered set $\mathbb{Q}$ of rational numbers. We consider a subgroup G of $\mathcal{A}(\mathbb{Q})$ which is formed by order automorphisms f of $\mathcal{A}(\mathbb{Q})$ such that

(a) $f(\xi + 1) = f(\xi) + 1$,

(b) on the interval $[\![\lambda, \mu]\!]$ the order automorphism f is a piecewise linear function.

It is obvious that the group G is right-orderable. Let GZ and GQ be subgroups of G consisting of the functions

$$l_q(\xi) = \xi + q$$

where $q \in \mathbb{Z}$ and $q \in \mathbb{Q}$ respectively. It is evident that GZ is a central subgroup of the group G.

Lemma 7.4.1. *Let H be a normal subgroup of G which is not contained in GZ. Then H contains the stabilizer G_0 of 0 in G.*

PROOF. Let D be a subgroup of G formed by the functions $f \in G$ for which there exist rational numbers α, β, $0 < \alpha < \beta < 1$, such that $f(\xi) = \xi$ for all ξ of $[\![0, \alpha]\!] \cup [\![\beta, 1]\!]$. It is easy to see that the group D is isomorphic to the Chehata group $D^0([\![0, 1]\!])$ (cf. 4.4.2).

We show that $H \cap D \neq E$. Let $f \in H$, $f \in GZ$. We set $g = f \cdot l_n^{-1}$, where n is an integer. Since g is a nonidentity automorphism, there is a rational number τ such that $0 < \tau, g(\tau) < 1$, $g(\tau) \neq \tau$, $g(\tau) \neq g^{-1}(1)$. We choose rational numbers α, β, γ such that $0 < \alpha < g(\tau) < \beta < \gamma < 1$ and the interval $[\![\alpha, \gamma]\!]$ does not contain τ and $g^{-1}(1)$. Let h be an arbitrary order automorphism of the group G such that $h(\alpha) = \alpha$, $h(g(\tau)) = \beta$, and h is a linear function on each of the intervals $[\![\alpha, g(\tau)]\!]$ and $[\![g(\tau), \gamma]\!]$; moreover, at the rest of the points ξ of the interval $[\![0, 1]\!]$ the equality $h(\xi) = \xi$ is fulfilled.

It is obvious that $p = g^{-1}h^{-1}gh \neq e$ and $p \in D$. Since $l_n \in GZ$, we have

$$f^{-1}h^{-1}fh = (gl_n)^{-1}h^{-1}(gl_n)h = g^{-1}h^{-1}gh = p$$

Hence $D \cap H \neq E$. By Theorem 4.4.3, the group D is simple. Consequently, $D \cap H \neq E$.

For the stabilizer G_0 of the point 0 in the group G we introduce the following notation: K is a subgroup of G formed by the order automorphisms $f \in G_0$ such that there exists a rational number α, $0 < \alpha < 1$, such that $f(\xi) = \xi$ on the interval $[\![0, \alpha]\!]$ and L is a subgroup formed by the functions $f \in G_0$ such that $f(\xi) = \xi$ on the interval $[\![\alpha, 1]\!]$ for some $\alpha \in \mathbb{Q}$, $0 < \alpha < 1$. In view of the condition (a), for any $f \in K \cup L$ there is a number $q \in \mathbb{Q}$ such that $l_q^{-1} \cdot f \cdot l_q \in D$. By virtue of the normality of H and the inclusion $D \subseteq H$, the subgroups K and L are contained in H.

Let $f \in G_0$. There are rational numbers α, β and an automorphism $h \in G_0$ such that $0 < \alpha$, $f(\alpha) < \beta < 1$ and $h(\alpha) = f(\alpha)$, $h(\beta) = \beta$. Let $g \in G$ such that $g(\xi) = f(\xi)$ for $0 \leqslant \xi \leqslant \alpha$ and $g(\xi) = h(\xi)$, $g(\xi) = \xi$ for $\beta \leqslant \xi \leqslant 1$. Then for $r = g^{-1}f$ the inclusion $r \in L$ is fulfilled. Therefore, $f = gr$, where $g \in K$, $r \in L$. Hence $G_0 \subseteq H$. □

Lemma 7.4.2. *Let H be an isolated subgroup of a group G and $H \neq G$. Then $H \cap GQ \neq E$.*

PROOF. Let $J(GZ)$ be the isolator of the subgroup GZ of G. Since GZ is a normal subgroup of G, we conclude that $J(GZ)$ is also a normal subgroup of G. Since $GQ \subseteq J(GZ)$, we have $J(GQ) = J(GZ)$.

By Lemma 7.4.1, $G_0 \subseteq J(GQ)$. For $f \in G$ there is an element $l_q \in GQ$ such that $f(0) = l_q(0)$. Hence $h = f \cdot l_q^{-1}$ is contained in G_0 and $f = h \cdot l_q$. Consequently, $G_0 \cdot GQ = G$ and $J(GQ) = G$.

Let H be an arbitrary isolated subgroup of G and $H \neq G$. In the case $H \cap GQ \neq E$ there are no nontrivial isolated subgroups. Therefore, $J(GQ) \subseteq H$ and, as above, we arrive at the equality $H = G$, which contradicts the inequality $H \neq GQ = E$. Hence $H \cap GQ = E$. □

Theorem 7.4.1 (Tararin [97]). *The group G of all order automorphisms of $\mathbb{Q}$ subject to the conditions* (a)–(b) *is right-orderable but not locally indicable.*

PROOF. We assume the contrary. Let a group G be locally indicable. By Corollary 6.1.3, there is a Conradian right order on G. All subgroups G_α of the system $\mathfrak{C}(G)$ formed by all convex subgroups of the group G are isolated.

By Lemma 7.4.2, $G_\alpha \cap GQ = E$ for any proper subgroup G_α of $\mathfrak{C}(G)$. Therefore, in $\mathfrak{C}(G)$ there is the greatest subgroup G_γ; moreover, its intersection with GQ is the trivial subgroup. Since the order on G is Conradian, G_γ is a normal subgroup of the group G and $G_0 \subseteq G_\gamma$, $G_0 \neq G_\gamma$.

Let $f \in G_\gamma \setminus G_0$. Then $f = h \cdot l_q$ for some $h \in G_0$ and $l_q \in GQ$, where $l_q \neq e$. Therefore, $l_q = h^{-1}f \in G_\gamma$. Hence $G_\gamma \cap GQ \neq E$, which contradicts the above assumption. Thus, the group G is not locally indicable. □

7.4.2. The Bergman example [89] of a finitely generated right-orderable group which is not indicable is based on ideas that are different from those in the example from Section 7.4.1.

As usual, we denote by $SL_2(\mathbb{R})$ the group of 2×2-matrices over the field $\mathbb{R}$ with the determinants 1. Let U be the universal covering group of $SL_2(\mathbb{R})$ (cf., for example, [98]). The elements of the group U can be regarded as linear transformations of the infinitely-sheeted plane P with the branch point 0. It is evident that the group $SL_2(\mathbb{R})$ acts exactly on the set of all rays X passing through 0 in P with the branch point 0. This set of rays can be identified with the real line $\mathbb{R}$. It is clear that the group U acts on this set in an obvious way and preserves the natural order on $\mathbb{R}$. The elements of U regarded as order automorphisms of the totally ordered set $\mathbb{R}$ have the periodicity property:

$$(\xi + \pi)f = (x)f + \pi$$

We observe the following fact. Let $h \in SL_2(\mathbb{R})$, and let a be a ray passing through the eigenvector of h. Then there exists exactly one element $g \in U$ covering h and leaving fixed all the rays of the plane P, which cover a.

We now consider the following parametrized families of elements of the group $SL_2(\mathbb{R})$:

$$a^r = \begin{bmatrix} 1 & -r \\ 0 & 1 \end{bmatrix}, \quad b^r = \begin{bmatrix} 1 & 0 \\ r & 1 \end{bmatrix}, \quad c_s = \begin{bmatrix} s^{1/2} & 0 \\ 0 & s^{-1/2} \end{bmatrix} \tag{7.4.1}$$

It is clear that

$$a^r \cdot a^{r'} = a^{r+r'}, \quad b^r \cdot b^{r'} = b^{r+r'} \quad c_s \cdot c_{s'} = c_{ss'} \tag{7.4.2}$$

Moreover,

$$c_s a^r c_s^{-1} = a^{rs}, \quad c_s^{-1} b^r c_s = b^{sr} \tag{7.4.3}$$

and for

$$d = aba = \begin{bmatrix} 0 & -1 \\ 1 & 0 \end{bmatrix} \tag{7.4.4}$$

we have

$$d^{-1} c_s d = c_{s^{-1}} \tag{7.4.5}$$

$$d^{-1} a^r d = b^r, \quad d^{-1} b^r d = a^r \tag{7.4.6}$$

Let r and s be real numbers. For a^r, b^r, and c_s there exist unique covers A^r, B^r, and C_s such that if a ray $u \in X$ covers a ray of proper subspace generated by a^r, b^r, and c_s, then it is a "fixed point" with respect to A^r, B^r, and C_s. We note that the corresponding equalities also hold in the universal covering group U:

$$A^r \cdot A^{r'} = A^{r+r'}, \quad B^r \cdot B^r \cdot B^{r'} = B^{r+r'}, \quad C_s \cdot C_{s'} = C_{s+s'}, \tag{7.4.2$'$}$$

$$C_s A^r C_s^{-1} = A^{sr}, \quad C_s^{-1} B^r C^s = B^{sr} \tag{7.4.3$'$}$$

From Eqs. (7.4.2) and (7.4.3), it follows that the left-hand and right-hand sides of each equality in (7.4.2′) and (7.4.3′) define elements that cover the same element of $SL_2(\mathbb{R})$. Therefore, it suffices to show that every side of all five equalities (7.4.2′) and (7.4.3′) determines an element corresponding to a fixed ray. The right-hand sides of the above-mentioned equalities determine elements corresponding to fixed rays by definition. For the left hand-side of the first formula such an assertion is true because all real powers of the transformation a have a common fixed ray. Hence a similar equality holds for the chosen coverings. An analogous reasoning can be applied to the rest of the formulas.

The left-hand sides of Eqs. (7.4.3′) have fixed rays, since they are conjugate to elements corresponding to fixed rays. Thus, the verification of the formulas (7.4.2′) and (7.4.3′) is complete.

By analogy with (7.4.4) we now introduce an element

$$D = ABA \tag{7.4.4'}$$

The validity of the equalities

$$D^{-1} C_s D = C_{s^{-1}} \tag{7.4.5'}$$

$$D^{-1} A^r D = B^r \tag{7.4.6'}$$

can be established as above. For any integer $n > 1$ we consider finitely generated subgroups $U_n = \operatorname{gp}(A, B, C_n)$ of the group U. By (7.4.3′) and (7.4.5′), the elements A, B, and C are conjugate to their powers. Hence any homomorphism of the group $U_n = \operatorname{gp}(A, B, C_n)$ onto an infinite cyclic group is trivial. Thus, we arrive at the following assertion.

Theorem 7.4.2 (Bergman [89]). *There is a finitely generated right-orderable group which is not indicable.*

It is easy to prove that the derived subgroup U_n' of the group U_n does not coincide with U_n. We now show that there exists a finitely generated right-orderable group that coincides with its derived subgroup. From (7.4.4′) it follows that

$$D \in \operatorname{gp}(A, B) \subseteq U$$

Conjugating the last relation by the element C_s and simplifying the result with the help of (7.4.3′) and (7.4.5′), we obtain

$$DC_{s^2} \in \operatorname{gp}(A^{s^{-1}}, B^s)$$

Setting $s = n$, where n is natural, we can conclude that the group $\operatorname{gp}(A^{n^{-1}}, B)$ contains the elements of D and D_{n^2}. Hence

$$C_{n^2} \in \operatorname{gp}(A^{n^{-1}}, B)$$

We consider the group $U(1/6) = \mathrm{gp}\,(A^{1/6}, B)$. Since

$$U(1/6) \supseteq \mathrm{gp}\,(A^{1/2}), \mathrm{gp}\,(A^{1/3}, B)$$

we have $C_4, C_9 \in U(1/6)$. By (7.4.3′), the generators $A^{1/6}, B$ are conjugate to their powers $(A^{1/6})^4$ and B^9 respectively. But since the numbers 3 $(= 4 - 1)$ and 8 $(= 9 - 1)$ are relatively prime, the quotient group $U(1/6)/U'(1/6)$ is trivial.

Corollary 7.4.1 (Bergman [89]). *There exists a finitely generated right-orderable group which coincides with its derived subgroup.*

We note that with the help of an embedding into the group U it is possible to establish the right-orderability of some specific groups (Bergman [89]), for example, for the group

$$G = \mathrm{gp}\,(x, y, z \mid x^2 = y^3 = z^7 = xyz)$$

7.5. The Local Indicability of Extensions

7.5.1. It was shown that the quasivarieties of locally indicable right-orderable groups were different. In this section, we prove that for a wide class of groups the existence of a right order on the group G provides its local indicability.

We denote by $\mathrm{r}(G)$ the locally nilpotent radical of a group G, i.e., its greatest locally nilpotent normal subgroup.

- A group is called *radical* if it has an increasing normal series $\mathrm{S}(G)$:

$$E = R_0(G) \subseteq R_1(G) \subseteq \ldots \subseteq R_\alpha(G) \subseteq R_{\alpha+1}(G) \subseteq \ldots \subseteq G$$

where $R_{\alpha+1}(G)/R_\alpha(G) = \mathrm{r}(G/R_\alpha(G))R_\alpha(G)$, and for the limit α

$$R_\alpha(G) = \bigcup_{\zeta<\alpha} R_\alpha(G)$$

We refer the reader to [45, 99] for the definition of a class that is radical in the Plotkin sense. It is evident that a subgroup of a radical group as well as the homomorphic image of a radical group are radical groups too.

For any subset X of a right-orderable group G we denote by $\mathrm{rc}\,(X)$ a right-relatively convex subgroup of G generated by X.

Theorem 7.5.1 (Tararin [77]). *If a right-orderable group G is an extension of a radical group by a periodic group, then it is locally indicable.*

PROOF. It suffices to consider the case of a finitely generated group G. Then the locally nilpotent radical $\mathrm{r}(G/H)$ of the quotient group G by a normal subgroup H is nontrivial provided that the quotient group G/H is torsion-free.

Let G be a finitely generated group subject to the hypotheses of the theorem. Then G has a maximal normal right-relatively convex subgroup. It suffices to consider only the case where G does not have proper normal right-relatively convex subgroups. We now show that G is an infinite cyclic group. Then the assertion of the theorem will follow.

For a radical A of G we have $A \neq E$ and $\mathrm{rc}\,(A) = G$. On G we fix a right order P. Let $\mathfrak{C}(A)$ be the system of convex subgroups A_α, $\alpha \in I$, of A with respect to the order induced on A by P. For every A_α of $\mathfrak{C}(A)$ we write B_α for the least convex subgroup of the group G such that $B_\alpha \cap A = A_\alpha$. Since G is finitely generated, in $\mathfrak{C}(A)$ there is the greatest subgroup A_0 different from A.

We show that $A_0 = B_0 = E$. To this end, we first show that A_0 is a normal subgroup of G. Since A is a locally nilpotent right-ordered group, by Theorem 2.4.2, A is Conradian. Hence A_0 is a normal subgroup of A and the quotient group A/A_0 is Archimedean. We note that $g^{-1}cg > e$ for any $c \in A \setminus A_0$, $c > e$, and $g \in G$. Since $\mathrm{rc}\,(A) = G$ and $c > e$, $c \notin A_0$, the set $\{c^k \mid k \in \mathbb{Z}\}$ is unbounded in G (cf. Corollary 5.1.7). Therefore, there are natural numbers r and k such that $g^{-1} > c^{-r}$ and $c^k > g^{-1}$. Consequently, $g^{-1}c^{r+k}g > c^k g > e$ and $g^{-1}cg > e$. Furthermore, $g^{-1}cg \notin A_0$ for every $g \in G$. Otherwise $g^{-1}cg > e$ and $g^{-1}cg \in A_0$ for some $g \in G$. However, in this case we must have $(g^{-1}cg)^t < c$ and $cg^{-1}c^{-t}g > e$ for any $t \in \mathbb{Z}$. Hence $g(cg^{-1}c^{-t}g)g^{-1} = gcg^{-1}c^{-t} > e$ and $gcg^{-1} > c^t$ for any integer t, i.e., the set $\{c^t \mid t \in \mathbb{Z}\}$ is bounded in G, which contradicts the assumptions $c \in A \setminus A_0$, $c > e$, $G = \mathrm{rc}\,A$. Thus, $(A \setminus A_0) \cap P$, and, consequently, A_0, is a normal subgroup of G. Hence $\mathrm{rc}\,(A_0)$ is a normal right-relatively convex subgroup of G and $E \subset A_0 \subset \mathrm{rc}\,(A_0) \subseteq B_0 \neq G$, which contradicts our assumptions on the group G and leads to the relation $A_0 = E$. The latter means that $B_0 = E$ and that A is an Archimedean group. Moreover, any element of A is unbounded in G and, since for every element of G some power of it belongs to A, any element of G is unbounded in G. Hence (cf. Corollary 5.1.7) the group G is Archimedean and, consequently, Abelian. Since G is finitely generated and does not have proper relatively convex subgroups, it is an infinite cyclic group. □

Corollary 7.5.1. *If a right-orderable group G is an extension of a radical group by a periodic group, then it admits a Conradian right order.*

The assertion follows from Theorem 7.5.1.

7.5.2. In the direction just discussed, Brodskiĭ [91] and Howie [100] have obtained the following result: *if $G = G_1 * G_2$ is a free product of locally indicable groups, and $r \in G$ is cyclically reduced and has length at least 2, then the quotient group $\overline{G}$ of the group G by a normal subgroup $(r)^G$ generated by r is locally indicable and the groups G_1 and G_2 are naturally embedded into $\overline{G}$.*

7.6. Lattice of Quasivarieties

It is easy to see that the set of all quasivarieties of right-orderable (orderable) groups $\mathrm{L}(\mathfrak{O}_r)$ ($\mathcal{L}_O\mathrm{L}(\mathfrak{O})$) is a complete by inclusion lattice. Moreover, the meet $\mathfrak{K}_1 \wedge \mathfrak{K}_2$ of two quasivarieties is the set-theoretical intersection of $\mathfrak{K}_1$ and $\mathfrak{K}_2$, and the join $\mathfrak{K}_1 \vee \mathfrak{K}_2$ is the least quasivariety containing both $\mathfrak{K}_1$ and $\mathfrak{K}_2$. The lattice $\mathrm{L}(\mathfrak{O}_r)$ ($\mathrm{L}(\mathfrak{O})$) contains the least element $\mathfrak{E}$, i.e., the quasivariety containing only the trivial group and the greatest element $\mathfrak{O}_r$ (($\mathfrak{O}$)), i.e., the quasivariety of all right-orderable (orderable) groups.

7.6.1. We show that the lattice $\mathrm{L}(\mathfrak{O}_r)$ ($\mathrm{L}(\mathfrak{O})$) contains the least nontrivial element.

Proposition 7.6.1. *The quasivariety $\mathfrak{A}$ of torsion-free Abelian groups is the least nontrivial quasivariety in the lattice* $\mathrm{L}(\mathfrak{O}_r)$ ($\mathrm{L}(\mathfrak{O})$).

Proof. Let $\mathfrak{K}$ be an arbitrary nontrivial quasivariety of right-orderable groups and $G \in \mathfrak{K}$, $G \neq E$. Then G contains an infinite cyclic subgroup (a). Therefore, $\mathfrak{K}$ contains the quasivariety $\mathrm{q}(a)$ generated by the infinite cyclic group (a). We show that $\mathrm{q}(a) = \mathfrak{A}$. Indeed, the reduced product

$$G = \bar{\prod_{n\in\mathbb{N}}}(a_n)/\mathcal{F}$$

of the infinite cyclic groups $\{(a_n) \mid n \in \mathbb{N}\}$ by an arbitrary nonprincipal ultrafilter contains divisible subgroups. Hence the quasivariety $\mathrm{q}(a)$ contains Abelian groups which are isomorphic to the additive group $\mathbb{Q}$ of rational numbers. Since any torsion-free Abelian group A is isomorphically embedded into a direct sum of groups isomorphic to $\mathbb{Q}$, we obtain $A \in \mathrm{q}(a)$. □

We indicate the quasivarieties of right-orderable groups covering $\mathfrak{A}$ in the lattices $\mathrm{L}(\mathfrak{O}_r)$ and $\mathrm{L}(\mathfrak{O})$. To this end, we introduce the notion of wreath product $A\mathrm{wr}B$ of groups A and B. This notion is close to that of groups of automorphisms of totally ordered sets (cf. 4.2). It can be utilized for any groups and corresponds to the groups A and B regarded as groups of permutations of the sets A and B respectively with their right regular representations. Let A and B be groups. For $b \in B$ we introduce a group A_b isomorphic to the group A and fix isomorphisms of the groups $\varphi_\alpha \colon A_\alpha \to A$. Let $\mathrm{fun}(B, A)$ be the direct product of groups A_b, $b \in B$.

- The semidirect product of the groups $\mathrm{fun}(B, A)$ and B is called the *wreath product* of groups A and B (denoted by $A\mathrm{wr}B$); moreover, $b^{-1}fb(x) = f(bx)$ for $f \in \mathrm{fun}(B, A)$, $x, b \in B$.

Every element $g \in A\mathrm{wr}B$ can be uniquely represented as $g = bf$, where $f \in \mathrm{fun}(B, A)$, $b \in B$; moreover, $b'f' \cdot bf = b'bf''$, where $f'' = b^{-1}f'bf = (f')^b f$.

Let F_2 be a free group of rank 2, N_2 be a free nilpotent group of class of rank 2, and $S = (a)\mathrm{wr}(b)$ be the wreath product of two infinite cyclic groups (a) and (b).

Proposition 7.6.2. *The quasivarieties* $\mathrm{q}(F_2)$, $\mathrm{q}(N_2)$, *and* $\mathrm{q}(S)$ *cover* $\mathfrak{A}$ *in the lattices* $\mathrm{L}(\mathfrak{O}_r)$ *and* $\mathrm{L}(\mathfrak{O})$.

PROOF. We consider the quasivariety $\mathrm{q}(F_2)$. Since any two noncommutative elements a and b of F_2 generate a free group of rank 2, in F_2 the quasi-identities

$$w(x, y) = e \Rightarrow (xy = yx)$$

hold for every nontrivial group word $w(x, y)$. Hence any non-Abelian group with two generators is free and, consequently, the quasivariety $\mathrm{q}\,(F_2)$ covers $\mathfrak{A}$ in the lattices $\mathrm{L}\,(\mathfrak{O}_{\mathrm{r}})$ and $\mathrm{L}\,(\mathfrak{O})$.

We consider the quasivariety $\mathrm{q}\,(N_2)$. As is known, any torsion-free nilpotent group of class 2 contains a subgroup isomorphic to N_2. Therefore, for any non-Abelian group G of $\mathrm{q}\,(N_2)$ the inclusions $\mathrm{q}\,(G) \supseteq \mathrm{q}\,(N_2)$ and $\mathrm{q}\,(G) \subseteq \mathrm{q}\,(N_2)$ hold. Consequently, the quasivariety $\mathrm{q}\,(N_2)$ covers $\mathfrak{A}$ in the lattices $\mathrm{L}\,(\mathfrak{O}_{\mathrm{r}})$ and $\mathrm{L}\,(\mathfrak{O})$.

Finally, we consider the quasivariety $\mathrm{q}\,((a)\mathrm{wr}(b))$. Let $S = (a)\mathrm{wr}(b)$. By definition,

$$S = \prod_{i=-\infty}^{+\infty} (a_{b^i})) \cdot (b)$$

where $b^{-1}a_{b^i}b = a_{b^{i+1}}$. We now totally order the group S by setting the element x as positive if $x = b^k f$, where

$$f \in \mathrm{fun}\,((b), (a)) = A = \prod_{i=-\infty}^{+\infty} (a_{b^i})$$

and $k > 0$ or $k = 0$, $f(b^i) = a_{b^i}^{k_i}$, $k_i > 0$, where i is the greatest integer such that $f(b^i) \neq e$. This totally ordered group is denoted by S^+. We note that in S^+ for any $f \in A$, $f \neq e$, and g, $g \notin A$, the following inequalities hold:

$$|g|^{-1}|f||g| \gg |f|$$

where $|x| = \max\{x, x^{-1}\}$. As usual, the expression $|b| \gg |a|$ stands for the infinite system of inequalities $\{|b| > |a|^k\}$, where k "runs through" the natural numbers.

We now consider the ultrapower

$$G = \bar{\prod_{i\in I}} S_i^+/\mathcal{F}$$

of the totally ordered group S^+ by an arbitrary ultrafilter $\mathcal{F}$. It is obvious that G is a totally ordered group and for any two noncommutative elements $x, y \in G$ at least one of the following relations holds:

$$|[x, y]|^{|x|} \gg |[x, y]| \qquad (7.6.1)$$

$$|[x, y]|^{|y|} \gg |[x, y]| \qquad (7.6.2)$$

But then the elements $\|[x,y]\|^{|x^n|}$ or $\|[x,y]\|^{|y^m|}$, where $n, m \in \mathbb{Z}$, are commutative and linearly independent. Therefore, for any two elements $x, y \in G$ one of the following assertions holds:

(a) the subgroup $\mathrm{gp}([x,y], x,)$ is isomorphic to the group S,

(b) the subgroup $\mathrm{gp}([x,y], x,)$ is isomorphic to the direct product of two infinite cyclic groups,

(c) the subgroup $\mathrm{gp}([x,y], x,)$ is isomorphic to an infinite cyclic group.

Let H be an arbitrary non-Abelian group of the quasivariety $\mathrm{q}(S)$. By Theorem 7.1.1, the group H is a subgroup of the Cartesian product of the groups each of which is some ultrapower of the group S. It is easy to check that for any two noncommutative elements $u, v \in H$ either $\mathrm{gp}(u, [u,v])$ or $\mathrm{gp}(v, [u,v])$ is isomorphic to S. □

Theorem 7.6.1 (Budkin [101]). *An arbitrary soluble cover of the quasivariety $\mathfrak{A}$ in the lattice $\mathrm{L}(\mathfrak{O})$ coincides with one of the quasivarieties $\mathrm{q}(N_2)$ and $\mathrm{q}(S)$.*

PROOF. If the quasivariety $\mathfrak{K}$ covers $\mathfrak{A}$ and contains the group S, then $\mathfrak{K} = \mathrm{q}(S)$. Therefore, we assume that $\mathfrak{K}$ covers $\mathfrak{A}$ but does not contain S. Since the group S is isomorphic to a subgroup of the free metabelian group of rank at least 2, the quasivariety $\mathfrak{K}$ does not have free metabelian groups. We denote by $\mathrm{var}\,\mathfrak{K}$ the group variety generated by $\mathfrak{K}$. Since free groups $\mathfrak{K}$ and $\mathrm{var}\,\mathfrak{K}$ coincide, the variety of metabelian groups $\mathfrak{A}^2$ is not contained in $\mathrm{var}\,\mathfrak{K}$. Kargapolov and Churkin [102] have shown that the variety $\mathrm{var}\,\mathfrak{K}$ is contained in the variety $\mathfrak{B}_n\mathfrak{N}_c\mathfrak{B}_n$, where $\mathfrak{B}_n$ is the variety of groups of period n and $\mathfrak{N}_c$ is the variety of nilpotent groups of nilpotent class at least c. Therefore, an arbitrary group G of $\mathfrak{K}$ is a periodic extension of a nilpotent group.

Since any orderable group is a group with unique roots, the locally nilpotent radical in the orderable group is an isolated subgroup (cf., for example, [45]). Thus, the group G is nilpotent. Hence $\mathfrak{K}$ coincides with the quasivariety $\mathrm{q}(N_2)$. □

Let p be a prime number. We consider the following group:

$$\begin{aligned} C_p = \mathrm{gp}\,(x_0, \ldots, x_{p-1}, y \mid\ & [x_i, x_j] = e \\ & y_{-1}x_iy = x_{1+1}, \quad 0 \leqslant i \leqslant p-2, \quad y^{-1}x_{p-1}y = x_0 \\ & [x_k, x_l] = e, \quad 0 \leqslant k, l \leqslant p-1, \quad x_0x_1 \ldots x_{p-1} = e) \end{aligned}$$

Proposition 7.6.3. *The group quasivariety* $\mathrm{q}(C_p)$ *covers* $\mathfrak{A}$ *in the lattice* $\mathrm{L}(\mathfrak{O}_{\mathrm{r}})$ *for an arbitrary prime number* p.

PROOF. It is easy to check that in the group C_p the following identities hold:

$$x^p, y^p] = e \tag{7.6.3}$$

$$[[x, y], z^p] = e \tag{7.6.4}$$

Therefore, by the above-mentioned result obtained by Kargapolov and Churkin, the variety $\mathrm{var}(\mathrm{q}(C_p))$ is contained in the variety $\mathfrak{B}_n\mathfrak{N}_c\mathfrak{B}_n$. Since the group C_p is a torsion-free group, $\mathrm{q}(C_p) \subseteq \mathfrak{N}_c\mathfrak{B}_n$. Since any torsion-free nilpotent group on which (7.6.3) and (7.6.4) are fulfilled is Abelian, any group of the quasivariety $\mathrm{q}(C_p)$ is an extension of a torsion-free Abelian group by a group of period p.

Let G be a non-Abelian group of $\mathrm{q}(C_p)$. Then there are elements $a, b \in G$ such that $[a, b] \neq e$. Since any nilpotent subgroup of G is Abelian, we have $[[a, b], a] \neq e$ or $[[a, b], b] \neq e$. For definiteness, we set $[[a, b], b] \neq e$ and $[[a, b], b^p] = e$. Let $[[a, b], b] = u$, and let (c) be an infinite cyclic group. Then the direct product $G \times (c)$ of the groups G and (c) is contained in $\mathrm{q}(C_p)$. We consider the elements $v = bc$ and u of $G \times (c)$ and the subgroup $L = \mathrm{gp}(u, v)$. It is easy to verify that L is isomorphic to the group C_p. □

Theorem 7.6.2 (Budkin [101]). *An arbitrary soluble cover of the quasivariety* $\mathfrak{A}$ *in the lattice* $\mathrm{L}(\mathfrak{O}_{\mathrm{r}})$ *coincides with one of the quasivarieties* $\mathrm{q}(N_2)$, $\mathrm{q}(S)$, *and* $\mathrm{q}(C_p)$.

PROOF. Let a quasivariety $\mathfrak{K}$ of soluble groups cover $\mathfrak{A}$ in the lattice $\mathrm{L}(\mathfrak{O}_{\mathrm{r}})$. We assume that $\mathfrak{K}$ does not contain non-Abelian nilpotent groups and the group $(a)\mathrm{wr}(b)$. We again apply the theorem by Kargapolov and Churkin [102] to conclude that the variety $\mathrm{var}(\mathrm{q}(C_p))$ is contained in the variety $\mathfrak{B}_n\mathfrak{N}_c\mathfrak{B}_n$. Since the quasivariety $\mathfrak{K}$ is contained in the quasivariety of torsion-free groups, we have $\mathfrak{K} \subseteq \mathfrak{N}_c\mathfrak{B}_n$. Since any torsion-free nilpotent group, on which (7.6.3) and (7.6.4) hold, is Abelian, we obtain $\mathfrak{K} \subseteq \mathfrak{N}_c\mathfrak{B}_n$. Therefore, we can assume that G is a non-Abelian metabelian group of the quasivariety $\mathfrak{K}$ and there are elements $a, b \in G$ such that $a \in G'$, $[a, b] \neq e$.

Let A be a normal Abelian subgroup of the group G and $G/A \in \mathfrak{B}_n$. If $[a, b^n] \neq e$, then, in view of $[a, b^n] \in A$ and $b^n \in A$, the subgroup $\mathrm{gp}(a, b^n)$ is a nilpotent non-Abelian group, which is impossible. Hence $[a, b^n] = e$. Taking the corresponding power of b instead of b, we can assume that the element b is chosen so as to satisfy the relations $[a,b] \neq e$ and $[a,b^p] = e$ for some prime number p. We denote by u the element $[a,b]$ for brevity. The group $G \times (c)$, where (c) is an infinite cyclic group, is contained in the quasivariety $\mathfrak{K}$. We consider in G the elements $v = bc$ and u and the subgroup $L = \mathrm{gp}\,(u,v)$. It is easy to check that the subgroup L is isomorphic to the group C_p. □

7.6.2. We show that not every quasivariety has a cover in the lattices $\mathrm{L}\,(\mathfrak{O}_{\mathrm{r}})$ and $\mathrm{L}\,(\mathfrak{O})$. We denote by $\mathfrak{P}^i_M$, $i = 1, 2$, the group quasivarieties defined by the following infinite systems of quasi-identities Σ_i:

$$(xy)^{-1}[x,y](xy) = [x,y]^{p^k} \Rightarrow [x,y] = e \tag{7.6.5}$$

$$(***)_i, \quad i = 1, 2$$

where $k \in \mathbb{N}$, $k \geqslant 2$, $p \in M$; M is a nonempty set of prime numbers and $(***)_i$ is a system of quasi-identities defining the quasivariety $\mathfrak{O}_{\mathrm{r}}$ of right-orderable groups for $i = 1$ and the quasivariety $\mathfrak{O}$ of orderable groups for $i = 2$.

Theorem 7.6.3 (Medvedev [88]). *The quasivariety $\mathfrak{P}^1_M$ does not have covers in the lattice $\mathrm{L}\,(\mathfrak{O}_{\mathrm{r}})$. The quasivariety $\mathfrak{P}^2_M$ does not have covers in the lattice $\mathrm{L}\,(\mathfrak{O})$.*

PROOF. We first consider the quasivariety $\mathfrak{P}^1_M$ and assume the contrary: a quasivariety $\overline{\mathfrak{P}}^1_M$ covers $\mathfrak{P}^1_M$ in the lattice $\mathrm{L}\,(\mathfrak{O}_{\mathrm{r}})$. Then there is a group G of $\overline{\mathfrak{P}}^1_M \setminus \mathfrak{P}^1_M$ on which some quasi-identity from (7.6.5) fails. For example, let the quasi-identity

$$(xy)^{-1}[x,y](xy) = [x,y]^{p^{k_0}} \Rightarrow [x,y] = e$$

fail for some numbers $k_0 \in \mathbb{N}$ and $p \in M$. Then there are elements $a, b \in G$ such that for $x = a$ and $y = b$ the left-hand side of the quasi-identity is true and the right-hand side is not true. Since the group G is right-orderable, the subgroup A of G generated by the elements $c = ab$ and $d = [a,b]$ is isomorphic to the group

$$G_{p^{k_0}} = \mathrm{gp}\,(c, d \mid c^{-1}dc = d^{p^{k_0}})$$

We consider the subgroup H of $G_{p^{k_0}}$ which is generated by the elements c^3 and d. It is clear that on H the quasi-identity

$$(xy)^{-1}[x,y](xy) = [x,y]^{p^{3k_0}} \Rightarrow [x,y] = e$$

fails for $x = c^3$ and $y = d$. Hence $H, G_{p^{k_0}} \notin P^1_M$. Let $\mathrm{q}(H)$ and $\mathrm{q}(G_{p^m})$ be the quasivarieties generated by the groups H and G_{p^m} respectively. Then

$$\overline{\mathfrak{P}}^1_M = \mathfrak{P}^1_M \vee \mathrm{q}(H) = \mathfrak{P}^1_M \vee \mathrm{q}(G_{p^m})$$

But the quasi-identity

$$(xy)^{-1}[x,y](xy) = [x,y]^{p^{2k_0}} \Rightarrow [x,y] = e$$

is valid on the quasivarieties $\mathfrak{P}^1_M$ and $\mathrm{q}(H)$ and fails on the group G_{p^m} for $x = c^2$ and $y = d$. Consequently,

$$G_{p^m} \notin \overline{\mathfrak{P}}^1_M = \mathfrak{P}^1_M \vee \mathrm{q}(H)$$

which contradicts the choice of the group G. Therefore, the group quasivariety $\mathfrak{P}^1_M$ does not have covers in the lattice $\mathrm{L}(\mathfrak{O}_\mathrm{r})$ of right-orderable groups.

The case of the lattice $\mathrm{L}(\mathfrak{O})$ of orderable groups is handled in a similar way. □

7.6.3. The lattices $\mathrm{L}(\mathfrak{O})$ and $\mathrm{L}(\mathfrak{O}_\mathrm{r})$ have a fairly complicated structure and until today have not been studied in detail. In this section, we establish that these lattices are not modular.

Lemma 7.6.1 (Budkin and Gorbunov [103]). *A finitely presented group H belongs to the group quasivariety $\mathrm{q}(G)$ generated by some group G if and only if the group H is isomorphically embedded into a suitable Cartesian power of the group G.*

PROOF. Let

$$H = \mathrm{gp}\,(h_1, \dots, h_n | r_1(h_1, \dots, h_n) = \dots = r_k(h_1, \dots, h_n) = e)$$

By the Mal'cev theorem (cf. Theorem 7.1.1 and [90]), we can assume that the group H is a subgroup of some reduced power $\overline{\prod\limits_{i\in I}} G_i/\mathcal{F}$ of the group G and

$$h_1 = (g_{1i})\mathcal{F}, \dots, h_n = (g_{ni})\mathcal{F}$$

By the definition of the reduced power, we have

$$I_0 = \{i \in I \mid r_1(g_{1i}, \dots, g_{ni}) = \dots = r_k(g_{1i}, \dots, g_{ni}) = e\} \in \mathcal{F}$$

We can check immediately that the elements

$$h_1' = (g_{1i}), \dots, h_n' = (g_{ni})$$

of the Cartesian power

$$\bar{\prod_{i \in I_0}} G_i$$

generate a subgroup isomorphic to the group. □

We now present the following groups studied by Budkin [96]:

$$B_4 = \mathrm{gp}\,(x_1, x_2, x_3, x_4, z \mid [x_1, x_3][x_2, x_4] = z, [x_i, z] = e,\ i = 1, 2, 3, 4)$$
$$G_4 = \mathrm{gp}\,(\overline{x}_1, \overline{x}_2, \overline{x}_3, \overline{x}_4 \mid [\overline{x}_1, \overline{x}_3][\overline{x}_2, \overline{x}_4] = e,\ [x_i, z] = e).$$

Let N_4 be a free nilpotent group of class 2 generated by the elements a_1, a_2, a_3, and a_4.

Lemma 7.6.2 (Baumslag [104]). *The group G_4 is approximated by free groups.*

Lemma 7.6.3 (Budkin [96]). *Any subgroup of the group G_4 generated by two elements is free.*

Let φ_1 denote the natural homomorphism of B_4 onto $B_4/(z)$ (obviously, $B_4/(z)$ is isomorphic to G_4), φ_2 stand for a homomorphism of B_4 onto N_4 defined by the relations

$$\varphi_2(x_i) = a_i,\ i = 1, 2, 3, 4; \quad \varphi_2(z) = [a_1, a_3][a_2, a_4]$$

and D be the kernel of the homomorphism φ_2. Then $D \cap (z) = E$. Since the group G_4 is approximated by free groups, it is orderable and $\mathrm{q}\,(G_4) = \mathrm{q}\,(F_2)$. In view of the equality $D \cap (z) = E$, the group B_4 is approximated by the orderable groups G_4 and N_4. Moreover, it is orderable and

$$B_4 \in \mathrm{q}\,(G_4) \vee \mathrm{q}\,(N_4)$$

in the lattice of group quasivarieties. Hence

$$\mathrm{q}\,(B_4) \vee \mathrm{q}\,(N_4) = \mathrm{q}\,(F_2) \vee \mathrm{q}\,(N_4)$$

Since there is an isomorphism between the groups $B_4/(z)$ and G_4, the group B_4 contains a subgroup isomorphic to the free group F_2. Therefore, $\mathrm{q}\,(B_4) \supseteq \mathrm{q}\,(F_2)$. We note that $B_4 \notin \mathrm{q}\,(F_2)$. Indeed, for $B_4 \in \mathrm{q}\,(F_2)$, in view of the finite presentability, the group B_4 is approximated by free groups. Hence there is a homomorphism φ of the group B_4 onto the free group F such that $z\varphi \neq e$. But in this case the element $z\varphi$ is contained in both the center and in the derived subgroup of the subgroup $B_4\varphi$ of the free group F. Hence $z\varphi = e$. The contradiction obtained means that $B_4 \notin \mathrm{q}\,(F_2)$.

We now show that in the lattice of group quasivarieties the following equalities hold:

$$\mathrm{q}\,(B_4) \wedge \mathrm{q}\,(N_4) = \mathrm{q}\,(F_2) \wedge \mathrm{q}\,(N_4) = \mathfrak{A}$$

We assume that there exists a 2-generated non-Abelian group G of the quasivariety $\mathrm{q}\,(B_4) \wedge \mathrm{q}\,(N_4)$. Since G is nilpotent, it is finitely presented (cf., for example, [53] or [3]). By Lemma 7.6.1, the group G is approximated by groups isomorphic to B_4.

We note now that any subgroup of B_4 generated by two elements is contained in the quasivariety $\mathrm{q}\,(F_2)$. Let A be a 2-generated subgroup of the group B_4. In the case of the trivial group $A\varphi_1$ we have $A \in \mathrm{q}\,(F_2)$. Therefore, we will assume that $A\varphi_1$ is not trivial. Then the group $A\varphi_1$ is free. Hence the group A is the central extension of an infinite cyclic group by the free group $A\varphi_1$. In this case the inclusion $A \in \mathrm{q}\,(F_2)$ is evident. Thus, G is a free and nilpotent non-Abelian group. Hence $\mathrm{q}\,(B_4) \wedge \mathrm{q}\,(N_4) = \mathfrak{A}$ and the sublattice of the lattice of the quasivarieties $\mathrm{L}\,(\mathfrak{O}_{\mathrm{r}})$ (respectively $\mathrm{L}\,(\mathfrak{O})$) generated by the quasivarieties $\mathrm{q}\,(B_4)$, $\mathrm{q}\,(N_4)$, and $\mathrm{q}\,(F_2)$ is isomorphic to the lattice L_2 mentioned in Chapter 1 (cf. Example 1.3.2). By Theorem 1.3.2 the following assertion is obvious.

Theorem 7.6.4 (Budkin [96]). *The lattices* $\mathrm{L}\,(\mathfrak{O}_{\mathrm{r}})$ *and* $\mathrm{L}\,(\mathfrak{O})$ *are neither modular nor distributive.*

Chapter 8

Semilinearly Ordered Groups

In this chapter, we consider a class of partially right-ordered groups (the so-called semilinearly ordered groups) which is of particular interest due to its closeness to the class of right-ordered groups. At the same time, the class under consideration has some specific properties that are "exotic" for partially ordered groups.

The notion of a semilinear order on a group was given by Frege in 1903. The introduction of this notion was stimulated by discussions on the role of an order in arithmetic as well as some questions on minimal axiomatic systems for ordered groups of different types.

The question on the independence of some axioms was open for a long time. In 1987, Adeleke, Dummett, and P. Neumann constructed in [105] the first example of a group which was ordered but not right-ordered. This paper initiated the development of the theory of semilinearly ordered groups.

8.1. Definitions

8.1.1. We begin with a definition.

- A directed-up partially right-ordered group is called *semilinearly ordered* if any two elements of it with a common lower bound are comparable.

In other words, a semilinearly ordered group $\{G; \cdot, \leqslant\}$ is a nonempty set G such that $\{G; \cdot\}$ is a group, $\{G; \leqslant\}$ is a partially ordered set, and the following axioms hold:

— if $x \leqslant y$, then $xz \leqslant yz$ for any $x, y, z \in G$,

— for any $x, y \in G$ there exists a $z \in G$ such that $x \leqslant z$, $y \leqslant z$,

— if $x \leqslant y$ and $x \leqslant z$ in G, then y and z are comparable in G, i.e., $y \leqslant z$ or $z \leqslant y$.

Since a partial right order on a group is stable under the actions of right translations, in a partially right-ordered group the last axiom is equivalent to the following relation:

— any two positive elements of the group are comparable.

For a semilinearly ordered group G, together with the positive cone $P = P(G)$ we consider the *negative cone* $P^{-1} = \{x \in G \mid x \leqslant e\}$.

In what follows, we use the notation

$$P^{\pm} = P \cup P^{-1}, \quad P_H = P \cap H, \quad P'_H = P \setminus P_H,$$

where $P = P(G)$ is the positive cone and H is a subgroup of a partially right-ordered group G.

In view of the ordinary properties of positive cones for partially right-ordered groups (cf. 1.5), a semilinearly ordered group can be defined in terms of positive and negative cones.

Theorem 8.1.1. *For a semilinearly ordered group G, the positive cone $P = \{x \in G \mid x \geqslant e\}$ possesses the properties* O.1, O.2, O.4 *(cf. Theorem* 1.5.1*) and the following property:*

O.7 $$P \cdot P^{-1} \subseteq P \cup P^{-1} = P^{\pm}$$

Conversely, if a subset P of a group G satisfies O.1, O.2, O.4, *and* O.7, *then the group G can be regarded as a semilinearly ordered group with positive cone P.*

The assertion directly follows from Theorem 1.5.1 and the above-mentioned axioms.

It is convenient to use another version of Theorem 8.1.1 which is based on the notion of the connectedness of a partial order (cf. 1.4 and 1.5). By Theorem 1.5.1, a partially right-ordered group is connected if and only if it is generated by its positive cone P (cf. Theorem 1.5.1).

Theorem 8.1.2. *A partially right-ordered group G is semilinearly ordered if and only if its positive cone P satisfies the conditions* O.1–O.3 *from Theorem* 1.5.1 *and* O.7 *from Theorem* 8.1.1.

PROOF. It suffices to prove that a connected partially ordered group any two of whose positive elements are comparable is directed, i.e., O.3 and O.7 guarantee the validity of O.4.

Indeed, if a group G satisfies the condition O.3, then any element $x \in G$ can be represented as follows:

$$x = p_1^{\varepsilon_1} p_2^{\varepsilon_2} \cdots x_k^{\varepsilon_k}$$

where $p_i \in P$, $\varepsilon_i = \pm 1$, $i = 1, 2, \ldots, k$, and $\varepsilon_i \varepsilon_{i+1} = -1$, $i = 1, 2, \ldots, k-1$. It is required to show that $x = p^{-1}q$ for suitable $p, q \in P$. We proceed by induction on k.

For $k = 1$ the assertion is obvious.

We assume that the assertion is true for all n, $1 \leqslant n \leqslant k-1$, $x = p_1^{\varepsilon_1} p_2^{\varepsilon_2} \cdots x_k^{\varepsilon_k}$. Then $x = p_1^{\varepsilon_1} y$, where $y = p_2^{\varepsilon_2} \cdots x_k^{\varepsilon_k} = u^{-1}v$, and $u, v \in P$, i.e., $x = p_1^{\varepsilon_1} u^{-1} v$. We consider all possible cases for ε_1.

If $\varepsilon_1 = -1$, then $p_1^{-1}u^{-1} = p^{-1}$, $p = up_1 \in P$, and $x = p^{-1}q$, where $q = v$.

If $\varepsilon_1 = +1$, then $x = p_1 u^{-1} v$. However, in view of O.7, we have $p_1 u^{-1} \in P^{\pm}$. For $p_1 u^{-1} \in P^{-1}$, we set $p = p_1 u^{-1}$ and $q = v$. If $r = p_1 u^{-1} \in P$, then we put $x = rv \in P$. This means that O.4 follows from O.3 and O.7. The application of Theorem 8.1.1 completes the proof. □

Proposition 8.1.1. *For a semilinearly ordered group G, the positive cone satisfies the following condition:*

O.8 $\qquad qPq^{-1} \subseteq P^{\pm}, \quad qP^{-1}q^{-1} \subseteq P^{\pm}, \quad qPq^{-1} \subseteq P^{\pm}$

for any $q \in P$.

PROOF. It suffices to prove the first inclusion in O.8. Let G be a semilinearly ordered group, and let $q, p \in P$. Then $q \geqslant e$, $p_1 = qp \geqslant$

$p \geqslant e$, and $qpq^{-1} = p_1 q^{-1}$. By O.7, we have the inclusion $p_1 q^{-1} \in P^{\pm}$, from which the proposition follows. □

We separate one subclass (left-invariant groups) from the class of all semilinearly ordered groups.

- A semilinearly order P on a group G is called *left-invariant* if the following relation holds:

O.9 $$qPq^{-1} \subseteq P, \quad q \in P$$

This notion was introduced by Adeleke, Dummett, and Neumann [105]. In the cited article, the definition of the left-invariant order with respect to positive elements was also given. This definition looks like the definition just introduced but with O.9 replaced by the following condition:

O.10 $$px \leqslant py \text{ for } p, x, y \in G \text{ such that } p \geqslant e,\ x \leqslant y$$

Proposition 8.1.2. *For any partially right-ordered group, the conditions* O.9 *and* O.10 *are equivalent.*

Proof. Let G be a partially right-ordered group satisfying O.9, and let $x \leqslant y$ in G. Then $yx^{-1} \in P$, $p \cdot yx^{-1} \cdot p^{-1} = py \cdot (px)^{-1} \in P$. Consequently, $py \leqslant px$.

Conversely, let the condition O.10 and the inclusion $q, p \in P$ hold in G. Multiplying the inequality $p \geqslant e$ from the left by the positive element q, we obtain the inequality $qp \geqslant q$, which implies $qpq^{-1} \in P$. The last inclusion shows the validity of O.9. □

It is clear that any right-ordered group is totally ordered. A group that is semilinearly ordered but not right-ordered is called *proper semilinearly ordered*, and the corresponding order is referred to as a *proper semilinear* order. The main purpose of this chapter is to study proper semilinearly ordered groups.

8.1.2. In this subsection, we give an example of a proper semilinearly ordered group.

Example 8.1.1 (Kopytov). We take an arbitrary integer p such that $p \neq 0$, $p \neq \pm 1$, and the group G of matrices

$$\begin{bmatrix} p^k & q \\ 0 & 1 \end{bmatrix}$$

where q is a rational number of the form $|p|^s m$ and k, s, and m are integers; moreover, p is not a divisor of m and $k, s, m \in \mathbb{Z}$, $q \in \mathbb{Q}_p$. We identify every matrix of G with the pair (k, q) of numbers. It is not hard to see that the group operations can be written as follows:

$$(k,q) \cdot (l,r) = (k+l, p^k r + q)$$
$$(k,q)^{-1} = (-k, -q \cdot p^{-k})$$

We consider the set P of matrices (k, q) such that $k \geqslant 1$, $q \in \mathbb{Z}$ or $k = 0$, $q \in \mathbb{Z}$, $q \geqslant 0$ and prove that P is the positive cone in the group G with respect to a proper semilinear order on G. To prove that P defines a semilinear order on G it suffices to verify O.1–O.3 and O.7. Let $(k,q),(l,p) \in P$. Then $(k,q)(l,r) = (k+l, p^k r + q)$, where $k,l,q,r \in \mathbb{Z}$, $k,l \geqslant 0$, $q \geqslant 0$ for $k = 0$ and $r \geqslant 0$ for $l = 0$. In the case $k + l > 0$ the number $p^k r + q$ is an integer. Consequently, $(k+l, p^k r + q) \in P$. If $k + l = 0$, then $k = 0$, $l = 0$, $r \geqslant 0$, $q \geqslant 0$, whence $(k+l, p^k r + q) \in P$, i.e., $P \cdot P \subseteq P$. Thus, P satisfies O.1.

Let $g = (k,q) \in P \cap P^{-1}$, i.e., $(k,q) \in P$, $(k,q)^{-1} = (-k, -qp^{-k}) \in P$. Then $k \geqslant 0$, $-k \geqslant 0$, i.e., $k = 0$, $g = (0,q)$. Since $g \in P$, $g^{-1} \in P$, we have $q \geqslant 0$, $-q \geqslant 0$, i.e., $q = 0$, $g = e$, and the condition O.2 is satisfied by P.

We note that the group G is generated by the elements $(1,0)$ and $(0,1)$, each of which is contained in P, i.e., the group G satisfies O.3.

Finally, let $g = (k,q)$, $h = (l,r)$, and $g,h \in P$. Then the element gh^{-1} assumes the form $gh^{-1} = (k-l, q-p^{k-l}r)$, where $k,l \in \mathbb{Z}$, $k,l \geqslant 0$, $q,r \in \mathbb{Z}$. If $k > l$, then $k-l > 0$, $p^{k-l} \in \mathbb{Z}$ and, consequently, $-q+p^{k-l}r \in \mathbb{Z}$, i.e., $gh^{-1} \in P$. If $k = l$, then $k - l = 0$ and $q - p^{k-l}r = q - r \in \mathbb{Z}$, and in this case we also have $gh^{-1} \in P^{\pm}$. For $k < l$ we consider the element $hg^{-1} = (gh^{-1})^{-1}$ and find, as above, $hg^{-1} \in P$, i.e., P satisfies O.7. The application of Theorem 8.1.2 completes the proof.

We remark that the semilinear order constructed in Example 8.1.1 is left-invariant for $p > 1$ and is not so for $p < -1$. Example 8.1.1 also shows that an order on a semilinearly ordered group cannot be isolated. Indeed, in the group from Example 8.1.1, the element $g = (0, p^{-1})$ is not comparable with the identity element, but $g^p = (0, p^{-1})^p = (0,1) > e = (0,0)$.

Proposition 8.1.3 (Adeleke, Dummett, and Neumann [105]). *If a semilinear order P on a group G is a partial order, then (G, P) is a totally ordered group.*

PROOF. Indeed, for any element g of the group (G, P) subject to the hypothesis of the proposition the equality $g = p^{-1}q$ holds with some $p, q \in P$. But $g = p^{-1}q = q \cdot q^{-1}p^{-1}q = qr^{-1}$, where $r = q^{-1}pq \in P$ and $r^{-1} \in P^{-1}$. By O.8, we obtain (cf. Proposition 8.1.1) $g \in P \cup P^{-1}$ and $G = P \cup P^{-1}$. Therefore, the group G is totally ordered. □

8.1.3. We indicate a universal method of constructing semilinear orders from a given right order.

Theorem 8.1.3 (Kopytov). *Let G be a right-ordered group with positive cone P and X be a system of generators of G contained in P. There exists the least semilinear order Q on the group G whose positive cone contains X and is contained in P.*

PROOF. We construct a semilinear order Q as the union of an increasing sequence of partial right orders P_n. We set $P_0 = \operatorname{sgr}(X)$. Since $X \subseteq P$, the semigroup P_0 is pure and is contained in P. If P_0 satisfies O.7, we set $P_0 = Q$. Otherwise we can consider the sequence

$$P_0 \subset P_1 \subset \cdots \subset P_n$$

of subsemigroups contained in P and such that

$$P_i \cdot P_i^{-1} \subseteq P_{i+1} \cup P_{i+1}^{-1}$$

for every i, $0 \leqslant i < n$. If for some i the order P_i satisfies the condition O.7, we set $Q = P_k$, where k is the least number i with the above-mentioned properties. If for some i the condition O.7 fails, then

$$P_n \cdot P_n^{-1} \not\subseteq P_n \cup P_n^{-1}$$

In this case we construct the set $R_n = \{x \in P \mid x = pq_{-1}\}$ for some $p, q \in P_n$. Since $P_n \cup R_n \subseteq P$, the set $P_{n+1} = \operatorname{sgr}\{P_n \cup R_n\}$ is a pure semigroup containing X and contained in P. Continuing the procedure, we arrive at either a semigroup P_n satisfying O.7 (in this case we set $Q = P_n$ and terminate the process) at some finite step or a strictly increasing sequence of semigroups P_n. In the second case we set $Q = \bigcup_{i=0}^{\infty} P_n$; moreover, Q satisfies O.7. Indeed, for $p, q \in Q$ we choose the least number i such that $p, q \in P_i$. Then $pq^{-1} \in P$ or $qp^{-1} = (pq^{-1})^{-1} \in P$, i.e., $pq^{-1} \in R_n$ or $(pq^{-1})^{-1} \in R_n$. This means that $pq^{-1} \in P_{n+1}$ or $pq^{-1} \in P_{n+1}^{-1}$. Consequently, $Q \cdot Q^{-1} \subseteq Q \cup Q^{-1}$, i.e., Q satisfies O.7. On the other hand, $X \subseteq Q$, i.e., Q generates the group G and Q satisfies the condition O.3.

By Theorem 8.1.2, we can assert that Q defines a semilinear order on G such that $Q \subseteq P$. From the construction of Q it follows that on the group G the order Q is the least semilinear order containing X. □

Unfortunately, the proposed procedure of constructing a semilinear order is not effective enough because it requires verification of the fact that the constructed semilinear order is proper.

8.2. Basic Properties

8.2.1. From Theorems 8.1.1, 8.1.2, and the ordinary properties of partially right-ordered groups, we obtain a number of assertions that are fundamental in the theory of semilinearly ordered groups.

Proposition 8.2.1. *A directed subgroup of a semilinearly ordered group is semilinearly ordered.*

PROOF. Let H be a directed subgroup and P be the positive cone for a semilinearly ordered group G. We assume that $P_H = P \cap H$. Since H is directed with respect to the order induced by a subgroup of the group G, we have $H = P_H^{-1} \cdot P_H$, i.e., the condition O.4 is satisfied by the partially ordered group H. Since $P_H \cdot P_H^{-1} \subseteq P \cup P^{-1}$ and $P_H \cdot P_H^{-1} \subseteq H$, we also have $P_H \cdot P_H^{-1} \subseteq (P \cup P^{-1}) \cap H = P_H \cup P_H^{-1}$, which proves the validity of O.7 for (H, P_H). By Theorem 8.1.1, (H, P_H) is a semilinearly ordered group. □

Corollary 8.2.1 (Medvedev). *Let G be a semilinearly ordered group and H be its normal subgroup. We assume that the quotient group G/H is a periodic group. Then the subgroup H is semilinearly ordered with respect to the induced order. In particular, any subgroup of a finite index of a semilinearly ordered group is a semilinearly ordered group with respect to the induced order.*

PROOF. By Proposition 8.2.1, it suffices to show that the subgroup H is directed. Let $h_1, h_2 \in H$. Since the group G is directed, there exists an element $w \in G$ such that $w \geqslant h_1$, $w \geqslant h_2$, $w \geqslant e$. It is obvious that

$$w^n \geqslant w^{n-1} \geqslant \cdots \geqslant w \geqslant e$$

in G for any natural number n. Since the quotient group G/H is periodic, we have $w^m \in H$ for some natural number m. Furthermore, $w^m \geqslant h_1$, $w^m \geqslant h_2$, which means that the subgroup H is directed with respect to the induced order. □

Proposition 8.2.2. *The quotient group G/N of a semilinearly ordered group G by a convex normal subgroup N is a semilinearly ordered group with respect to the natural order on the quotient group G/N.*

PROOF. If P is the positive cone for a semilinearly ordered G, then the semigroup $\overline{P} = \{pN \mid p \in P\} = P \cdot N$ is the positive cone with respect to the natural order on the quotient group $\overline{G} = G/N$. Therefore,

$$\overline{P}^{-1} \cdot \overline{P} = (PN)^{-1} \cdot PN = P^{-1}P \cdot N = G \cdot N$$

Consequently, $\overline{P}^{-1} \cdot \overline{P} = \overline{G}$. □

8.2.2. For any subset X of a partially ordered set Y we denote by $U(X)$ the set of all upper bounds of the set X in Y.

Lemma 8.2.1. *For any elements g of a semilinearly ordered group (G, P) and $h \in U(h) \cap P = U(h, e)$ the following relation holds:*

$$hgh^{-1} \in P \cup P^{-1}$$

PROOF. If $g \in G$ and $h \geqslant e$, $h \geqslant g$, then $hg^{-1} \geqslant e$ and $hg^{-1}h^{-1} \geqslant h^{-1}$. At the same time, we have $e \geqslant h^{-1}$. Since the order P on the group G is semilinear and the elements $hg^{-1}h^{-1}$ and e have a common bound, these elements are comparable, i.e., $hg^{-1}h^{-1} \geqslant e$ or $hg{-}1h^{-1} \leqslant e$. This means that $hg^{-1}h^{-1} \in P^{\pm}$ and $hgh^{-1} \in P^{\pm}$. □

Proposition 8.2.3. *For any semilinearly ordered group (G, P) the following equality holds:*

$$G = \bigcup_{q \in P^{-1}} q(P^{\pm})q^{-1} = \bigcup_{g \in G} g(P^{\pm})g^{-1}$$

Proposition 8.2.4. *A left-invariant semilinear order on a group can be extended to a total order on the same group.*

PROOF. It is evident that for any element g of a semilinearly ordered group G with order P the order gPg^{-1} defines a semilinear order on G; moreover, if the order P is left-invariant, then gPg^{-1} is also left-invariant.

The union of an increasing sequence of left-invariant semilinear orders is a semilinear left-invariant order too. Therefore, in the set of all left-invariant orders P_α that are extensions of the order P there is the maximal element Q. Hence $pQp^{-1} \subset Q$ for any $p \in Q$. Moreover, $pQp^{-1} = Q$ for any $p \in Q$ since Q is the maximal left-invariant order. Consequently, $p_{-1}Qp = Q$ for any $p \in Q$. Taking into account that $G = Q^{-1}Q$, we can conclude that $gQg^{-1} = Q$ for any $g \in G$, i.e., the semilinear order Q is a partial order. By Proposition 8.1.3, Q is a total order on the group G. □

Theorem 8.2.1 (Petrova). *Any semilinearly ordered group is right-orderable.*

PROOF. We employ the Conrad criterion for right-orderability of a group (cf. Theorem 3.1.1). Let $x_1, x_2, \ldots, x_n$ be a finite collection of nontrivial elements of a semilinearly ordered group G. Since the group G is directed, there is an element $h \in G$ such that $h \geqslant e$, $h \geqslant x_1, x_2, \ldots, x_n$. By Lemma 8.2.1, $hx_ih^{-1} \in P^{\pm}$ for each $i = 1, 2, \ldots, n$. We now choose numbers $\varepsilon_1, \varepsilon_2, \ldots, \varepsilon_n$ equal to ± 1 to satisfy the inclusion $hx_ih^{-1} \in P$ for every i. Since the semigroup P is pure and all the elements x_i differ from e, we obtain

$$e \notin \mathrm{sgr}(hx_1^{\varepsilon_1}h^{-1} \cdot hx_2^{\varepsilon_2}h^{-1} \cdots hx_n^{\varepsilon_n}h^{-1}) = h \cdot \mathrm{sgr}(x_1^{\varepsilon_1} \cdot x_2^{\varepsilon_2}\ cdots x_n^{\varepsilon_n}) \cdot h^{-1}$$

But in this case the following relation is fulfilled:

$$e \notin \mathrm{sgr}(x_1^{\varepsilon_1} \cdot x_2^{\varepsilon_2} \cdots x_n^{\varepsilon_n})$$

The application of Theorem 3.1.1 completes the proof. □

Corollary 8.2.2 (Rachunek [106] and Petrova). *A semilinearly ordered group is torsion-free.*

We recall that $c(X)$ stands for the centralizer of a set X.

Lemma 8.2.2. *If a is incomparable with the identity element of a group G, then*

$$U(a) \cap P \cap c(a) = \varnothing$$

PROOF. Assume by way of contradiction that in G there are elements a and h such that $a \notin P^{\pm}$, $h \geqslant e$, $h \geqslant a$, and $ha = ah$. Then $ha^{-1} \geqslant e$ and $ha^{-1}h^{-1} \geqslant h^{-1}$. But $e \geqslant h^{-1}$. Consequently, the elements $ha^{-1}h^{-1}$ and e are comparable. However, in view of $h \in \mathfrak{C}(a)$, the equality $ha^{-1}h^{-1} = a$ holds and the element a^{-1} is comparable with the identity element, $a^{-1} \in P^{\pm}$, $a \in P^{\pm}$, which contradicts the choice of the element a. □

8.3. Convex Subgroups

In Section 8.2, it was shown (cf. Proposition 8.2.1) that any directed subgroup H of a semilinearly ordered group (G, P) is semilinearly ordered itself and the following relations hold: $H = P_H^{-1} \cdot P_H$, $P_H \cdot P_H^{-1} \subseteq P_H^{\pm}$, where $P_H = P \cap H$. Thereby, we have established an order isomorphism between partially-ordered-by-inclusion sets, i.e., the sets of all directed subgroups of a semilinearly ordered group G and the set of all those subsemigroups Q of the positive cone P of the group G subject to O.7. In this section, we study the properties of convex subgroups as well as convex directed subgroups of semilinearly ordered groups and some properties of extensions of such groups.

8.3.1. We first consider partially ordered groups.

Lemma 8.3.1. *For any subgroup H of a partially right-ordered group (G, P) the following conditions are equivalent:*

$$H \quad \textit{is a convex subgroup} \tag{8.3.1}$$

$$P_H = P \cap H \quad \textit{is a convex subset of } P \tag{8.3.2}$$

$$P \cdot P'_H \subseteq P'_H, \; P'_H = P \setminus H \tag{8.3.3}$$

PROOF. Let H be a convex subgroup of a partially ordered group (G, P). It is obvious that P_H is a convex subset of P. Hence (8.3.1) implies (8.3.2).

Conversely, we assume that (8.3.2) holds, $a \leqslant b \leqslant c$ for $a, c \in H$. Multiplying the last inequality from the right by a^{-1}, we obtain $e \leqslant ba^{-1} \leqslant ca^{-1}$, where $ba^{-1} \in P$, $ca^{-1} \in P_H$ and, in view of (8.3.2), $ba^{-1} \in P_H$, $b \in H$.

Let H be a convex subgroup, $p \in P'_H$, and $q \in P$. Then $q \geqslant e$ and $qp \geqslant p > e$. Since P_H is a convex subset of P, the element qp cannot belong to P_H because $p \notin H$. Hence $qp \in P'_H$ and $P \cdot P'_H \subseteq P'_H$. Thus, we have proved that the convexity of H implies (8.3.3).

Let (8.3.3) hold and $e \leqslant a \leqslant h$ for some elements $a, h \in G$ and $h \in P_H$. It is required to prove that $a \in P_H$. It is clear that under our assumptions we have $ha^{-1} \in P$ and $h = ha^{-1} \cdot a$. For $a \in P'_H$ we have $h \in P'_H$ because of (8.3.3), which contradicts the choice of $h \in P_H$. Thus, in view of (8.3.3), we conclude that H is a convex subgroup of G. □

Lemma 8.3.2. *If H is a convex directed subgroup of a semilinearly ordered group G and $p \in P'_H$, then $p > x$ for any $x \in H$.*

Proof. Let $p \in P'_H$, $x \in H$, $x = a^{-1}b$, where $a, b \in P_H$. Then $px^{-1} = pb^{-1}a$, where $pb^{-1} \in P \cup P^{-1}$. If the inequality $pb^{-1} < e$ holds, then the following relations must be valid: $e < p < b$ and $p \in P \cap H = P_H$, which contradicts the choice of p. Consequently, $pb^{-1} \in P$, but $pb^{-1} \notin H$, otherwise $p \in H$. Thus, $pb^{-1} = p_1 \in P'_H$. But then $p_1 a \in P$, $px^{-1} = p_1 a > e$, $p > x$. □

8.3.2. We are now in a position to describe the structure of convex directed subgroups and their location in a semilinearly ordered group.

Proposition 8.3.1 (Kopytov). *The set of all convex directed subgroups of a semilinearly ordered group (G, P) is totally ordered by inclusion and order-isomorphic to the set of all those convex subgroups of the positive cone which contain the identity element of the group.*

Proof. If H is a convex directed subgroup of a semilinearly ordered group G with positive cone P, then H is semilinearly ordered (cf. Proposition 8.2.1). Consequently, $H = P_H^{-1} \cdot P_H$, where $P_H = P \cap H$ is a convex subsemigroup contained in P and containing e. Thus, we have a mapping φ of the system $\mathfrak{C}_{\mathrm{cd}}(G)$ constituted by all convex directed subgroups of the group G into the system $\mathfrak{C}_{\mathrm{cd}}(P)$ constituted by all those convex subsemigroups of a totally ordered semigroup P which contain e. It is obvious that $\mathfrak{C}_{\mathrm{cd}}(P)$ is totally ordered by inclusion. In view of the relations $H = P_H^{-1} \cdot P_H$ and $P_H = P \cap H$, the mapping φ is monotone and one-to-one. Its inverse mapping φ^{-1} is also monotone, i.e., φ is an order isomorphism. □

Corollary 8.3.1 (Kopytov). *The set of all directed subgroups of a semilinearly ordered group is a complete totally ordered system of subgroups.*

Theorem 8.3.1 (Kopytov). *In any semilinearly ordered group G there exists the greatest convex right-ordered subgroup G_0.*

PROOF. Any convex right-ordered subgroup H of a semilinearly ordered group G is a convex directed subgroup, i.e., it belongs to $\mathfrak{C}_{\mathrm{cd}}(G)$. Since the union of an increasing sequence of right-ordered groups of $\mathfrak{C}_{\mathrm{cd}}(G)$ is a right-ordered subgroup of $\mathfrak{C}_{\mathrm{cd}}(G)$, in the system $\mathfrak{C}_{\mathrm{cd}}(G)$ the subgroup G_0 is the greatest one among convex right-ordered subgroups of the group G. □

Theorem 8.3.2 (Rachunek [106]). *A proper convex directed subgroup of any semilinearly ordered group is right-ordered.*

PROOF. Let $H \in \mathfrak{C}_{\mathrm{cd}}(G)$, and let H be a normal subgroup of a group G. By Lemma 8.3.2, $p > x$ for any $p \in P'_H$ and $x \in H$. If $h \in H$ and $p \in P$, then $p > e$, $p^{-1}h^{-1}p \in H$, and $p^{-1}h^{-1}p < p$. Consequently, $p^{-1}h^{-1} < e$ and $p^{-1} < h$. The last inequality and the inequality $e < p$ imply that the elements e and h are comparable, i.e., H is right-ordered. □

Corollary 8.3.2 (Rachunek [106]). *In any semilinearly ordered group G there exists the greatest right-ordered convex subgroup G_n containing all convex normal directed subgroups of the group G; moreover, G is the lexicoextension of G_n by the semilinearly ordered group $\overline{G} = G/G_n$ without proper normal convex directed subgroups.*

PROOF. The existence of a subgroup G_n follows from Theorem 8.3.2, and the fact that G is the lexicoextension of its subgroup G_n is obtained from Lemma 8.3.2. □

Corollary 8.3.3. *If $A \prec B$ is a subnormal jump in the system of convex directed subgroups of a semilinearly ordered group G, then $A \subseteq G_0$.*

8.3.3. We describe one more convex directed subgroup of a semilinearly ordered group G, which is naturally connected with the corresponding positive cone. We consider the subset P^* of $P^{\pm}$ constituted by elements comparable with any element of $P^{\pm}$:

$$\begin{aligned} P^* &= \{x \in P^{\pm} \mid xy^{-1} \in P^{\pm},\ \ y \in P^{\pm}\} \\ &= \{x \in P^{\pm} \mid xy \in P^{\pm},\ \ y \in P^{\pm}\}. \end{aligned}$$

It is easy to see that $y \in P^*$ for any $x \in P^*$ and $y \in G$, $y \geqslant x$. It is obvious that $P \subseteq P^*$. Let

$$G_r = P^* \cap P^{*-1}$$

Theorem 8.3.3 (Kopytov). *In any semilinearly ordered group G the set G_r is a convex right-ordered subgroup; moreover,*

$$G_n \subseteq G_r \subseteq G_0$$

PROOF. We first show that P^* is a subgroup. Let $a, b \in P^*$, i.e., $ax \in P^{\pm}$, $bx \in P^{\pm}$, for any $x \in P^{\pm}$. Then $ab \cdot x = a \cdot bx = ay$, where $y \in P^{\pm}$. Hence $ay \in P^{\pm}$, i.e., $ab \in P^*$. Consequently, G_r is a subgroup. Furthermore, G_r, being an intersection of convex sets, is convex. It is obvious that the group G_r is right-ordered, hence $G_r \subseteq G_0$. On the other hand, G is the lexicoextension of its normal right-ordered subgroup G_n. This means that any element of G_n is comparable with any element of $P^{\pm}$, i.e., $G_n \subseteq G_r$. □

We now indicate some properties of an arbitrary convex normal subgroup of a semilinearly ordered group. Let M be a subset of a group G.

- A subgroup N of the group G is called *infrainvariant* (M*-infrainvariant*) if $gHg^{-1} \subseteq H$ or $gHg^{-1} \supseteq H$ for any $g \in G$ ($gHg^{-1} \subseteq H$ for any $g \in M$).

Theorem 8.3.4 (Kopytov). *If H is a proper convex normal subgroup of a semilinearly ordered group G with positive cone P, then H contains the convex directed subgroup H_{cd}, $H_{\mathrm{cd}} \neq E$, such that*

(a) $H_{\mathrm{cd}} = P_H^{-1} \cdot P_H$,

(b) H_{cd} *is P-infrainvariant,*

(c) $pH_{\mathrm{cd}}p^{-1} \subseteq P_H \cup P_H^{-1}$ *for* $p \in P \setminus H$,

(d) $H = \bigcup_{p \in P} p^{-1} H_{\mathrm{cd}} p$

PROOF. As was mentioned, for any element $g \in G$ there is an element $p \in P$ such that $pgp^{-1} \in P \cup P^{-1}$. Therefore, a nontrivial normal subgroup H of the group G has a nonidentity intersection with P_H. Since H is convex, P_H is a convex subgroup of the semigroup P. By Proposition 8.3.1, $P_H^{-1} \cdot P_H$ is a convex directed subgroup of the group G. The assertion (a) is proved.

Let $H_{\text{cd}} = P_H^{-1} P_H$. In view of O.8 (cf. Proposition 8.1.1), $pqp^{-1} \in P^{\pm}$ for any $p \in P$ and $q \in P_H$. Since H is a normal subgroup of G, we have $pP_Hp^{-1} \subseteq P_H^{\pm}$, $p \cdot P_H^{\pm} \cdot p^{-1} \subseteq P_H^{\pm}$. Consequently, $pHp^{-1} \subseteq H$ for any $p \in P$, i.e., H is a P-infrainvariant subgroup of the group G. Thus, the assertion (b) is proved.

Let $p \in P'_H = P \setminus H$, $a \in H_{\text{cd}}$. By Lemma 8.3.2, $px > e$ for any $x \in H_{\text{cd}}$. Hence $e < pa^{-1} = (ap^{-1})^{-1} = (p^{-1} \cdot pap^{-1})^{-1}$, i.e., $p^{-1} \cdot pap^{-1} < e$, $p^{-1} < pa^{-1}p^{-1}$. But, at the same time, $p^{-1} < e$. By the axioms formulated at the beginning of the chapter, the elements $pa^{-1}p^{-1}$ and e are comparable, $pap^{-1} \in P_H^{\pm}$. Thus, the assertion (c) is proved.

By the normality of the semigroup P, from Lemma 8.2.1 we obtain (d). □

We note that in Theorem 8.3.4 it is impossible to assert that the subgroup H_{cd} is right-ordered. Corresponding examples will be presented later.

8.3.4. On the basis of the results just obtained, we are able to indicate some classes of groups on which every semilinear order is right. One class of such groups can be described with the help of Proposition 8.1.3.

Proposition 8.3.2 (Adeleke, Dummett, and Neumann [105]). *Any semilinear order on an Abelian group is total.*

Theorem 8.3.5 (Varaksin [107]). *Any semilinear order on a nilpotent group is right.*

PROOF. We proceed by induction on the nilpotency class of the group. If G is a semilinearly ordered nilpotent group of the nilpotency class 1, then it is Abelian and, by Proposition 8.3.2, is totally ordered. We suppose that every semilinearly ordered group of the nilpotency class k, where $1 \leqslant k \leqslant n-1$, is a totally ordered group. Let a group G be a nilpotent group of the nilpotency class n. Furthermore, we assume that it is semilinearly ordered. Let Z be the center of the group G. By Lemma 8.2.2, every element $z \in Z$ is comparable with the identity element of the group G. We denote by H the convex closure in G of the subgroup Z, i.e.,

$$H = \{x \in G \mid a \leqslant x \leqslant b\}$$

for some elements $a, b \in Z$. It is clear that

$$H = \{x \in G \mid a^{-1} \leqslant x \leqslant a^{-1}\}$$

for some element $a \in Z$, and $H = H^{-1}$.

Let $x, e \in H$. Then there is an element $a \in Z$ such that $a^{-1} \leqslant x, y \leqslant a$. Since $x \leqslant a$, we obtain $xy \leqslant ay = ya$. But $y \leqslant a$ and $ya \leqslant a^2$. Consequently, $xy \leqslant a^2$.

We can deduce that $a^{-1} \leqslant x$ and $a^{-1}y \leqslant xy$ in a similar way. Moreover, in view of the inequality $a^{-1} \leqslant y$, we have the equality $a^{-2} \leqslant ya^{-1} = a^{-1}y$, from which we obtain the following inequality: $a^{-2} \leqslant xy$, i.e., $a^{-1} \leqslant xy \leqslant a^2$, $a^2 \in Z$. Together with the equality $H^{-1} = H$, the last relation means that H is a subgroup of the group G. Furthermore, by construction, H is a convex directed subgroup of the semilinearly ordered group G. It is obvious that the subgroup H is normal. By Corollary 8.3.2, G is the lexicoextension of the right-ordered group H by the semilinearly ordered group G/H. Since $Z \subseteq H$, the quotient group G/H is nilpotent with the nilpotency class strictly less than n. By the induction assumption, the group G/H is right-ordered. Since the lexicoextension of a right-ordered group by a right-ordered group is also a right-ordered group, a semilinear order on the group G is a right order. □

Corollary 8.3.4. *Any semilinearly ordered group in which any two elements generate a nilpotent group is a right-ordered group. In particular, any locally nilpotent semilinearly ordered group is a right-ordered group.*

Proof. Let G be a semilinearly ordered group with order P, and let $g \in G$. Then $g = a^{-1}b$, where $a, b \in P$. We consider the subgroup $H = \mathrm{gp}(a, b)$ of G and put $P_H = P \cap H$. Since $a, b \in P_H$, the set P_H generates the subgroup H and

$$P_H \cdot P_H^{-1} \subseteq (P \cup P^{-1}) \cap H = P_H \cup P_H^{-1}$$

Consequently, P_H is a semilinear order on the subgroup H. Since the subgroup H is nilpotent, P_H is a right order on H, i.e., the element g is comparable with e in H and, consequently, in G. □

The following example shows that there exist nonnilpotent 2-generated groups on which any semilinear order is a right one.

EXAMPLE 8.3.1. Let $G = \mathrm{gp}(a, b \mid b^{-1}ab = a^{-1})$, and let P be an arbitrary semilinear order on the group G. A subgroup $A = \mathrm{gp}(a, b^2)$ is a normal Abelian subgroup of the group G and the quotient group G/A is a cyclic group of order 2. By Corollary 8.2.1 and Theorem 8.3.5, the subgroup A is a semilinearly ordered group with respect to the induced order. Every element $g \in G$ can be represented as $g = a^n b^k$, $n, k \in \mathbb{Z}$. We assume that the element g is incomparable with e. Then $g \in G \setminus A$. Consequently, $g = a^n b^{2p+1}$. Since the group G is directed, there exists its element w such that $w = a^m b^l$, $w \geqslant g = a^n b^{2p+1}$, $w \geqslant e$, $w \geqslant b^{-1}$, $w \geqslant a^n b^{2k} = h$.

In the case $l = 2t$, we multiply the inequality $w \geqslant h$ from the right by b and obtain $a^m b^{2t+1} \geqslant a^n b^{2p+1} = g$. From the inequality $a^m b^{2t} \geqslant b^{-1}$ we deduce $a^m b^{2t+1} \geqslant e$. Then for $u = a^m b^{2t+1}$ we have $u^2 = (a^m b^{2t+1})^2 = b^{4t+2} \geqslant u \geqslant g$, $u^2 \geqslant e$, and $u \in U(g) \cap P \cap c(g)$, $b^{4t+2} \neq e$, which contradicts Lemma 8.2.2.

The case $l = 2t + 1$ is considered in a similar way. Therefore, in the group G there are no elements incomparable with e. Consequently, P is a right order on the group G. □

8.4. Constructions

In this section, we present some methods of constructing semilinearly ordered groups from given right-ordered ones.

8.4.1. We first discuss the question on lexicoextensions of semilinear groups (cf. 8.3).

Proposition 8.4.1 (Rachunek [106]). *Let a partially right-ordered group G be a lexicoextension of its normal subgroup H by the partially right-ordered group $\overline{G} = G/H$. The group G is semilinearly ordered if and only if H is proper semilinearly ordered and $G = H$, or $\overline{G}$ is semilinearly ordered and H is right-ordered.*

PROOF. Let G be a semilinearly ordered group and G be the lexicoextension of H by $\overline{G}$. Since the quotient group $\overline{G} = G/H$ is partially ordered with respect to the natural order and H is a convex normal subgroup of G, in view of Proposition 8.2.2, we conclude that $\overline{G}$ is semilinearly ordered. In the case $H \neq G$ there exists an element g such that $g > e$ and $g \notin H$. Let a be an arbitrary element of H and $b = g^{-1}ag$, $b \in H$. Since G is the lexicoextension of H, we find that $gb^{-1} > e$ and $gbg^{-1} > g^{-1}$. However, $e > g^{-1}$ in G. In view of the semilinearity of the order, the elements e and $a = gbg^{-1}$ are comparable, i.e., the group H is right-ordered.

Conversely, if G is the lexicoextension of a right-ordered group H by the semilinearly ordered group $\overline{G} = G/H$, then the order P on the group G is defined by the relation

$$P = \{x \in G \mid xH > H \text{ в } G/H \text{ или } x \in H, x \geqslant e \text{ в } H\}$$

It is easy to check that P is a semilinear order on the group G. □

8.4.2. In this subsection, we indicate how to construct strict semilinear orders on groups starting from given semilinearly ordered groups, in particular, from right-ordered groups.

Theorem 8.4.1 (Kopytov). *Let H, N, and C be subgroups of a group G such that the following conditions are satisfied:*

(a) *N is a normal subgroup of G and $G = N \cdot H$;*

(b) *H is a semilinearly ordered group with order P_H; moreover, $N \cap H$ is a convex directed subgroup of H;*

(c) *G is P_H-infrainvariant and $C \neq N$, $N \cap H \subseteq C$;*

(d) *$N = \bigcup_{h \in H} hCh^{-1}$;*

(e) *C is a right-ordered group with order P_C.*

Then on G it is possible to introduce a proper semilinear order Q such that N is convex with respect to this order, semilinearly ordered groups $(G, Q)/N$ and $(H, P_H)/(H \cap N)$ are isomorphic, and $Q \cap C = P_C$.

Proof. As an order Q we take the set of elements $g \in G$ such that $g \in P_C$ or $g \notin N$, $g = pa$, where $p \in P'_H = P_H \setminus N$, $a \in p^{-1}Cp = C^p$, i.e., $pap^{-1} \in C$. We show that the definition of the set Q is correct. To prove this fact, it suffices to show that $q \in P'_H$ and $b \in C^q$ for $pa = qb$, where $p, q \notin N$, $a, b \in N$, $p \in P'_H$, $a \in G^p$. Indeed, under the above assumptions we have $pa = qb$, $q^{-1}p = ba^{-1} = c \in H \cap N \subseteq C$,

$q = p \cdot p^{-1}c^{-1}p = pc'$, where $c' = p^{-1}c^{-1}p$ and $c' \in N \cap H$. Since (H, P_H) is semilinearly ordered and $N \cap H$ is a convex directed subgroup of the group H, in view of Lemma 8.3.2, we have $q = pc' \in P'_H$. Further,

$$\begin{aligned} qbq^{-1} &= pc' \cdot q \cdot (c')^{-1}p^{-1} = pc'ca(c')^{-1}p^{-1} \\ &= p \cdot p^{-1}c^{-1}p \cdot ca \cdot p^{-1}cp \cdot p^{-1} = c^{-1} \cdot pcp^{-1} \cdot pap^{-1} \cdot c \end{aligned}$$

here $c \in H \cap N \subseteq C$ and $pcp^{-1} \in C$, since C is a P_H-infrainvariant subgroup, $p \in P_H$, and $pap^{-1} \in C$ by the above hypothesis. This means that $qbq^{-1} \in C$, and the definition of the set Q is correct.

We now show that Q satisfies all the hypotheses of Theorem 8.1.2. Let $x, y \in Q$. We study all possible cases of the location of elements x and y.

If $x, y \notin N$, then $x = pa$ and $y = qb$, where $p, q \in P'_H$, $pap^{-1}, qbq^{-1} \in C$. Then $x \cdot y = pa \cdot qb = pq \cdot q^{-1}aq \cdot b$; moreover, $pq \in P'_H$ and $q^{-1}aq \cdot b \in N$. Therefore, to prove the relation $xy \in Q$, it suffices to obtain the formula $pq \cdot q^{-1}aq \cdot b \cdot (pq)^{-1} \in C$. We have

$$\begin{aligned} pq \cdot q^{-1}aq \cdot b \cdot (pq)^{-1} &= p \cdot qq^{-1}aqq^{-1} \cdot qbq^{-1} \cdot p^{-1} \\ &= pap^{-1} \cdot p \cdot qbq^{-1} \cdot p^{-1} \end{aligned}$$

By the above hypothesis, $pap^{-1} \in C$ and $qbq^{-1} \in C$. Since C is P_H-infrainvariant, the inclusion $p \cdot qbq^{-1} \cdot p^{-1} \in C$ holds. Consequently,

$$pq \cdot q^{-1}aq \cdot b \cdot (pq)^{-1} \in C, \quad xy \in Q$$

Let $x \notin N$ and $y \in N$. Then $x = pa$, $p \in P'_H$, $pap^{-1} \in C$, $y \in P_C \subseteq C$. Therefore, $xy = p \cdot ay$, $p \in P'_H$, $ay \in N$, and $p \cdot ay \cdot p^{-1} = pap^{-1} \cdot pyp^{-1} \in C$ because $pap^{-1} \in C$ by the above hypothesis and $pyp^{-1} \in C$ in view of the inclusion $y \in C$ and the P_H-infrainvariance of C.

For $x \in N$ and $y \notin N$ we have $x \in P_C \subseteq C$, $y = qb$, $q \in P'_H$, $qbq^{-1} \in C$. Hence $xy = x \cdot qb = q \cdot q^{-1}xq \cdot b$; moreover, $q \in P'_H$, $q^{-1}xq \cdot b \in N$. Furthermore, $q \cdot q^{-1}xq \cdot b \cdot q^{-1} = x \cdot qbq^{-1} \in C$ since $x \in C$, $qbq^{-1} \in C$ by the above assumption.

For $x, y \in N$ we have $x, y \in P_C$ and $xy \in P_C$, $xy \in Q$ since P_C is a semigroup. Thereby, we have proved that the condition O.1 from Theorem 8.1.2 is satisfied.

We now show that Q is a pure semigroup. Let $x \in Q$. For $x \notin N$ we have $x = pa$, where $p \in P'_H$, $pap^{-1} \in C$. Then $x^{-1} = p^{-1} \cdot pa^{-1}p^{-1} \notin Q$; consequently, $Q \cap Q^{-1} \subseteq N$. If $x \in Q \cap N$, then $x \in P_C$ and for $x \neq e$, we have $x \notin P_C^{-1}$, i.e., $x \notin Q$. This means that $Q \cap Q^{-1} = \{e\}$. The condition O.2 is satisfied.

It is not hard to see that P_C generates C. Therefore, $P'_H \cup P_C$ generates a subgroup containing H. Moreover, by the condition (d), N is contained in the subgroup generated by the set $P'_H \cup P_C$. Hence the group G is generated by the set $P'_H \cup P_C$ and, moreover, by the set Q, i.e., the condition O.3 is satisfied.

It remains to verify the condition O.7 for the group G:

$$Q \cdot Q^{-1} \subseteq Q \cup Q^{-1}$$

Let $x, y \in Q$. We show that $xy^{-1} \in Q^{\pm}$. As above, we consider all possible cases of the location of elements x and y.

Let $x, y \in Q \setminus N$, $x = pa$, $y = qb$, $p, q \in P'_H$, $pap^{-1} \in C$, $qbq^{-1} \in C$. Three cases are possible.

CASE 1: $\overline{p} > \overline{q}$ in the quotient group $\overline{H} = H/N \cap H$. If $\overline{p} > \overline{q}$, then $xy^{-1} = (pa)(qb)^{-1} = pq^{-1} \cdot qaq^{-1} \cdot qb^{-1}q^{-1} = r \cdot c$; here $r = pq^{-1} \in P'_H$ and $c = qaq^{-1} \cdot qb^{-1}q^{-1} = qaq^{-1} \cdot b' \in N$, where $b' = qb^{-1}q^{-1} \in C$, by condition. Hence

$$rcr^{-1} = r \cdot qaq^{-1} \cdot b' \cdot r^{-1} = rqaq^{-1}r^{-1} \cdot rb'r^{-1}$$

However,

$$rqaq^{-1}r^{-1} = pq^{-1} \cdot qaq^{-1} \cdot qp^{-1} = pap^{-1} \in C$$

by condition and $rb'r^{-1} \in C$ since $b' \in C$, $r \in P'_H$, and C is a P_H-infrainvariant subgroup. Thus, we have $xy^{-1} \in Q$.

CASE 2: $\overline{q} > \overline{p}$ in $\overline{H}$. Arguing as in Case 1, we find $yx^{-1} \in Q$, i.e., $xy^{-1} \in Q^{-1}$.

CASE 3: $\overline{p} = \overline{q}$. We have $pq^{-1} = c \in H \cap N \subset C$ and $pap^{-1} \in C$, $qbq^{-1} \in C$. Moreover, $q = c^{-1}p$. For xy^{-1} it is clear that

$$xy^{-1} = pa \cdot b^{-1}q^{-1} = pq^{-1} \cdot qaq^{-1} \cdot qb^{-1}q^{-1} = c \cdot c^{-1}pap^{-1}c \cdot qb^{-1}q$$

By condition, $pap^{-1} \in C$, $qbq^{-1} \in C$, $c \in C$, and the group C is right-ordered. Consequently,

$$xy^{-1} \in C = P_C \cup P_C^{-1} \subset Q \cup Q^{-1}$$

This completes the consideration of Case 3.

Thus, we have proved that $xy^{-1} \in Q^{\pm}$ provided that $x, y \in Q \setminus N$.

Let $x \in Q \setminus N$ and $y \in Q \cap N$. Then $x = pa$, $p \in P'_H$, $a \in N$, $pap^{-1} \in C$, and $y = b$, $b \in P_C$. In this case $xy^{-1} = pa \cdot b^{-1} = p \cdot ab^{-1}$; moreover, $p \in P'_H$ and $ab^{-1} \in N$. In addition,

$$p \cdot ab^{-1} \cdot p^{-1} = pap^{-1} \cdot pbp^{-1}$$

where $pap^{-1} \in C$ by condition and $pb^{-1}p^{-1} \in C$ because C is P_H-infrainvariant. This means that $xy^{-1} \in Q$.

In the case $x \in Q \cap N$ and $y \in Q \setminus N$, arguing as above, we conclude that $yx^{-1} \in Q$, i.e., $xy^{-1} \in Q^{-1} \subset Q^{\pm}$.

For $x, y \in Q \cap N$ we have $x, y \in P_C$ and $xy^{-1} \in P_C^{\pm}$ since P_C is a right order on C.

Thus, we have proved that Q is a positive cone with respect to the semilinear order on G, which is an extension of the order P_C; moreover, the group N satisfies the condition (8.3.3) and, consequently, is convex with respect to the order Q. The validity of (8.3.3) immediately follows from the definition of the order Q. It is also obvious that the semilinearly ordered quotient group $(G, Q)/N$ is order-isomorphic to a semilinearly ordered quotient group $(H, P_H)/(N \cap H)$.

To complete the proof, it suffices to remark that $Q \cap N = P_C$ and $(Q \cap N) \cup (Q \cap N)^{-1} \neq N$ because $C \neq N$. In other words, Q is not a right order, i.e., it is a proper semilinear order on the group G. □

8.4.3. Theorem 8.4.1 does not give any specific recommendations for the choice of the group $G = N{\cdot}H$. However, in some cases it is possible to indicate standard methods that allow us to construct proper semilinearly ordered groups from given semilinearly ordered or right-ordered groups N and H. We demonstrate them for $G = N{\cdot}H$ and $N \cap H = E$. First, we introduce a definition. Let $\varphi : H \to \mathrm{Aut}(N)$ be a homomorphism of a group H into the group of automorphisms of the group N.

- The set G of all pairs (u, a), $u \in H$, $a \in N$, with the multiplication $(u, a) \cdot (v, b) = (uv, \varphi a \cdot b)$ for $u, v \in H$ and $a, b \in N$ is called the *splitting extension* of the group N by the group H, corresponding to the homomorphism φ.

It is well known that the set G so constructed is a group; moreover, the mappings

$$\nu : N \to G, a\nu = (e, a),$$
$$\mu : H \to G, u\mu = (u, e)$$

are isomorphisms of N and H into the group G respectively. Furthermore, $G = N\nu \cdot H\mu$ and $N\nu \cap H\mu = E$. The constructed group G is also called the *semidirect product* of its normal subgroup N and the complement factor H. In similar cases we will identify N with $N\nu$ and H with $H\mu$.

The assertions below follow immediately from Theorem 8.4.1.

Theorem 8.4.2 (Kopytov). *Let $G = N \cdot H$ be a group such that it is the semidirect product of its normal subgroup N and the complement factor H. We suppose that* (d), (e) *and the following conditions are satisfied:*

(f) *H is a semilinearly ordered group with order P_H,*

(g) *there exists a P_H-infrainvariant subgroup C of the group N such that $C \neq N$ and $N \cap H \subseteq C$.*

Then a proper semilinear order Q can be introduced on G so that N is convex, the semilinearly ordered groups $(G, Q)/N$ and $(H, P_H)/(H \cap N)$ are isomorphic, and $Q \cap C = P_C$.

Theorem 8.4.3 (Kopytov). *The wreath product $G = A\mathrm{wr}B$ of a right-ordered group $A \neq E$ and a semilinearly ordered group $B \neq E$ admits a proper semilinear order that extends given orders on A and B under the canonical embeddings of the groups A and B into G.*

PROOF. Indeed, if $G = A\mathrm{wr}B$, then the group G is a semidirect product of its normal subgroup $N = \prod_{x \in B} A_x = \mathrm{fun}(B, A)$ and the subgroup $H = B$, acting on N as follows: if $A_u = \{f \in N \mid f(x) = e$ for $x \neq u,\ u, x \in B\}$, then $vA_uv^{-1} = A_{uv}$ for any $u, v \in B$.

We put $C = \prod_{x \in P_B} A_x$, where P_B is a given semilinear order on the group B. On the group C we fix an arbitrary right order P_C that extends a given right order P_A on the group A, where the latter is identified, as usual, with the subgroup A_e of N (such an order exists on C). The application of Theorem 8.4.2 completes the proof. □

Theorem 8.4.4 (Kopytov). *Let $\overline{G} = F/U$ be a group such that it is the quotient group of a free group F with a free basis X by its normal subgroup U. We suppose that the group $\overline{G}$ admits a semilinear order $\overline{P}$ such that, under the natural homomorphism of F onto $\overline{G}$, the images $\overline{x}$ of the generators $x \in X$ are contained in $\overline{P}$. Then, on the quotient group $G = F/U'$ of the group F by the derived subgroup $N' = [N, N]$ of the subgroup N, a proper semilinear order can be introduced.*

PROOF. Let F be a free group with the free basis $X = \{x_i \mid i \in I\}$, and let the subgroup U (regarded as a normal subgroup) of F be generated by w_j, $j \in J$. We consider a free Abelian group A with the free basis $\{a_i \mid i \in I\}$ and the wreath product $G^* = A \operatorname{wr} \overline{G}$. On the group G^*, acting as in Theorem 8.4.3, we construct a proper semilinear order P^* such that it extends the given semilinear order $\overline{P}$ on the group $\overline{G}$. For this order all the elements $c_i = (x_i, a_i)$, $i \in I$, are positive.

Further, we consider the subgroup G^* of G generated by (x_i, a_i), $i \in I$, where $x_i \in \overline{G}$, $a_i \in A$. Then G is generated by positive elements and, consequently, is a directed subgroup of the semilinearly ordered group G^*. By Proposition 8.2.1, the group G is semilinearly ordered. We note that this subgroup is isomorphic to the group F/N' (cf., for example, [108]). To complete the proof, it suffices to show that the semilinear order just constructed is proper. Under the standard embedding of the group G into the wreath product $A \operatorname{wr} \overline{G}$ the subgroup N/N' goes to the basis subgroup $\overline{A} = \prod_{g \in \overline{G}} A_g$. Since $P^* \cap \overline{A} \subset \prod_{g \in \overline{P}} A_g$, the element $a = c_{g_1} c_{g_2} \cdots c_{g_k}$, where $c_{g_i} \in A_{\underline{g_i}}$ for $1 \leqslant i \leqslant k$, is contained in P^* only if all the elements g_i belong to $\overline{P}$. But it is evident that in the subgroup N/N' there are elements a that are not subject to this condition. Therefore, in G there are elements incomparable with the identity element. □

Theorem 8.4.5 (Kopytov). *The free product of nonidentity right-ordered groups admits a proper semilinear order that extends the given orders on the factors.*

PROOF. It suffices to consider the case of two factors. Let $G = G_1 * G_2$ be the free product of right-ordered groups G_1 and G_2, and let $\overline{G} = G_1 \operatorname{wr} G_2$ be the wreath product of the groups G_1 and G_2. By the definition of the free product of groups, there exists a homomorphism φ of the group G onto the group $\overline{G}$. Let K be the kernel of this homomorphism. We fix a right order P_K on the group K and a proper semilinear order $\overline{P}$ on the group $\overline{G}$. Such an order exists on $\overline{G}$ by Theorem 8.4.3. Then it is possible to construct a partial right order P on G such that the subgroup K will be convex and the group G will be the lexicoextension of the right-ordered group (K, P_K) by the semilinearly ordered group $(\overline{G}, \overline{P})$ (cf. Theorem 1.6.2). Applying Proposition 8.4.1, we conclude that the group (G, P) is semilinearly ordered. □

References

1. BIRKHOFF, G., *Lattice Theory*, Am. Math. Soc., Providence, Rhode Island, 1967.
2. STONE, M. N., The theory of representations for Boolean algebras, *Trans. Am. Math. Soc.*, **40**, No. 1, 37–111, 1936.
3. HALL, M., JR., *The Theory of Groups*, Macmillan, New York, 1959.
4. LEVI, F. W., Ordered groups, *Proc. Indian Acad. Sci. A.*, **16**, 256–263, 1942.
5. SMIRNOV, D. M., Right-ordered groups, *Algebra Logika*, **5**, No. 6, 41–59, 1966.
6. MCCLEARY, S. H., An even better representation for free lattice-ordered groups, *Trans. Am. Math. Soc.*, **290**, No. 1, 81–100, 1985.
7. DROSTE, M., Representations of free lattice-ordered groups, *Order*, **10**, No. 2, 375–381, 1993.
8. CONRAD, P. F., Right-ordered groups, *Mich. Math. J.*, **6**, No. 3, 267–275, 1959.
9. ZAĬTSEVA, M. I., On the family of orders on an Abelian group, *Usp. Mat. Nauk*, **8**, No. 11, 135–137, 1953.
10. HÖLDER, O., Die Axiome Quantität und die Lehre vom Mass., *Ber. Verh. Sächs. Wiss. Leipzig, Math.-Phis. Cl.*, **53**, 1–64, 1901.

11. IWASAWA, K., On linearly ordered groups, *J. Math. Soc. Japan*, **1**, No. 1, 1–9, 1948.
12. MAL'CEV, A. I., On ordered groups, *Izv. Akad. Nauk SSSR, Ser. Mat.*, **13**, No. 6, 473–482, 1949.
13. RHEMTULLA, A. H., Right-ordered groups, *Can. J. Math.*, **24**, No. 5, 891–895, 1972.
14. AULT, J. S., Right-ordered locally nilpotent groups, *J. London Math. Soc.*, **2**, No. 4, 662–666, 1972.
15. TEH, H. H., Construction of orders in Abelian groups, *Proc. Cambridge Philos. Soc.*, **57**, No. 3, 476–482, 1961.
16. TREVISAN, G., Classificazione dei semplici ordenamenti di un gruppo libero commutativo con *n* generatori, *Rend. Semin. Mat. Univ. Padova*, **22**, 143–156, 1953.
17. HION, YA. V., Archimedean ordered rings, *Usp. Mat. Nauk*, **9**, No. 4, 237–242, 1954.
18. MATSUSITA, S., Sur la puissance des ordres dans un groupe libre, *Proc. Koninkl. Nederl. Akad. Wet. A.*, **56**, 15–16, 1953.
19. FUCHS, L., *Partially Ordered Algebraic Systems*, Pergamon Press, New York, 1963.
20. OHNISHI, H., Linear order on a group, *Osaka J. Math.*, **4**, 17–18, 1952.
21. LOŚ, J., On the existence of linear order in a group, *Bull. Acad. Pol. Sci.*, **2**, No. 1, 21–33, 1954.
22. LORENZEN, P., Über halbgeordneter Gruppen, *Arch. Math.*, **52**, No. 1/2, 66–70, 1949.
23. MAL'CEV, A. I., On the full ordering of groups, *Tr. Mat. Inst. Steklova*, **38**, 173–175, 1951.
24. NEUMANN, B. H., An embedding theorem for algebraic systems, *Proc. London Math. Soc.*, **4**, No. 1, 138–153, 1954.
25. ZAĬTSEVA, M. I., Right-ordered groups, *Uchen. Zap. Shuisk. Gos. Ped. Inst.*, **6**, 215–226, 1958.
26. BURNS, R. G. and HALE, V. W., A note on group rings of certain torsion-free groups, *Can. Math. Bull.*, **15**, No. 3, 441–445, 1972.
27. WIGOLD, J., Semigroups covering of groups (II), *Mat.-Fiz. Čas.*, **12**, 217–223, 1962.
28. SMIRNOV, D. M., One-side orders in groups with increasing central series, *Algebra Logika*, **6**, No. 2, 77–88, 1967.
29. PODDERYUGIN, V. D., Conditions of orderability of a group, *Izv. Akad. Nauk SSSR, Ser. Mat.*, **21**, No. 2, 199–208, 1957.
30. RIEGER, L. S., On the ordered and cyclically ordered groups (I–III), *Věstn. Kral. Čs. Společ Nauk*, **6**, 1–31, 1946; **1**, 1–33, 1947; **1**, 1–26, 1948.
31. KOKORIN, A. I., Intersection and join of relatively convex subgroups of orderable groups, *Algebra Logika*, **7**, No. 3, 48–50, 1968.

32. KOKORIN, A. I., Γ-fully orderable and relatively convex subgroups of orderable groups, *Sib. Mat. Zh.*, **7**, No. 3, 713–717, 1966.
33. DARNEL, M. R., GLASS, A. M. W., and RHEMTULLA, A. H., Groups in which every right order is two-sided, *Arch. Math.*, **53**, 538–542, 1989.
34. KIM, Y. K. and RHEMTULLA, A. H., Orderable groups satisfying an Engel condition, *Ordered Algebraic Structures* (Gainesville, FL, 1991), Kluwer Academic Publishers, Dordrecht, 1993, pp. 73–79.
35. COHN, P. M., Group of order automorphisms of ordered sets, *Mathematika*, **4**, No. 1, 41–50, 1957.
36. KOPYTOV, V. M. and MEDVEDEV, N. YA., *The Theory of Lattice-Ordered Groups*, Kluwer Academic Publishers, Dordrecht–Boston–London, 1994.
37. KOPYTOV, V. M., *Lattice-Ordered Groups*, Nauka, Moscow, 1984.
38. HOLLAND, CH. W., The lattice-ordered group of an ordered set, *Mich. Math. J.*, **10**, No. 4, 399–408, 1963.
39. NEUMANN, H., *Varieties of Groups*, Springer Verlag, Berlin, 1967.
40. DANWOODY, M. Y., On verbal subgroups of free groups, *Math. Z.*, **104**, 91–97, 1968.
41. TODORINOV, S. A., Right preorderable groups, *Compt. Rend. Acad. Bulg. Sci.*, **35**, No. 10, 1323–1325, 1982.
42. MURA, R. B. and RHEMTULLA, A. H., *Notes on Orderable Groups*, University of Alberta, Edmonton, 1975.
43. PIERCE, K. R., Extensions of right orders in solvable groups, *J. London Math. Soc.*, **11**, No. 1, 81–87, 1975.
44. OHNISHI, H., On linearisation of ordered groups, *Osaka J. Math.*, **2**, 161-164, 1950.
45. KUROSH, A. G., *The Theory of Groups*, II, Chelsea Publishing Co., New York, 1956.
46. KARGAPOLOV, M. I., Fully orderable groups, *Algebra Logika*, **2**, No. 6, 5–14, 1963.
47. KOKORIN, A. I., On fully orderable groups, *Dokl. Akad. Nauk SSSR*, **151**, 31–33, 1963.
48. KOKORIN, A. I. and KOPYTOV, V. M., *Fully Ordered Groups*, Nauka, Moscow, 1972.
49. HOLLAND, CH. W., Extensions of ordered algebraic structures, Doctoral Thesis, Tulane University, 1961.
50. FUCHS, L. and SASIADA, E., Note on orderable groups, *Ann. Univ. Sci. Budap., Sect. Math.*, **7**, 13–17, 1965.
51. KOPYTOV, V. M., On the theory of fully orderable groups, *Algebra Logika*, **5**, No. 6, 27–31, 1966.
52. HOLLAND, CH. W., Transitive lattice-ordered groups, *Math. Z.*, **87**, No. 3, 420–433, 1965.
53. KARGAPOLOV, M. I. and MERZLJAKOV, YU. I., *Fundamentals of*

the Theory of Groups, Springer Verlag, New York, 1979.

54. HOLLAND, CH. W. and MCCLEARY, S. H., Wreath products of ordered permutation groups, *Pac. J. Math.*, **31**, No. 3, 703–716, 1969.
55. MCCLEARY, S. H., o-Primitive ordered permutation groups, *Pac. J. Math.*, **40**, No. 2, 349–372, 1972.
56. MCCLEARY, S. H., o-Primitive ordered permutation groups (II), *Pac. J. Math.*, **49**, No. 2, 431–445, 1973.
57. KOPYTOV, V. M., Free lattice-ordered groups, *Algebra Logic*, **18**, No. 4, 259–270, 1979.
58. VARAKSIN, S. V., Lattice-ordered groups constructed from right-ordered groups, *Algebra Logika*, **28**, No. 5, 524–533, 1989.
59. DARNEL, M. R., Special-valued l-groups and Abelian covers, *Order*, **4**, No. 2, 191–194, 1987.
60. DARNEL, M. R., The free lattice-ordered group over a nilpotent group, *Proc. Am. Math. Soc.*, **111**, No. 2, 301–307, 1991.
61. HOLLAND, CH. W., MEKLER, A. H., and SHELAH, S., Lawless order, *Order*, **1**, No. 3, 383–397, 1985.
62. HIGMAN, G., On infinite simple permutation groups, *Publ. Math.*, Debrecen, **3**, No. 3/4, 221–226, 1955.
63. CHEHATA, C. G., An algebraically simple ordered group, *Proc. London Math. Soc.*, **2**, 183–197, 1952.
64. DLAB, V., On a family of simple ordered groups, *J. Aust. Math. Soc.*, **8**, No. 3, 591–608, 1968.
65. HOLLISTER, H., Contribution to the theory of partially ordered groups, Doctoral Thesis, Univ. of Michigan, 1965.
66. CONRAD, P. F., Free lattice-ordered groups, *J. Algebra*, **16**, 190–203, 1970.
67. MAL'CEV, A. I., On partially ordered nilpotent groups, *Algebra Logika*, **1**, No. 2, 5–9, 1962.
68. TARARIN, V. M., On groups of automorphisms of totally ordered sets, Dep. VINITI, No. 677–684, Moscow, 1982.
69. TODORINOV, S. A., On relatively convex subgroups, *God. VTUZ*, **16**, No. 2, 75–83, 1980.
70. PLOTKIN, B. I., On the theory of noncommutative torsion-free groups, *Mat. Sb.*, **30**, 197–212, 1952.
71. HALL, P., Nilpotent groups, *Notes of Lectures given at the Canadian Mathematical Congress Summer Seminar University of Alberta*, 12–30 *August*, 1957. Queen Mary College, London, 1969 (Math. Notes).
72. TODORINOV, S. A. and PETROVA, N. L., Groups possessing unique right order, *God. VTUZ*, **16**, No. 2, 55–60, 1980.
73. TARARIN, V. M., On groups having a finite number of right orders, Dep. VINITI, No. 627–687, Moscow, 1987.
74. KOPYTOV, V. M., On linearly ordered soluble groups, *Algebra Logic*,

12, No. 6, 374–380, 1973.

75. BLUDOV, V. V., Uniquely orderable groups, *Algebra Logic*, **13**, No. 6, 343–360, 1974.
76. BUTTSWORTH, R. N., A family of groups with a countable infinite number of full orders, *Bull. Aust. Math. Soc.*, **4**, No. 1, 97–104, 1971.
77. TARARIN, V. M., On radically right-ordered groups, *Sib. Mat. Zh.*, **32**, No. 5, 203–204, 1991.
78. FOX, C., An embedding theorem for ordered groups, *Bull. Aust. Math. Soc.*, **12**, No. 3, 321–335, 1975.
79. GURCHENKOV, S. A., On the completion of normal nilpotent subgroups in totally ordered groups, *Mat. Zametki*, **51**, No. 2, 35–39, 1992.
80. KOPYTOV, V. M., On the completion of the center of orderable groups, *Ural. Gos. Univ. Mat. Zap.*, **4**, No. 3, 20–24, 1963.
81. MEDVEDEV, N. YA., Relative convexity of subgroups of finite rank, *Mat. Zametki*, **46**, No. 4, 34–37, 1989.
82. MURA, R. B. and RHEMTULLA, A. H., Ordered solvable groups satisfying the maximal condition on isolated subgroups and groups with finitely many relatively convex subgroups, *J. Algebra*, **36**, No. 1, 38–45, 1975.
83. VINOGRADOV, A. A., On the free product of ordered groups, *Mat. Sb.*, **25**, 163–168, 1949.
84. BERGMAN, G. M., Ordering coproducts of groups and semigroups, *J. Algebra*, **133**, No. 2, 313–339, 1990.
85. STROJNOWSKI, A. and WOJAS, W., Some remarks on *o*-groups, *Demonstr. Math.*, **23**, No. 3, 799–801, 1990.
86. MAGNUS, W., KARRASS, A., and SOLITAR, D., *Combinatorial Group Theory: Presentations of Groups in Terms of Generators and Relations*, Wiley, New York, 1966.
87. PASSMAN, D. S., *Infinite Crossed Products*, Academic Press, Boston, 1989.
88. MEDVEDEV, N. YA., Quasivarieties of *l*-groups and groups, *Sib. Math. J.*, **26**, No. 5, 717–722, 1985.
89. BERGMAN, G. M., Right-orderable groups that are not locally indicable, *Pac. J. Math.*, **147**, No. 2, 243–248, 1991.
90. MAL'CEV, A. I., *Algebraic Systems*, Springer Verlag, Berlin, 1976.
91. BRODSKIĬ, S. D., Equations over groups, and groups with one defining relation, *Sib. Mat. Zh.*, **25**, No. 2, 235–250, 1984.
92. MOSTOWSKI, A. and EHRENFEUCTH, A., A model of axiomatic theories admitting automorphisms, *Fundam. Math.*, **43**, 40–68, 1956.
93. RABIN, M. O., Universal groups of automorphisms of models, *Theory of Models* (Proc. 1963 Internat. Sympos. Berkeley), North-Holland, Amsterdam, 1965, pp. 274–284.
94. ERSHOV, YU. L. and PALYUTIN, E. A., *Mathematical Logic*, Nau-

ka, Moscow, 1987.

95. FEDOROV, YU. G., On infinite groups of which all nontrivial subgroups have a finite index, *Usp. Mat. Nauk*, **6**, No. 1, 187–189, 1951.
96. BUDKIN, A. I., Quasi-identities in a free group, *Algebra Logic*, **15**, No. 1, 25–33, 1976.
97. TARARIN, V. M., On the theory of right-orderable groups, *Mat. Zametki*, **54**, No. 2, 96–98, 1993.
98. MASSEY, W., *Algebraic Topology: An Introduction*, Harcourt, Brace & World, New York–Chicago–San Francisco–Atlanta, 1967; Massey, W. and Stallings, J., *Algebraic Topology: An Introduction* [Russian translation], Mir, Moscow, 1977.
99. ROBINSON, D. J. S., *Finiteness Conditions and Generalized Soluble Groups* (Parts 1, 2), Springer Verlag, Berlin, 1972.
100. HOWIE, J., On locally indicable groups, *Math. Z.*, **180**, No. 4, 445–461, 1982.
101. BUDKIN, A. I., Axiomatic rank of a quasivariety of right-ordered groups, *Algebra Logic*, No. 5, 313–318, 1986.
102. KARGAPOLOV, M. I. and CHURKIN, V. A., On varieties of solvable groups, *Algebra Logic*, **10**, No. 6, 395–398, 1971.
103. BUDKIN, A. I. and GORBUNOV, V. A., Quasivarieties of algebraic systems, *Algebra Logic*, **14**, No. 2, 73–84, 1975.
104. BAUMSLAG, G., On generalized free products, *Math. Z.*, **78**, 423–438, 1962.
105. ADELEKE, S. A., DUMMETT, M. A. E., and NEUMANN, P. M., On a question of Frege's about right-ordered groups, *Bull. London Math. Soc.*, **19**, No. 6, 513–521, 1987.
106. RACHUNEK, J., Convex directed subgroups of right-ordered free groups, *Czech. Math. J.*, **41**, No. 1, 99–103, 1991.
107. VARAKSIN, S. V., Semilinear orderings for solvable and nilpotent groups, *Algebra Logic*, **29**, No. 6, 415–417, 1990.
108. SHMEL'KIN, A. L., On free products of groups, *Mat. Sb.*, **79**, No. 4, 616–620, 1979.

Subject Index

Algebra, Boolean 11

Amalgamation property 178

Automorphism 5

Basis 59

Bound
- greatest lower 4, 5
- least upper 4, 5
- lower 4
- upper 4

Centralizer 36
- of jump 36

Class
- locally closed 63
- superaccessible 184
- ultraclosed 182

Complement 11

Completion
- Dedekind 77
- of group 157

Component
- global 86
- local 86
- of group vector 118

Cone
- α-positive 122
- negative 15, 218
- positive 15

Convex G-congruence 74

Diagonal 173

Dominant component 65

Dominion 175

Element
- α-positive 122
- negative 15
- positive 15
- strictly negative 15
- strictly positive 15

Elements
- comparable 2
- incomparable 2

Extension
- lexicographic 26
- of order 2
- splitting 236

Filter 181
- Fréchet 182

Group
- G-torsion-free 54
- Γ-orderable 59
- o-2-transitive 82
- o-primitive 81
- $\mathfrak{R}$-projectable 184
- Archimedean 33
- Conradian 32
- automorphisms periodic 95
- connected 12
- directed 12
- divisible 58
- fully orderable 67
- fully right-orderable 62
- indicable 50
- lattice-ordered 12
- locally $\mathfrak{R}$-projectable 184
- locally nilpotent 40
- locally projectable 50
- nilpotent 40
- of order automorphisms 73
- order-simple 23
- orderable 45
- partially left-ordered 12
- partially ordered 12
- partially right-ordered 12
- pathological 95
- radical 205
- right-orderable 45
- right-ordered 12
- semilinearly ordered 218
- totally ordered 12
- transitive 78
- universal 190
- vector 118

Homomorphism 22
- natural 75
- of Boolean algebra 11
- of lattices 7
- strict order 22

Ideal 20
- of totally ordered set 76

Isolator 51

Isomorphism 22
- Boolean 11
- of lattices 7
- of *po*-sets 7

Join of elements 4

Jump 31

Kernel of congruence 75

Lattice 4
- Boolean 11
- Brouwerian 10
- complete 6
- completely distributive 10
- conditionally complete 6
- distributive 8
- dually Brouwerian 10
- modular 8

Lattices, isomorphic 7
Length of word 166

Mapping
coordinate 87
order 7
Meet 4
Module
G-ordered 59
Γ-ordered 59
free 59

Nilpotency class 40
Normalizer 36
Null element 6

Order
coordinate-wise 2
left-invariant 220
lexicographic 2
partial 1
proper semilinear 220
strongly isolated 119
trivial 19

Pair, covering 84
Permutation 5
Point, fracture 103
Principle, Archimedean 33
Product
cardinal (of *po*-sets) 3
Cartesian 24
Cartesian (of lattices) 8
Cartesian (of *po*-sets) 3
direct 24
free 165, 169
lexicographic 25
lexicographic (of *po*-sets) 3
reduced 182
semidirect 27, 236
wreath 208
wreath, generalized 89

Quasi-identity 181
Quasivariety 181
Quotient group 21, 75

Rank, axiomatic 194
Representation
faithful 73
orderable 122
partially ordered 122
totally ordered 122
transitive 78
Ring, totally ordered 167

Separation property 81
Series
central 40
rational 134
Set
G-invariant 75
normal 17
partially ordered 1
totally ordered 2
transitivity 78
unstable 100
well-ordered 2
Sets, order-isomorphic 7
Skolemization 192
Stabilizer 75
point-wise 75

Subgroup
- G-fully orderable 68
- M-infrainvariant 229
- Cartesian 169
- convex 20
- equalizer 175
- infrainvariant 229
- isolated 50
- relatively convex 153
- right-relatively convex 127

Sublattice 7

Subsemigroup, α-closed 123

Subset
- convex 7, 29
- isolated 29
- strongly isolated 54

System
- complete 31
- infrainvariant 31
- normal 30
- soluble 31
- subnormal 31

Transversal 91

Ultrafilter 182

Ultraproduct 182

Unit 6

Wreath product
- Cartesian 86
- direct 86

(ζ, θ)-component 82

A-order 52

G-block 74

G-congruence trivial 81

G-module 59

G-orbit 78

V-formation 178

α-cone local 122

l-group
- l-simple 23
- normal-valued 96
- with subnormal jumps 96

l-homomorphism 23

l-isomorphism 23

l-subgroup 19

o-group 12

o-homomorphism 22

o-homomorphism, natural 23

o-isomorphism 22

po-group 12

po-set 1
- connected 4
- directed down 4
- directed up 4

pro-group 12

ro-group 12

Author Index

Adeleke 217, 220, 221, 230

Ault 41

Baumslag 214

Bergman 166, 167, 172, 173, 176, 177, 178, 204, 205

Birkhoff 11

Bludov 147

Brodskiĭ 188

Budkin 196, 197, 210, 211, 213, 214, 215

Burns 50

Chehata 102, 112

Cohn 57

Conrad 32, 34, 37, 38, 46, 57, 113, 114, 117

Darnel 57, 97

Dlab 113, 147

Droste 32

Dummett 217, 220, 221, 230

Ehrenfeucth 190, 192

Fedorov 195

Fox 160

Fuchs 47, 72

Glass 57

Gorbunov 213

Gurchenkov 160

Hölder 33

Hale 50

Higman 100

Hion 42

Holland 58, 71, 77, 80, 82, 83, 84, 85, 87, 91, 100

Hollister 113

Iwasawa 36

Kargapolov 70, 72

Kim 57

Kokorin 56, 57, 69, 70, 153, 154, 155, 156, 160, 161

Kopytov 27, 72, 97, 142, 143, 144, 153, 154, 156, 160, 161, 220, 222, 227, 228, 229, 233, 237, 238

Levi 27

Loś 48

Lorenzen 48

Mal'cev 36, 49, 52, 54, 99, 100, 119, 183, 213

Matsusita 43

McCleary 32, 80, 82, 83, 84, 85, 87, 91, 94, 95

Medvedev 163, 178, 212, 223

Mekler 100

Mostowski 190, 192

Mura 66, 152, 164

Neumann, B. 41, 49

Neumann, P. 217, 220, 221, 230,

Ohnishi 48, 68

Passman 178

Petrova 130, 225

Plotkin 128

Podderyugin 52, 53, 54, 68

Rabin 190, 192

Rachunek 225, 228, 232

Rhemtulla 41, 51, 57, 63, 64, 66, 152, 164

Rieger 52, 54

Sasiada 72

Shelah 100

Smirnov 31, 32, 52, 58, 59, 60, 61

Stone 11

Strojnowski 166

Tararin 122, 123, 124, 125, 126, 128, 129, 130, 131, 132, 134, 138, 148, 151, 202, 206

Teh 42

Todorinov 62, 63, 127, 129, 130

Trevisan 42

Varaksin 97, 230

Vinogradov 165, 168

Wigold 50

Wojas 166

Zaĭtseva 32, 42, 49, 57, 98, 100

MIX
Papier aus verantwortungsvollen Quellen
Paper from responsible sources
FSC® C105338

If you have any concerns about our products,
you can contact us on
ProductSafety@springernature.com

In case Publisher is established outside the EU,
the EU authorized representative is:
Springer Nature Customer Service Center GmbH
Europaplatz 3, 69115 Heidelberg, Germany

Printed by Libri Plureos GmbH
in Hamburg, Germany